Contents

How to use this book

Welcome to your BTEC National Applied Science course.

You are joining a course that has a 30-year track record of learner success, with the BTEC National widely recognised within the industry and in higher education as the signature vocational qualification. Over 62 per cent of large companies recruit employees with BTEC qualifications and 100,000 BTEC learners apply to UK universities every year.

A BTEC National in Applied Science qualification will give you the opportunity to develop a range of specialist skills that will prepare you for the world of work, or for continued scientific study at a higher level.

BTEC Applied Science is a vocational course, available at a range of sizes, which is recognised and respected by employers and higher education institutions alike. Its flexible, unit-based structure allows you to choose the areas you wish to study, and focus on the aspects of science that interest you most.

In your BTEC course, you will not only get a solid grounding in scientific theories and concepts, but also develop the practical, investigative skills that underpin this sector. You will have the opportunity to focus on more specialist areas, such as forensics, genetics, material science, molecular biology and cryogenics. In addition to gaining science-specific skills, throughout your BTEC course you will develop more generic skills such as team-working, presentational skills and research strategies. These will ensure that you are ready to meet the demands of the modern workplace.

Scientific developments help to shape our world, and provide a huge range of employment opportunities. The field of genetics and genetic engineering help us to understand human diseases and how we can control food production. Forensics sheds light on how crimes are committed and how accidents can be investigated in a methodical and effective way. Material science provides an understanding of how materials behave and what uses they can be put to. Biomedical science opens the door to careers in health care and related industries, giving you the choice to follow your interests and realise your ambitions. Most importantly, science is an area that is continually changing, and a BTEC Applied Science course reflects these developments and allows you to keep pace with the exciting innovations that emerge from scientific study.

How your BTEC is structured

Your BTEC National is divided into **mandatory units** (the ones you must do) and **optional units** (the ones you can choose to do). The number of units you need to do and the units you can cover will depend on the type and size of qualification you are doing.

This book covers **units 1, 2, 3, 4, 8, 9, 10 and 11**. If you are doing the **Certificate**, **Extended Certificate** or **Foundation Diploma** in **Applied Science**, you will find all the mandatory units you need in this book. If you are taking the **Foundation Diploma,** there are all the mandatory units and enough optional units here for you to choose from to complete your course. If you are studying the **Diploma or Extended Diploma**, this book is designed to be used together with the *Pearson BTEC National Applied Science Student Book 2,* which includes further mandatory and optional units for these larger sizes of qualification. The table below shows how each unit in this book maps to the BTEC National Applied Science qualifications.

endorsed for
BTEC

Pearson
BTEC National
Applied
Science

Student Bo

Frances Annets
Joanne Hartley
Sue Hocking
Roy Llewellyn
Chris Meunier
Catherine Parmar
Alison Peers

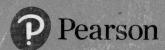
Pearson

Published by Pearson Education Limited, 80 Strand, London, WC2R 0RL.

www.pearsonschoolsandfecolleges.co.uk

Copies of official specifications for all Edexcel qualifications may be found on the website: www.edexcel.com

Text © Pearson Education Limited 2016
Page design by Andy Magee
Typeset by Tech-Set Ltd
Original illustrations © Pearson Education Ltd
Illustrated by Tech-Set Ltd
Cover design by Vince Haig
Picture research by Cristina Lombardo
Cover photo © Poprotskiy Alexey / Shutterstock.com

First published 2016

19 18 17 16
10 9 8 7 6 5 4 3 2 1

British Library Cataloguing in Publication Data
A catalogue record for this book is available from the British Library

ISBN 978 1 292 13409 3

Printed in the UK by Bell & Bain Ltd, Glasgow

Acknowledgements
We would like to thank Chongwei Chua, Neil Harris and Jennifer Smith for their invaluable help in reviewing this book.

The publisher would like to thank the following for their kind permission to reproduce their photographs:

(Key: b-bottom; c-centre; l-left; r-right; t-top)

123RF.com: 479, lightpoet 141, Stephen Coburn 1; **Alamy Images:** Art Directors & TRIP 95, Blend Images 185, Derek Meijer 285, Douglas Lander 200, Geoffrey Robinson 354, Lebrecht Music & Arts 71, OJO Images Ltd 286, PCN Photography 212 (c), Phanie 488, Phototake 38l, 39, sciencephotos 182, Suzanne Long 193t, Tengku Mohd Yusof 190t, Yon Marsh 358; **Courtesy Golden Rice Humanitarian Board.** www.goldenrice.org: 496; **Dr H. Jastrow - www.drjastrow.de:** 40; **eiscolabs.com:** 263c; **Fotolia.com:** Alexey Brin 395, apfelweile 212 (d), Dirk Vonten 212 (e), gerasimenuk 212 (b), karelnoppe 489, kristo74 190b, laviejasirena 84, micromonkey 89, Monkey Business 499, moonrise 487, Susan Flashman 375, Syda Productions 474bl, vadymvdrobot 490, WavebreakMediaMicro 441, 443; **Getty Images:** Dimitri Otis 442, Fritz Goro 263t, Photofusion 193b, Ulrich Baumgarten 119; **GNU Free Documentation License: Wikipedia:** Thenickman100 495r; **Health & Safety Executive:** Contains public sector information published by the Health and Safety Executive and licensed under the Open Government License 232; **Imagemore Co., Ltd:** 108; **Pearson Education Ltd:** Gareth Boden 394, Trevor Clifford 93, 100, 130, Adam Hale 362 (a), 362 (b), 362 (c), 362 (d), 362 (e), 362 (f), 374 (a), 374 (b), 374 (c), 374 (d), HL Studios 474tl, Coleman Yuen. Pearson Education Asia Ltd 64; **PhotoDisc:** Karl Weatherly 165; **Science Photo Library Ltd:** 315, 456, Adrian T Sumner 461, AJ Photo 491, Alfred Pasieka 454, Andrew Lambert Photography 62, 70, Biophoto Associates 38r, 469t, CNRI 463tl, Dept. of Clinical Cytogenetics, Addenbrookes Hospital 463tr, Dr Alexey Khodjakov 469tc, 469b, Dr Keith Wheeler 324, Dr Gladden Willis, Visuals Unlimited 300, Edward Kinsman 63, ER Degginger 125, Frans Lanting, Mint Images 495l, Gene Cox 472, Look at Sciences 463b, M I Walker 470, Marek Mis 271, Martin Shields 143, 494, Martyn F Chillmaid 104, 242, Mauro Fermariello 263b, Nancy Kedersha 337, Prof G Gimenez-Martin 469bc, Professor T Naguro 473, Ted Kinsman 146; **Shutterstock.com:** 192, Andrey Pavlov 212 (a), AVAVA 500, Denis Pepin 331, Felix Mizioznikov 142, george green 189, jadimages 199, Karuka 474tc, l i g h t p o e t 335, Marcin Balcerzak 225, martin33 212 (f), michaeljung 336, Minerva Studio 281, Rawpixel 474br, Sebastian Kaulitzki 288, Stuart Jenner 218, sylv1rob1 393, Zurijeta 474tr; **Wikimedia Commons:** Sparkla 77

All other images © Pearson Education

The publisher would like to thank the following for their kind permission to reproduce their materials:

Figures 1.12, 1.13, 1.14 http://www.bbc.co.uk/bitesize/higher/chemistry/energy/bsp/revision/1/, BBC; Figure 1.8 www.chemguide.co.uk/atoms/bonding/dative.html, with the permission of Jim Clark; Figure 1.11 www.

bbc.co.uk/schools/gcsebitesize/science/add_ocr_gateway/periodic_table/metalsrev2.shtml, BBC; Figure 1.16 www.peoi.org/Courses/Coursessp/chemintro/ch/ch10a.html, © David W Ball; Figure 1.62 www.chemguide.co.uk/atoms/bonding/vdwstrengths.html, with the permission of Jim Clark; Figure 1.28 adapted from www.phschool.com/science/biology_place/biocoach/plants/walls.html, Pearson Education, Inc., Used by permission of Pearson Education, Inc. All Rights Reserved; Figure 1.29 adapted from *Biology*, 7 ed (P Raven, 2005) Figure 5.5 p.84, McGraw-Hill Education; Figure 1.30 adapted from http://cell-specialisation-jesse.wikispaces.com, Jesse Claire, 26 February 2012; Figure 1.38 adapted from, Elaine N Marieb *Essentials of Human Anatomy and Physiology* 7th ed ©2003. Reprinted and electronically reproduced by permission of Pearson Education, Inc., New York, NY; Figure 2.21 from 'Thin layer chromatography' Croatian-English Chemistry Dictionary & Glossary, 15 December 2015. KTF-Split. 8 February 2016 http://glossary.periodni.com Copyright © 2015 by Eni Generalic. All rights reserved; Figure 4.3 from http://www.cleapss.org.uk/secondary/secondary-science/hazcards, CLEAPSS. The information it contains must not be used directly to inform any risk assessment as users looking for this information must use the most up to date version available via the CLEAPSS website www.cleapss.org.uk; Figure 4.5 from http://www.hse.gov.uk/pubns/books/lawposter-a2.htm, contains public sector information published by the Health and Safety Executive and licensed under the Open Government Licence; Figures 8.7 and 8.8 adapted from 'How Skeletal Muscles Work', *Biological Sciences Review* Vol. 22, No. 4, p.11 (David Jones 2010); Figure 8.10 adapted from *Textbook of Medical Physiology*, 10 ed, Saunders (A Guyton and J Hall, 2000) p.171, copyright Elsevier (2000); Figure 8.15 after *Mosby's Medical, Nursing and Allied Health Dictionary*, 6 ed, Elsevier Health (Mosby 2001) p.821, copyright Elsevier (2000); Figures 9.34, 9.35, 9.36, 9.37, 9.38, 9.39, 9.40 from *Advanced Biology for You* by Gareth Williams (2e, OUP, 2015), copyright © Williams Services Ltd 2015, reproduced by permission of Oxford University Press; Figure 10.1 adapted from www.columbia.edu/cu/biology/courses/c2005/images/protolec2.98.html, © Larry Chasin, by permission of Larry Chasin; Figure 10.40 adapted from https://commons.wikimedia.org/wiki/File%3AThylakoid_membrane.svg, By Yikrazuul (Own work) [CC BY-SA 3.0 (http://creativecommons.org/licenses/by-sa/3.0)], via Wikimedia Commons; Figure 10.43 after http://users.rcn.com/jkimball.ma.ultranet/BiologyPages/C/Chlorophyll.html, Kimball's Biology Pages © John W Kimball, distributed under a Creative Commons Attribution 3.0 Unported (CC BY 3.0) license; Figure 10.44 from *OCR Biology A2 Heinemann*, Heinemann (S Hocking, 2008) p.62, Pearson Education Limited; Figure 10.46 adapted from *OCR Biology A2*, Heinemann (S Hocking, 2008) p.64, Pearson Education Limited.

Websites
Pearson Education Limited is not responsible for the content of any external internet sites. It is essential for tutors to preview each website before using it in class so as to ensure that the URL is still accurate, relevant and appropriate. We suggest that tutors bookmark useful websites and consider enabling students to access them through the school/college intranet.

A note from the publisher
In order to ensure that this resource offers high-quality support for the associated Pearson qualification, it has been through a review process by the awarding body. This process confirms that this resource fully covers the teaching and learning content of the specification or part of a specification at which it is aimed. It also confirms that it demonstrates an appropriate balance between the development of subject skills, knowledge and understanding, in addition to preparation for assessment.

Endorsement does not cover any guidance on assessment activities or processes (e.g. practice questions or advice on how to answer assessment questions), included in the resource nor does it prescribe any particular approach to the teaching or delivery of a related course.

While the publishers have made every attempt to ensure that advice on the qualification and its assessment is accurate, the official specification and associated assessment guidance materials are the only authoritative source of information and should always be referred to for definitive guidance.

Pearson examiners have not contributed to any sections in this resource relevant to examination papers for which they have responsibility.

Examiners will not use endorsed resources as a source of material for any assessment set by Pearson.

Endorsement of a resource does not mean that the resource is required to achieve this Pearson qualification, nor does it mean that it is the only suitable material available to support the qualification, and any resource lists produced by the awarding body shall include this and other appropriate resources.

Unit title	Mandatory	Optional
Unit 1 Principles and Applications of Science	All sizes	
Unit 2 Practical Scientific Procedures and Techniques	All sizes	
Unit 3 Science Investigation Skills	All sizes except Certificate	
Unit 4 Laboratory Techniques and their Application	All sizes except Certificate and Extended Certificate	
Unit 8 Physiology of Human Body Systems		All sizes except Certificate
Unit 9 Human Regulation and Reproduction		All sizes except Certificate
Unit 10 Biological Molecules and Metabolic Pathways		All sizes except Certificate
Unit 11 Genetics and Genetic Engineering		All sizes except Certificate

Your learning experience

You may not realise it but you are always learning. Your educational and life experiences are constantly shaping you, your ideas, your thinking, and how you view and engage with the world around you.

You are the person most responsible for your own learning experience so it is really important you understand what you are learning, why you are learning it and why it is important both to your course and your personal development.

Your learning can be seen as a journey which moves through four phases.

Phase 1	Phase 2	Phase 3	Phase 4
You are introduced to a topic or concept; you start to develop an awareness of what learning is required.	You explore the topic or concept through different methods (e.g. research, questioning, analysis, deep thinking, critical evaluation) and form your own understanding.	You apply your knowledge and skills to a task designed to test your understanding.	You reflect on your learning, evaluate your efforts, identify gaps in your knowledge and look for ways to improve.

During each phase, you will use different learning strategies. As you go through your course, these strategies will combine to help you secure the core knowledge and skills you need.

This student book has been written using similar learning principles, strategies and tools. It has been designed to support your learning journey, to give you control over your own learning and to equip you with the knowledge, understanding and tools to be successful in your future studies or career.

Features of this book

In this student book there are lots of different features. They are there to help you learn about the topics in your course in different ways and understand it from multiple perspectives. Together these features:

▶ explain what your learning is about

▶ help you to build your knowledge

▶ help you understand how to succeed in your assessment

▶ help you to reflect on and evaluate your learning

▶ help you to link your learning to the workplace.

In addition, each individual feature has a specific purpose, designed to support important learning strategies. For example, some features will:

▶ get you to question assumptions around what you are learning

▶ make you think beyond what you are reading about

▶ help you make connections across your learning and across units

▶ draw comparisons between your own learning and real-world workplace environments

▶ help you to develop some of the important skills you will need for the workplace, including team work, effective communication and problem solving.

Features that explain what your learning is about

Getting to know your unit

This section introduces the unit and explains how you will be assessed. It gives an overview of what will be covered and will help you to understand *why* you are doing the things you are asked to do in this unit.

Getting started

This appears at the start of every unit and is designed to get you thinking about the unit and what it involves. This feature will also help you to identify what you may already know about some of the topics in the unit and acts as a starting point for understanding the skills and knowledge you will need to develop to complete the unit.

Features that help you to build your knowledge

Research

This asks you to research a topic in greater depth. Using these features will help to expand your understanding of a topic as well as developing your research and investigation skills. All of these will be invaluable for your future progression, both professionally and academically.

Worked example

Our worked examples show the process you need to follow to solve a problem, such as a maths or science equation or the process for writing a letter or memo. This will also help you to develop your understanding and your numeracy and literacy skills.

Theory into practice

In this feature you are asked to consider the workplace or industry implications of a topic or concept from the unit. This will help you to understand the close links between what you are learning in the classroom and the affects it will have on a future career in your chosen sector.

Discussion

Discussion features encourage you to talk to other students about a topic in greater detail, working together to increase your understanding of the topic and to understand other people's perspectives on an issue. This will also help to build your team working skills, which will be invaluable in your future professional and academic career.

Features connected to your assessment

Your course is made up of a series of mandatory and optional units. There are two different types of mandatory unit:

▶ externally assessed
▶ internally assessed.

The features that support you in preparing for assessment are below. But first, what is the difference between these two different types of units?

Externally assessed units

These units give you the opportunity to present what you have learned in the unit in a different way. They can be challenging, but will really give you the opportunity to demonstrate your knowledge and understanding, or your skills in a direct way. For these units you will complete a task, set directly by Pearson, in controlled conditions. This could take the form of an exam or it could be another type of task. You may have the opportunity in advance to research and prepare notes around a topic, which can be used when completing the assessment.

Internally assessed units

Most of your units will be internally assessed. This involves you completing a series of assignments, set and marked by your tutor. The assignments you complete could allow you to demonstrate your learning in a number of different ways, from a written report to a presentation to a video recording and observation statements of you completing a practical task. Whatever the method, you will need to make sure you have clear evidence of what you have achieved and how you did it.

Assessment practice

These features give you the opportunity to practise some of the skills you will need when you are assessed on your unit. They do not fully reflect the actual assessment tasks, but will help you get ready for doing them.

Plan – Do – Review

You'll also find handy advice on how to plan, complete and evaluate your work after you have completed it. This is designed to get you thinking about the best way to complete your work and to build your skills and experience before doing the actual assessment. These prompt questions are designed to get you started with thinking about how the way you work, as well as understand why you do things.

Getting ready for assessment

For internally assessed units, this is a case study from a BTEC National student, talking about how they planned and carried out their assignment work and what they would do differently if they were to do it again. It will give you advice on preparing for the kind of work you will need to for your internal assessments, including 'Think about it' points for you to consider for your own development.

Getting ready for assessment

This section will help you to prepare for external assessment. It gives practical advice on preparing for and sitting exams or a set task. It provides a series of sample answers for the types of questions you will need to answer in your external assessments, including guidance on the good points of these answers and how these answers could be improved.

Features to help you reflect on and evaluate your learning

PAUSE POINT

Pause points appear after a section of each unit and give you the opportunity to review and reflect upon your own learning. The ability to reflect on your own performance is a key skill you'll need to develop and use throughout your life, and will be essential whatever your future plans are.

Hint
Extend

These also give you suggestions to help cement your knowledge and indicate other areas you can look at to expand it..

Reflect

This allows you to reflect on how the knowledge you have gained in this unit may impact your behaviour in a workplace situation. This will help not only to place the topic in a professional context, but also help you to review your own conduct and develop your employability skills.

Features which link your learning with the workplace

Case study

Case studies are used throughout the book to allow you to apply the learning and knowledge from the unit to a scenario from the workplace or the industry. Case studies include questions to help you consider the wider context of a topic. This is an opportunity to see how the unit's content is reflected in the real world, and help you to build familiarity with issues you may find in a real-world workplace.

THINK ▶FUTURE

This is a special case study where someone working in the industry talks about the job role they do and the skills they need. This comes with a *Focusing your skills* section, which gives suggestions for how you can begin to develop the employability skills and experiences that are needed to be successful in a career in your chosen sector. This is an excellent opportunity to help you identify what you could do, inside and outside of your BTEC National studies, to build up your employability skills.

Principles and Applications of Science I 1

Getting to know your unit

All scientists and technicians need to understand core science concepts. Chemists need to understand atoms and electronic structure to predict how a range of chemical substances will react to make useful products. Medical professionals need to understand the structure and workings of cells when they think about how the body stays healthy as well as when diagnosing and treating illness.

Scientists working in the communication industry need a good understanding of waves.

How you will be assessed

The external paper for this unit will be split into three sections, each worth 30 marks.

▶ **Section A** – Chemistry (Structure and bonding in applications of science, Production and uses of substances in relation to properties)

▶ **Section B** – Biology (Cell structure and function, Cell specialisation, Tissue structure and function)

▶ **Section C** – Physics (Working with waves, Waves in communication, Use of electromagnetic waves in communication)

The paper will contain a range of question types, including multiple choice, calculations, short answer and open response. These question types, by their very nature, generally assess discrete knowledge and understanding of content in this unit.

You need to be able to apply and synthesise knowledge from this unit. The questions on the paper will be contextualised in order for you to show you can do this.

There will be two opportunities each year to sit this paper: January and May/June.

Throughout this chapter, you will find assessment practices that will help you prepare for the exam. Completing each of these will give you an insight into the types of questions that will be asked and, importantly, how to answer them.

Unit 1 has four Assessment Outcomes (AO) which will be included in the external examination. These are:

▶ **AO1**: Demonstrate knowledge of scientific facts, terms definitions and scientific formulae
  · Command words: give, label, name, state
  · Marks: ranges from 12 to 18 marks
▶ **AO2**: demonstrate understanding of scientific concepts, procedures, processes and techniques and their application
  · Command words: calculate, compare, discuss, draw, explain, state, write
  · Marks: ranges from 30 to 45 marks
▶ **AO3**: Analyse, interpret and evaluate scientific information to make judgements and reach conclusions
  · Command words: calculate, compare, comment complete, describe, discuss, explain, state
  · Marks: ranges from 18 to 24 marks
▶ **AO4**: Make connections, use and integrate different scientific concepts, procedures, processes or techniques
  · Command words: compare, comment, discuss, explain
  · Marks: ranges from 9 to 12 marks

Here are some of the command words. The rest are found in the specification .

Command word	Definition – what it is asking you to do
Analyse	Identify several relevant facts of a topic, demonstrate how they are linked and then explain the importance of each, often in relation to the other facts.
Comment	Requires the synthesis of a number of variables from data/information to form a judgement. More than two factors need to be synthesised.
Compare	Identify the main factors of two or more items and point out their similarities and differences. You may need to say which are the best or most important. The word *Contrast* is very similar.
Define	State the meaning of something, using clear and relevant facts.
Describe	Give a full account of all the information, including all the relevant details of any features, of a topic.
Discuss	Write about the topic in detail, taking into account different ideas and opinions.
Evaluate	Bring all the relevant information you have on a topic together and make a judgement on it (for example, on its success or importance). Your judgement should be clearly supported by the information you have gathered.
Explain	Make an idea, situation or problem clear to your reader, by describing it in detail, including any relevant data or facts.

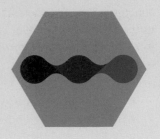

Getting started

Scientists working in a hospital laboratory use a range of core scientific principles. Write a list of core scientific principles you think they might need and why they are useful. Remember these may be to do with physics, chemistry or biology. When you have completed this unit, see if you can add any more principles to your list.

 A Periodicity and properties of elements

A1 Structure and bonding in applications in science

The electronic structure of atoms

You should already know about the structure of an atom. The nucleus contains positive protons and neutral neutrons. Surrounding the nucleus are energy shells containing negative electrons. You should also know that protons and neutrons both have a relative mass of 1 and that the relative mass of an electron is almost 0.

Lab technicians need to understand the electronic structure of atoms. They can use this knowledge to predict how chemical substances will behave and react.

The protons and the neutrons are found in the nucleus at the centre of an atom. The electrons are in shells or energy levels surrounding the nucleus. Each shell can hold electrons up to a maximum number. When the first shell is full, electrons then go into the second shell and so on. The maximum number of electrons in each shell is shown in Table 1.1.

▸ **Table 1.1:** Maximum number of electrons for each electron shell

Electron shell	Maximum number of electrons
1	2
2	8
3	18
4	32
5	50

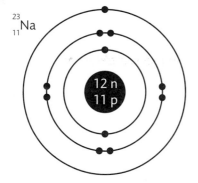

$^{23}_{11}$Na

▸ **Figure 1.1:** Simple atomic structure of sodium

A sodium atom containing 11 electrons has an electron arrangement of 2, 8, 1.

This can be represented by a simple Bohr diagram, as shown in Figure 1.1.

This is the simple version of electron structure you will have seen at Key Stage 4.

Under Bohr's theory, an electron's shells can be imagined as orbiting circles around the nucleus.

However, it is more complicated than this. Electrons within each shell will not have the same amount of energy and so the energy levels or shells are broken down into sub-shells called **orbitals**. These are called s, p, d and f orbitals. The orbitals have different energy states.

The Aufbau principle states that electrons fill the orbital with the lowest available energy state in relation to the proximity to the nucleus before filling orbitals with higher energy states. This gives the most stable **electron configuration** possible.

> **Key term**
>
> **Orbitals** – regions where there is a 95% probability of locating an electron. An orbital can hold a maximum of two electrons.

Electrons have the same charge and so repel each other, so if there is more than one orbital in an energy level (sub-shell) they will fill them singly until all the orbitals in that sub-shell have an electron in them and then they will pair up.

Figure 1.2 shows the energy levels of the shells, sub-shells and orbitals for an atom.

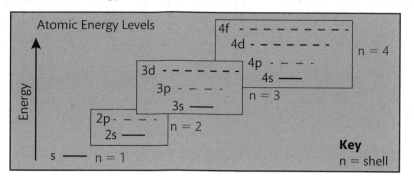

▶ **Figure 1.2:** Energy levels of the shells subshells and orbitals for an atom

Step by step: Electron structures

8 Steps

When writing out electron structures, you should follow these rules.

Half arrows are used to represent each electron in the orbitals. They are drawn facing up and down as each electron in an orbital will have a different **spin**.

1 The electrons sit in orbitals within the shell. Each orbital can hold up to two electrons.

▼

2 The first shell can hold two electrons in an *s*-type orbital.

▼

3 The second shell consists of one *s*-type orbital and three *p*-type orbitals. This diagram represents lithium.

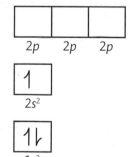

▼

4 The third shell consists of one *s*-type orbital, three *p*-type orbitals and five *d*-type orbitals.

▼

5 Electrons fill the lowest energy level orbitals first.

▼

6 Where there are several orbitals of exactly the same energy, for example, the three *2p* orbitals in the second shell, then the electrons will occupy different orbitals wherever possible.

▼

7 So the electronic structure of nitrogen (which has 7 electrons) is:

↑	↑	↑
$2p^1$	$2p^1$	$2p^1$

↑↓
$2s^2$

↑↓
$1s^2$

▼

8 and the electronic structure of a sodium atom (which has 11 electrons) becomes:

↑
$3s^1$

↑↓	↑↓	↑↓
$2p^2$	$2p^2$	$2p^2$

↑↓
$2s^2$

↑↓
$1s^2$

Assessment practice 1.1

Copy out the following table and complete the electronic structures for the elements. Three have been done for you.

Element	Number of electrons	Electron structure
Hydrogen	1	$1s^1$
Helium		
Lithium		
Boron		
Carbon	6	$1s^2\ 2s^2\ 2p^2$
Oxygen	8	$1s^2\ 2s^2\ 2p^4$
Magnesium		
Chlorine		
Calcium		

Ⅱ PAUSE POINT Try explaining what you have learned so far.

Hint   Close the book and write out all the key concepts you have learned so far. What do you know about electronic structure? Could you draw the electronic structure for calcium?

Extend   What is new compared to what you learned at level 2 about electronic structure?

One of the tasks of a lab technician is to make up solutions ready for experiments or for making products. Different types of compounds dissolve in different types of solvents depending on what type of bonding is in the compound. The lab technician must know what type of compound they are using in order to select the correct solvent.

Ionic bonding

Noble gases (elements in group 0 of the periodic table) have a stable electronic configuration. They have full outer shells. This means they do not react easily and most do not react at all. Elements in the other groups do not have full outer shells. This means that they react to gain stable electronic configurations.

Ionic bonding occurs when an atom of an element loses one or more electrons and donates it to an atom of a different element. The atom that loses electrons becomes positively charged and the atom that gains electron(s) become negatively charged because of the imbalance of protons and electrons.

For example, the bonding in sodium chloride is ionic. This means that the sodium atom loses the electron in its outer shell to become the positively charged sodium ion, Na^+, with the same electron configuration as neon. Chlorine gains an electron to become the negatively charged chloride ion, Cl^-, with the same electron configuration as argon. This means that both the sodium ion and the chloride ion have a full outer shell and become stable. The positive charge on the sodium ion and the negative charge on the chloride ion are attracted.

Key term

Ionic bonding – electrostatic attraction between two oppositely charged ions.

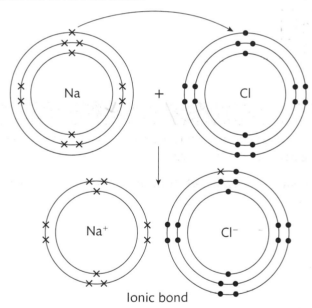

▶ **Figure 1.3:** Electron transfer and bonding in sodium chloride

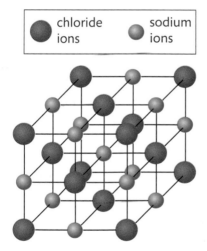

chloride ions   sodium ions

▶ **Figure 1.4:** Lattice structure of sodium chloride

Figure 1.3 shows bonding using a dot and cross diagram. The dots and crosses represent electrons in the shells.

Ions containing more than one element can also be formed. For example, in sodium hydroxide, Na^+ bonds with the hydroxide ion $(OH)^-$.

The opposite charges on the ions are what hold them together. This is **electrostatic attraction**.

The opposite charged ions in sodium chloride form a **giant ionic lattice** (see Figure 1.4) where the ions are arranged in a regular pattern.

Key terms

Electrostatic attraction – the force experienced by oppositely charged particles. It holds the particles strongly together.

Giant ionic lattice – a regular arrangement of positive ions and negative ions, for example, in NaCl.

The strength of the electrostatic force and, therefore, of the ionic bond is dependent on the ionic charge and the ionic radii of the ions. The more electrons a positive ion has, the more shells it will have. If an ion has more shells, then its radius will be bigger than an ion with fewer shells.

The electrostatic force is stronger when the ionic charge is higher. However, the force becomes weaker if the ionic radii are bigger. This is because, when the ionic radius is bigger, the ionic charge is spread over a larger surface area.

Covalent bonding

Covalent bonding usually occurs between atoms of two non-metals. A covalent bond forms when an electron is shared between the atoms. These electrons come from the top energy level of the atoms.

A chlorine molecule has a covalent bond (see Figure 1.5). The highest shell in each chlorine atom contains seven electrons. One electron from the highest shell in each atom is shared to give each chlorine atom the electron configuration of argon with a stable full outer shell.

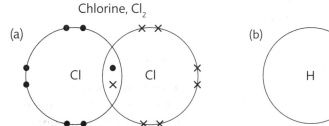
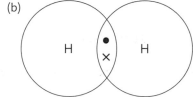
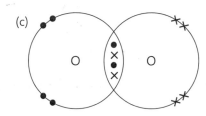

▶ **Figure 1.5:** Covalent bonding in (a) a chlorine molecule, (b) a hydrogen molecule and (c) oxygen moelcule

Multiple bonds

In some covalent molecules, both sharing electrons come from one atom. This is called a dative (coordinate) covalent bond (see Figure 1.6).

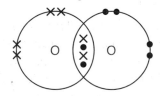

▶ **Figure 1.6:** The double bonds between the oxygen are formed by two shared pairs of electrons.

If three pairs of electrons are shared, then a triple covalent bond is formed. A triple bond is present in a nitrogen molecule (see Figure 1.7).

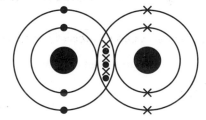

▶ **Figure 1.7:** Triple bond in a nitrogen molecule

Single bonds have a greater length than double bonds and double bonds have a greater length than triple bonds. The shorter the length of the bond, the stronger the bond is. Therefore, triple bonds are stronger than double or single bonds. A single bond between carbon atoms has a length of 154 pm and a bond energy of

347 kJ mol⁻¹. A double bond between carbon atoms has a length of 134 pm and a bond energy of 612 kJ mol⁻¹. A triple bond between atoms has a bond length of 120 pm and a bond energy of 820 kJ mol⁻¹.

An ammonium ion contains a dative bond (see Figure 1.8). When ammonia reacts with hydrochloric acid, a hydrogen ion from the acid is transferred to the ammonia molecule. A **lone pair** of electrons on the nitrogen atom forms a dative covalent bond with the hydrogen ion.

Key term

Lone pair – a non-binding pair of electrons.

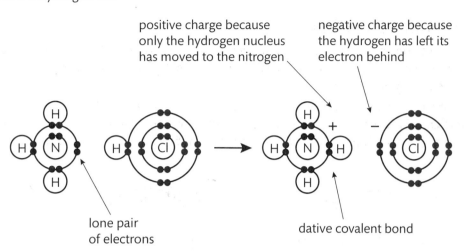

positive charge because only the hydrogen nucleus has moved to the nitrogen

negative charge because the hydrogen has left its electron behind

lone pair of electrons

dative covalent bond

▶ **Figure 1.8:** Dative bond formation in reaction between ammonia and hydrogen chloride

Covalent bonding in organic molecules

Carbon makes four covalent bonds so it forms many compounds which are called **organic compounds**.

Methane has the formula CH_4. Each carbon atom bonds covalently with four hydrogen atoms. The carbon gains the stable electron structure of neon and hydrogen gains the stable electron structure of helium.

These four bonds mean that methane is not a flat molecule. It has a tetrahedral structure (see Figure 1.9). This is because the bonds are as separated from each other as possible, because the negative electron pairs repel each other, with each bond angle being 109.5°. If you were to build a model of a methane molecule, it would have a 3D shape with a hydrogen pointing down towards you, one pointing down away from you, one pointing down to the side and one pointing up, all connected to the carbon in the centre.

Key term

Organic compound – a compound that contains one or more carbons in a carbon chain.

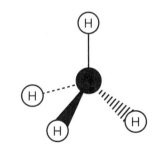

▶ **Figure 1.9:** Tetrahedral structure of methane

Step by step: Building models of organic compounds   3 Steps

1 Use molecular model kits to build models of the following organic compounds.
- methane CH_4
- ethane CH_3CH_3
- propane $CH_3CH_2CH_3$.

2 Write down what you notice about the structure of these molecules.

3 Look at one of the carbons in each molecule and the atoms bonded to it. Write down what you notice about the shape.

Organic compounds with three or more carbons in a chain cannot be linear because of the tetrahedral structure around each central carbon (see Figure 1.10).

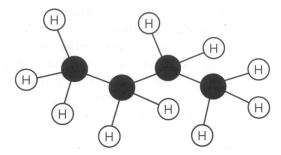

▶ **Figure 1.10:** A butane model

Metallic bonding

Metals are giant structures of atoms held together by metallic bonds. The metal structure is a regular lattice (see Figure 1.11).

Metallic bonding is caused because the electrons in the highest energy level of a metal atom has the ability to become **delocalised**. They are free to move through the metal in a 'sea' of electrons. This gives the metal nuclei a positive charge, which is attracted to the negative charge on the delocalised electrons. There is a very strong force of attraction between the positive metal nuclei and the negative delocalised electrons. However, the forces in metallic bonding are not as strong as in covalent or ionic bonding.

free electrons from higher energy
level of metal atoms

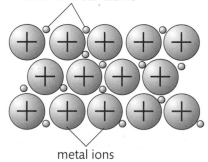

metal ions

▶ **Figure 1.11:** Metallic structure

The metal structure is a lattice of positive ions with electrons flowing between these ions.

Key term

Delocalised electrons – electrons that are free to move. They are present in metals and are not associated with a single atom or covalent bond.

Ⅱ PAUSE POINT What have you learned about bonding?

Hint Describe the differences between ionic, covalent and metallic bonding.

Extend Give two examples of elements, compounds or molecules with each type of bond.

The **electronegativity** of two atoms will determine what type of bond will form between them.

Key term

Electronegativity – the tendency of an atom to attract a bonding pair of electrons.

Atoms that have similar electronegativities form covalent bonds.

There is a strong electrostatic attraction between the two nuclei and the shared pair(s) of electrons between them. This is the covalent bond. Both atoms have the same electronegativity, and so the electrons are equally shared. The molecule is **non-polar** (see Figure 1.12). Hydrogen only has one shell containing one electron. This electron from each hydrogen is shared to give each atom the electronic configuration of helium. Oxygen only has six electrons in its highest energy shell. Each oxygen atom shares two of its electrons with another oxygen atom, giving both eight electrons in their outer shell. This makes the atoms in the oxygen molecule stable.

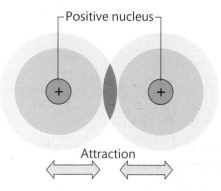

▶ **Figure 1.12 :** Non-polar covalent bond

In most covalent compounds, the bonding is **polar** covalent (see Figure 1.13). The shared electrons are attracted more to one nucleus in the molecule than the other. The atom with the higher electronegativity will attract the electrons more strongly. This gives the atom a slight negative charge. The other atom in the molecule will have a slight positive charge.

Key terms

Non-polar molecule – a molecule where the electrons are distributed evenly throughout the molecule.

Polar molecule – a molecule with partial positive charge in one part of the molecule and similar negative charge in another part due to an uneven electron distribution.

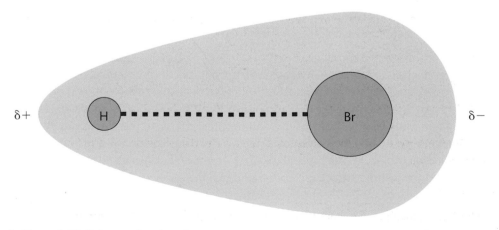

▶ **Figure 1.13:** Polar covalent bond

As the difference in electronegativity between the atoms increases, the bond will become more polar. See Figure 1.14.

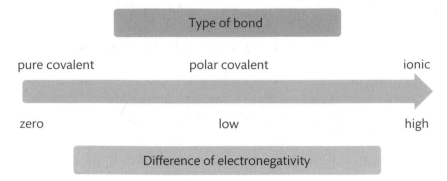

▶ **Figure 1.14:** Electronegativity spectrum

The electronegativities of some of the common elements you will use are shown in Table 1.2. It increases across a period and decreases down a group.

▶ **Table 1.2:** Electronegativities of elements

Element	Electronegativity
Fluorine	3.98
Oxygen	3.44
Nitrogen	3.04
Carbon	2.55
Chlorine	3.16
Hydrogen	2.20
Lithium	0.98
Sodium	0.93

Intermolecular forces

Intermolecular forces also affect how chemical substances behave. A laboratory technician must know where these are present and understand how they will affect the behaviour and reactions of chemical substances they are working with.

London dispersion forces

One type of intermolecular force is called London dispersion forces (also called temporary **dipole** – induced dipole forces). They are weak forces present between non-polar covalent molecules. They are less than 1% of the force of a covalent bond (see Figure 1.15).

When the electron distribution in a molecule becomes non-symmetrical (i.e. there are more electrons at one end of the molecule than the other), then one end of the molecule can become more positive and one end can become more negative. This causes a temporary dipole. The positive and negative charge in the dipole can disturb the electrons in a nearby molecule, repelling the electrons and so causing (inducing) a dipole in that molecule. The molecule with the temporary dipole and the molecule with the induced dipole attract each other and pull the molecules together. The forces are temporary because the electrons are constantly moving, so electron density in any part of a molecule is constantly changing. Larger molecules have more electrons which can move further so more temporary dipoles can form, meaning the force is bigger.

more electrons → more movement → bigger dipoles → stronger attraction

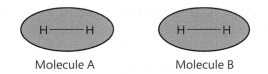

Even distribution of electrons throughout both molecules.

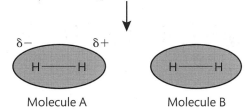

Uneven distribution of electrons in Molecule A causes a temporary dipole in the molecule. This will induce a dipole in molecule B as the electrons in Molecule B will be attracted to the positive end of Molecule A.

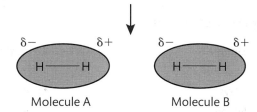

This forms a temporary dipole – induced dipole.

▶ **Figure 1.15:** London dispersion forces

London dispersion forces are the only forces that exist between noble gases and non-polar molecules.

Assessment practice 1.2

Pentane (C_5H_{12}) boils at 309 K and ethane (C_2H_6) boils at 185 K. This means that pentane is a liquid at room temperature (293 K) and ethane is a gas.

Explain why pentane is a liquid at room temperature but ethane is a gas.

Dipole-dipole forces

Another form of **van der Waals forces** are dipole-dipole forces. These are permanent forces between polar molecules (see Figure 1.16). Polar molecules have a permanent negative end and a permanent positive end. These oppositely charged ends attract each other. Dipole-dipole forces are slightly stronger than London dispersion forces but are still weak in comparison to a covalent bond. The force is about 1% of the strength of a covalent bond. Molecules that have permanent dipole-dipole forces include hydrogen chloride, HCl, and iodine monochloride, ICl. In both cases, the chlorine atom in the molecule is slightly negative. The hydrogen and iodine atoms are slightly positive.

Key term

Van der Waals forces – all intermolecular attractions are van der Waals forces.

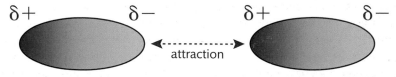

▶ **Figure 1.16:** Dipole-dipole forces

There are dipole-dipole forces between molecules of iodine monochloride (ICl).

Hydrogen bonding

The strongest form of intermolecular force is a hydrogen bond. These are a special type of dipole-dipole bond and are forces that are about 10% of the strength of a covalent bond. Hydrogen bonds will form when compounds have hydrogen directly bonded to fluorine, oxygen or nitrogen. This is because there is a large difference in electronegativity between hydrogen and these three atoms. This large difference means that very polar bonds are formed so the molecules have permanent dipoles. When two of these molecules are close together, there will be an attraction between the positive end of one and the lone pair of electrons of the other. This is a hydrogen bond.

This is different to other dipole-dipole forces because there are inner bonding electrons. The single electron in the hydrogen atom is drawn to the nitrogen (see Figure 1.17), oxygen or fluorine atom. There are no non-bonding electrons shielding the nucleus of the hydrogen. The hydrogen proton is strongly attracted to the lone pair of electrons on the nitrogen atom of another molecule.

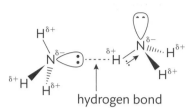

▶ **Figure 1.17:** Hydrogen bond in ammonia

hydrogen bond

Discussion

Hydrogen bonding in water is the reason why water has such unusual properties. For example, solid water is less dense than liquid water, it has a higher boiling point than expected, it is a good solvent for many chemical substances.

Research how hydrogen bonding is caused in a water molecule. Work in pairs to list properties of water due to the hydrogen bonding. In groups, explain the properties to other pairs of learners.

Ⅱ PAUSE POINT Try to describe all the different types of intermolecular forces to a partner.

Hint Draw a table showing the different types of intermolecular bonding and their properties.

Extend Explain how each type of intermolecular bond affects the properties of the molecules.

Quantities used in chemical reactions

Balancing equations

All chemical reactions can be written as a balanced equation using the chemical formulae for the reactants and the products involved in the reaction. Symbols for elements can be found in the periodic table. The numbers in the formulae show how many atoms of each element there are. You can use the periodic table to predict whether the compound is covalent or ionic. The group numbers will show you how many electrons the atom needs to lose or gain or share to form a bond.

The equation must balance like a maths equation. There should be the same number and types of atoms on both sides of the equation.

Step by step: Writing a balanced equation   5 Steps

1 Write the equation as a word equation, including all the reactants and all the products.

▼

2 Write out the formulae for each substance in the reaction. Note that gaseous elements (except those in group 0) like hydrogen and oxygen are diatomic (molecules with two atoms), so they must be written as H_2 and O_2. Metal elements and the noble gases are monatomic (one atom).

▼

3 Write out the number of each element on both sides.

▼

4 Make the number of each atom equal on each side. Remember that you cannot change the formula of the compounds. To increase the number of atoms of a particular element, you must place a number in front of the compound it is in. This will affect the number of atoms of all the other elements in the compound.

▼

5 Check that there is the same number of atoms of each element on both sides.

Worked Example

1 Write a balanced equation for the following reaction.

Ethanol + oxygen $\rightarrow$ carbon dioxide + water

Step 1: Write out the formulae for each substance in the reaction.

$C_2H_5OH + O_2 \rightarrow CO_2 + H_2O$

Step 2: Write out the number of each element on both sides.

left-hand side	right-hand side
C 2	C 1
H 6	H 2
O 3	O 3

Step 3: Make the number of each atom equal on each side.

In this case, start by putting a 2 in front of the carbon dioxide to equal out the carbons. This will also add two more oxygens to the right-hand side.

$C_2H_5OH + O_2 \rightarrow 2CO_2 + H_2O$

left-hand side	right-hand side
C 2	C ~~1~~ 2
H 6	H 2
O 3	O ~~3~~ 5

Put a 3 in front of the water to balance the hydrogens. Remember to add to the oxygens again.

$C_2H_5OH + O_2 \rightarrow 2CO_2 + 3H_2O$

left-hand side	right-hand side
C 2	C ~~1~~ 2
H 6	H ~~2~~ 6
O 3	O ~~3~~ ~~5~~ 7

The carbons and hydrogens are now equal on both sides, so you must multiply the oxygens on the left-hand side to finish balancing the equation.

$C_2H_5OH + 3O_2 \rightarrow 2CO_2 + 3H_2O$

left-hand side	right-hand side
C 2	C ~~1~~ 2
H 6	H ~~2~~ 6
O ~~3~~ 7	O ~~3~~ ~~5~~ 7

This equation is now balanced.

ⓘPAUSE POINT

Write a balanced equation for the following reaction:
butanol (C_4H_9OH) + water $\rightarrow$ carbon dioxide + water

Hint   Remember you can only change the number of moles of each substance, you cannot change the formula.

Extend   Now write a balanced equation for:
magnesium carbonate + hydrochloric acid $\rightarrow$ magnesium chloride + water + carbon dioxide

Assessment practice 1.3

Write balanced equations for the following reactions.

1   methane (CH_4) + oxygen $\rightarrow$ carbon dioxide + water
2   calcium carbonate ($CaCO_3$) + hydrochloric acid (HCl) $\rightarrow$ calcium chloride + carbon dioxide + water
3   calcium hydroxide ($CaOH)_2$ + hydrochloric acid $\rightarrow$ calcium chloride + water

Moles, molar masses and molarities

Chemical equations allow you to work out the masses of the reactants you need to use in order to get a specific mass of product. Chemists never use one molecule of a substance because that would be too small. Even 0.1 g of hydrochloric acid will contain millions of molecules of the acid. These numbers are very big and difficult to work with so chemists use a quantity called a **mole** with the symbol mol.

Do not let the idea of a mole confuse you. It is just a number. One mole of a chemical means there are 6.023×10^{23} particles (Avogadro's constant).

6.023×10^{23} is a number is standard form. This is a simple way of showing a very large number. 1×10^3 is how you would write 1000 in standard form. The 10^3 means that if you write the number out in full, it will have 3 zeroes at the end. So 6.023×10^{23} is a simple way to write 6023 with 20 zeroes at the end.

A mole is the amount of a substance which has the same number of particles as there are atoms in 12 g of carbon-12.

So one mole of carbon dioxide has the same number of particles as one mole of gold. The **molar mass** of a substance is equal to the mass of one mole of a substance.

It is useful to be able to convert masses into moles and moles into masses.

Mass (g) = molar mass × number of moles

▶ The relative atomic mass (Ar) of an element on the periodic table tells you how much mass there is in one mole of the element. The relative atomic mass is the average mass of an atom of an element compared to one twelfth of the mass of an atom of carbon–12. The relative atomic mass of hydrogen is 1.0. The relative atomic mass of oxygen is 16.0.

▶ The relative formula mass is the sum of all the relative atomic masses of all the atoms in the empirical formula (simplest formula) of a compound (Mr).
The relative formula mass of water, H_2O, is $(1 \times 2) + 16 = 18$
Relative atomic and formula masses do not have any units, as they are only relative to carbon–12.

Key terms

Mole – a unit of substance equivalent to the number of atoms in 12 g of carbon-12. One mole of a compound has a mass equal to its relative atomic mass expressed in grams.

Molar mass – the mass of one mole of a substance.

Assessment practice 1.4

What is the relative formula mass for these molecules? You will need to use the periodic table to find the relative atomic masses.

1 CO_2

2 NaOH

3 H_2SO_4

4 $Ca(OH)_2$

5 Fe_2O_3

The following worked examples show how to convert masses to moles.

Worked Example

1 What is the number of moles in 136.5 g of potassium?

Number of moles of an element = mass/A_r

For potassium $A_r = 39$

Number of moles $= \dfrac{136.5}{39}$

$= 3.5$ moles

2 What is the number of moles in 20 g of sodium hydroxide, NaOH?

Number of moles = mass/M_r

For sodium hydroxide $M_r = 23 + 16 + 1 = 40$

Number of moles $= \dfrac{20}{40}$

$= 0.5$ moles

Empirical formula

This shows the ratio between elements in a chemical compound. It is useful when discussing giant structures such as sodium chloride. The empirical formula of a compound can be calculated from the masses of each element in the compound. These masses are worked out through experimental analysis of the compound.

Step by step: Empirical formula   `3 Steps`

1 Divide the mass of each element present in the compound by its molar mass to get its molar ratio.

▼

2 Divide the answer for each element by the smallest molar ratio calculated. This gives you a ratio of 1:x for each element present.

▼

3 If the answers are not all whole numbers, multiply them all by the same number to get whole numbers. e.g. if the ratio is 1:1.5:3 then multiplying all the numbers by 2 will give you an answer with all whole numbers: 2:3:6.

Molecular formula

Molecular formulae are used for simple molecules. To work out the molecular formula you need to know the empirical formula and the relative molecular mass. For example, a compound has the empirical formula CH_2. This has an empirical formula mass of $12 + (1 \times 2)$ It has a relative molecular mass of 42. To work out its molecular formula, you first divide its relative molecular mass by the empirical mass: $42/14 = 3$. You write out the formula multiplying each part of the CH_2 unit by 3. This gives C_3H_6. This is the molecular formula.

Reacting quantities

When carrying out **titrations**, a chemist has to use **solutions** of a known concentration. These are called **standard solutions**. They have been prepared and tested to ensure they are of the specific concentration needed.

The number of moles of **solute** in a given volume of **solvent** tells you how concentrated the solution is.

When 1 mole of solute is dissolved in 1 cubic decimetre of solution, its concentration is written as:

$$1 \, mol \, dm^{-3}$$

This can be written as 1M for short. This is the molarity of the solution.

I mole of HCl has a mass of $1 + 35.5 = 36.5 \, g$.

$36.5 \, g$ of HCl in $1 \, dm^3$ of solution has a concentration of $1 \, mol \, dm^{-3}$ or 1M or $36.5 \, g \, dm^{-3}$.

Key terms

Titration – a method of volumetric analysis used to calculate the concentration of a solution.

Solution – a liquid mixture where a solute is dissolved in a solvent.

Standard solution – a solution of known concentration used in volumetric analysis.

Solute – the substance dissolved in a solvent to form a solution.

Solvent – a liquid that dissolves another substance.

Worked Example

1   How many moles of hydrochloric acid are there in $100 \, cm^3$ of 1M hydrochloric acid solution?

Number of moles (N) = molarity (C) × volume of solution (V) (dm^3)

$$N = CV$$

The volume is given in cm^3 so this needs to be converted into dm^3 by dividing by 1000. (Remember $1 \, dm^3 = 1000 \, cm^3$)

$$\text{Number of moles} = \frac{100}{1000} \times 1$$

$$= 0.1 \, mol$$

2 What is the concentration of a sample of sodium hydroxide solution if 10 dm³ contains 0.5 mol?

Number of moles (N) = molarity (C) × volume of solution (V) (dm³)

$$N = CV$$
$$0.5 = C \times 10$$
$$C = \frac{0.5}{10} = 0.05M$$

3 What volume in cm³ of 2M sulfuric acid solution would you need to ensure you had a sample containing 0.05 mol?

Number of moles (N) = molarity (C) × volume of solution (V) (dm³)

$$N = CV$$
$$0.05 = 2 \times V$$
$$V = \frac{0.05}{2} = 0.025 \text{ dm}^3$$

Multiply by 1000 to give answer in cm³

$$0.025 \times 1000 = 25 \text{ cm}^3$$

4 Calculate the number of moles of HCl in 20 cm³ of a 2 mol dm⁻³ solution of HCl(aq).

Convert 20 cm³ to dm³ by dividing by 1000.

$$\frac{20}{1000} = 0.02 \text{ dm}^3$$

Use the equation

Number of moles (N) = molarity (C) × volume of solution (V) (dm³)

$$N = CV$$
$$0.02 \times 2 = 0.04 \text{ mol of HCl in solution}$$

Using a chemical equation to calculate the quantities of reactants and products

Chemical equations can be used to calculate the quantities of reactants and products. Here is an example. Calcium chloride can be produced by reacting calcium carbonate with hydrochloric acid. This is the equation for the reaction.

$$CaCO_3 (s) + 2HCl (aq) \rightarrow CaCl_2 (aq) + CO_2 (g) + H_2O (l)$$

Note that the equation includes state symbols. A solid substance is indicated by (s), a solution is indicated by (aq), a liquid is indicated by (l) and a gas is indicated by (g). The equation shows that one mole of calcium carbonate reacts with two moles of hydrochloric acid. One mole of calcium chloride is produced as well as one mole each of carbon dioxide and water. This is an example of **stoichiometry**.

> **Key term**
>
> **Stoichiometry** – involves using the relationships between the reactants and the products in a chemical reaction to work out how much product will be produced from given amounts of reactants.

Worked Example

1. Calculate the expected mass of calcium chloride produced when 50 g of calcium carbonate is reacted with excess hydrochloric acid.

$A_r(H) = 1$, $A_r(C) = 12$, $A_r(O) = 16$, $A_r(Cl) = 35.5$, $A_r(Ca) = 40$

One mole of $CaCO_3$ produces one mole of $CaCl_2$.

You know this from the balanced equation $CaCO_3 + 2HCl \rightarrow CaCl_2 + CO_2 + H_2O$. This shows a one-to-one (1:1) ratio.

Add up the relative atomic masses for each compound.

$40 + 12 + (3 \times 16)\,g = 100\,g$ of $CaCO_3$ produces $40 + (35.5 \times 2)\,g = 111\,g$ of $CaCl_2$.

As one mole of $CaCO_3$ produces one mole of $CaCl_2$ then

100 g $CaCO_3$ produces 111 g $CaCl_2$

In this case, only 50 g of $CaCO_3$ was used, so

50 g $CaCO_3$ produces $\dfrac{111}{100} \times 50\,g$ $CaCl_2$

50 g $CaCO_3$ produces 55.5 g $CaCl_2$

You could say that only $\frac{1}{2}$ a mole of $CaCO_3$ was used so therefore only half the amount of $CaCl_2$ would be produced and this would give the same answer of 55.5 g.

This is the theoretical mass.

2. Calculate the expected mass of water if 10 g of oxygen is reacted with excess hydrogen.

$A_r(H) = 1$, $A_r(O) = 16$

Use a balanced equation to find out the ratio between oxygen and water.

$2H_2 + O_2 \rightarrow 2H_2O$

So one mole of oxygen gives 2 moles of water. This is a 1:2 ratio.

Add up the relative atomic masses for each substance. Remember there will be two lots of water.

$2 \times 16\,g = 32\,g$ of O_2 produces $\mathbf{2} \times (2 \times 1) + 16\,g = 36\,g$ of H_2O

32 g O_2 produces 36 g H_2O

So

10 g O_2 produces $\dfrac{36}{32} \times 10\,g$ of H_2O

10 g O_2 produces 11.25 g H_2O

Assessment practice 1.5

Silver iodide is used in the manufacture of photographic paper.

Calculate the theoretical yield of silver iodide for 34 g of silver nitrate reacting with excess sodium iodide. The equation for the reaction is as follows.

$AgNO_3\,(aq) + NaI\,(aq) \rightarrow AgI\,(s) + NaNO_3\,(aq)$

$A_r(N) = 14$, $A_r(O) = 16$, $A_r(Ag) = 108$, $A_r(I) = 127$

Percentage yields

The **theoretical mass** is the amount of product you can produce in a reaction. In most reactions it is unlikely that the total amount of product possible is made.

Some may be lost in transferring product from one vessel to another. Some of the reactants or products may react with impurities. In **reversible reactions**, products react to become the reactants and so are not all extracted from the reaction system. Chemists need to know how efficient their reaction process is, so they calculate the **percentage yield**.

The percentage yield is the actual mass compared to the theoretical mass. An efficient process would give a percentage yield as close to 100% as possible.

The formula for calculating percentage yield is:

$$\text{Percentage yield} = \frac{\text{actual number of moles}}{\text{expected number of moles}} \times 100\%$$

It can also be calculated as:

$$\text{Percentage yield} = \frac{\text{actual mass}}{\text{theoretical mass}} \times 100\%$$

The first step in working out percentage yield is to measure accurately the mass of product that you have obtained. How accurate your measurements are may depend on the equipment you have, but the mass should be measured to at least two decimal places. You should be able to use a top pan balance for this. If you are using small quantities, or if you want more accurate measurements, you may use a chemical balance that measures to 3 decimal places.

Once you have measured the mass of your product, you can work out how many moles you have produced. You can then divide this by the number of moles you were expecting to obtain and multiply by 100.

If you are using solutions, then you will need to calculate the number of moles for the volume of solution used. Calculating concentration times volume, CV, will give you the number of moles in the volume of solution used. This equation can be rearranged to find out what volume of a known concentration of solution is needed in a reaction.

> **Key terms**
>
> **Reversible reaction** – a reaction where the reactants react to form products and the products simultaneously react to re-form the reactants, for example, in $NaCl$.
>
> **Percentage yield** – the actual amount of mass worked out as a percentage of the theoretical mass.

Worked Example

1 When 50 g of calcium carbonate is reacted with excess hydrochloric acid solution to make calcium chloride, the theoretical yield is 55 g. When the reaction was carried out, only 44 g of calcium chloride was produced.

Calculate the percentage yield of calcium chloride.

$$\text{Percentage yield} = \frac{\text{actual mass}}{\text{theoretical mass}} \times 100\%$$

$$\text{Percentage yield} = \frac{44}{55} \times 100\%$$

Percentage yield is 80%

2 When 1 mole of oxygen reacts with excess hydrogen, 2 moles of water should be produced.

When this reaction was carried out, the actual number of moles was 1.8 moles of water.

Calculate the percentage yield.

$$\text{Percentage yield} = \frac{\text{actual number of moles}}{\text{expected number of moles}} \times 100\%$$

$$\text{Percentage yield} = \frac{1.8}{2} \times 100\%$$

Percentage yield is 90%

A2 Production and uses of substances in relation to properties

The periodic table

Key:
mass number
atomic symbol
name
atomic (proton) number

H hydrogen 1

1	2											3	4	5	6	7	0
																	4 **He** helium 2
7 **Li** lithium 3	9 **Be** beryllium 4											11 **B** boron 5	12 **C** carbon 6	14 **N** nitrogen 7	16 **O** oxygen 8	19 **F** fluorine 9	20 **Ne** neon 10
23 **Na** sodium 11	24 **Mg** magnesium 12											27 **Al** aluminium 13	28 **Si** silicon 14	31 **P** phosphorus 15	32 **S** sulfur 16	35.5 **Cl** chlorine 17	40 **Ar** argon 18
39 **K** potassium 19	40 **Ca** calcium 20	45 **Sc** scandium 21	48 **Ti** titanium 22	51 **V** vanadium 23	52 **Cr** chromium 24	55 **Mn** manganese 25	56 **Fe** iron 26	59 **Co** cobalt 27	59 **Ni** nickel 28	64 **Cu** copper 29	65 **Zn** zinc 30	70 **Ga** gallium 31	73 **Ge** germanium 32	75 **As** arsenic 33	79 **Se** selenium 34	80 **Br** bromine 35	84 **Kr** krypton 36
85 **Rb** rubidium 37	88 **Sr** strontium 38	89 **Y** yttrium 39	91 **Zr** zirconium 40	93 **Nb** niobium 41	96 **Mo** molybdenum 42	[98] **Tc** technetium 43	101 **Ru** ruthenium 44	103 **Rh** rhodium 45	106 **Pd** palladium 46	108 **Ag** silver 47	112 **Cd** cadmium 48	115 **In** indium 49	119 **Sn** tin 50	122 **Sb** antimony 51	128 **Te** tellurium 52	127 **I** iodine 53	131 **Xe** xenon 54
133 **Cs** caesium 55	137 **Ba** barium 56	139 **La*** lanthanum 57	178 **Hf** hafnium 72	181 **Ta** tantalum 73	184 **W** tungsten 74	186 **Re** rhenium 75	190 **Os** osmium 76	192 **Ir** iridium 77	195 **Pt** platinum 78	197 **Au** gold 79	201 **Hg** mercury 80	204 **Tl** thallium 81	207 **Pb** lead 82	209 **Bi** bismuth 83	[209] **Po** polonium 84	[210] **At** astatine 85	[222] **Rn** radon 86
[223] **Fr** francium 87	[226] **Ra** radium 88	[227] **Ac*** actinium 89															

▶ **Figure 1.18:** A section from the periodic table

Periods 1, 2, 3 and 4

The periodic table (see Figure 1.18) shows all the chemical elements arranged in order of increasing **atomic number**. Chemists can use it to predict how elements will behave, or what the physical or chemical properties of the element may be.

A laboratory technician needs to be very familiar with the periodic table. It is an information sheet on all the elements and their properties.

The elements on the periodic table are organised into groups (vertical columns) and periods (horizontal rows). Chemical properties are similar for elements in the same group. The atomic number increases as you move from left to right across a period. This is because each successive element has one more proton than the one before.

▶ **Table 1.3:** Characteristics of each period

Period	Characteristics
1	Contains hydrogen and helium. Both are gases. The electrons in these two elements fill the 1s orbital. Helium only has two electrons and, chemically, helium is unreactive. Hydrogen readily loses or gains an electron, and so can behave chemically as both a group 1 and a group 7 element. Hydrogen can form compounds with most elements and is the most abundant chemical element in the universe.
2	Contains eight elements: lithium, beryllium, boron, carbon, nitrogen, oxygen, fluorine and neon. The outer electrons in these elements fill the 2s and 2p orbitals. Nitrogen, oxygen and fluorine can all form diatomic molecules. Neon is a noble gas. Carbon is a giant molecular structure.
3	Contains eight elements: sodium, magnesium, aluminium, silicon, phosphorus, sulfur, chlorine and argon. The outer electrons in these elements fill the 3s and 3p orbitals.
4	Contains 18 elements, from potassium to krypton. The first row of the transition elements is in this period. The outer electrons on these elements fill the 4s, 4p and 3d orbitals.

Key term

Atomic number – the number of protons in an atom. (This is the same as the number of electrons in the atom.)

Discussion

Look at the periodic table and write down five key features of the periodic table. Work in pairs and try to list the names of any groups in the periodic table as you can. Discuss any facts you know about the elements in the groups you have listed. These may be properties of the elements or trends within groups.

Groups – s block, p block, d block

The periodic table is also organised by element blocks. An element block is a set of elements in groups that are next to each other. Element blocks are named for the orbital that the highest energy electrons are in for that set of elements. Groups 1 and 2 of the periodic table are in s block. Groups 3 to 7 and group 0 make up p block. This block contains all the non-metals except for hydrogen and helium. The transition metals are in the d block.

For example, carbon had electronic structure of $1s^2 2s^2 2p^2$. The highest energy electron in carbon is in a p orbital and therefore carbon is a p block element.

Assessment practice 1.6

Explain why calcium is an s block element.

Ⅱ **PAUSE POINT** Summarise what you have learned about the periodic table.

Hint Consider what you know about groups, periods and trends.

Extend Choose three elements in different areas of the table, and explain why their atomic structure and properties means they are in the position they are in.

Physical properties of elements

Atomic radius

The radius of an atom changes depending on what is around it. The only way to measure the radius is to measure the distance between the nuclei of two touching atoms and divide by two.

The atomic radius decreases across the period from left to right. Across the group, more protons and electrons are added. However, the extra electrons are added to the same s and p sub-shells and so the size does not increase. The extra protons increase nuclear charge. The increased nuclear charge attracts the extra electrons and pulls them closer to the nucleus. This leads to a decrease in atomic radius.

As you go down a group, the atomic radii increases. This is because the extra electrons are added to additional shells and so the radius increases. Although nuclear charge increases, the number of inner shells increases and so the nuclear charge is shielded more. This means that the atomic radius increases.

The trend is slightly different for the transition metals. The atomic radii get slightly smaller as you go across the start of the transition metals but then the radii stay very similar. This is because the additional nuclear charge is balanced by the extra shielding by the 3d electrons of the outer 4s sub-shell.

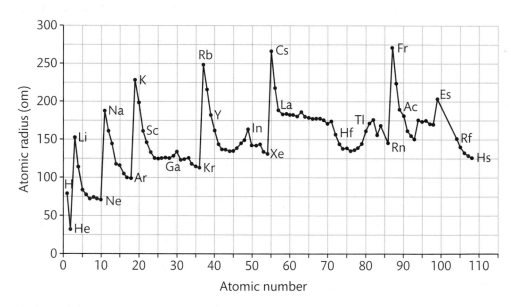

▶ **Figure 1.19**: Periodic trends in atomic radii

Ionic radius

The trends in ionic radius down a group follow a similar pattern to the trend for atomic radius down a group. This is because extra electrons are added to extra shells as you go down the group, therefore giving a larger size.

Cations have a smaller radius than their corresponding atom. As you go across a period, the cations all have the same electronic structure. They are **isoelectronic**, therefore although the number of electrons remains the same, the nuclear charge increases, for example, Na^+, Mg^{2+}, Al^{3+}. However, the number of protons increases across the period. This pulls the electrons more strongly to the centre of the ion, so the ionic radii of the cations decreases as you go across the period.

Key terms

Cations – ions with a positive charge.

Isoelectronic – having the same numbers of electrons.

Anions have a larger radius than the corresponding atom because there is more repulsion between the extra electrons. As you go across the period, the anions are all isoelectronic, for example, N^{3-}, O^{2-}, F^-. They have more electrons, not fewer. The number of protons still increases as you go across the period while the number of shells and electrons stays the same, so the ionic radius of the anions also decreases as you go across the period.

> **Key term**
>
> **Anions** – ions with a negative charge.

Electronegativity

Electronegativity is a measure of the tendency of an atom to attract a bonding pair of electrons. It increases as you go across a period. It decreases as you go down a group. This means that fluorine is the most electronegative element. The Group 0 gases such as argon that do not form bonds do not have electronegativity that can be reliably determined, because they do not form compounds/bonds.

Electronegativity depends on the number of protons in the nucleus, the distance from the nucleus of the bonding pair of electrons and how much shielding there is from inner electrons.

As you go across the period, the bonding pair of electrons will be shielded by the same number of electrons. However, the number of protons will increase, so the group 7 element will be more electronegative than the group 1 element.

As you go down a group, there is more shielding from inner electrons and the bonding pair of electrons are further from the nucleus. This adds up to less pull on the bonding pair from the positive charge of the nucleus and so electronegativity decreases.

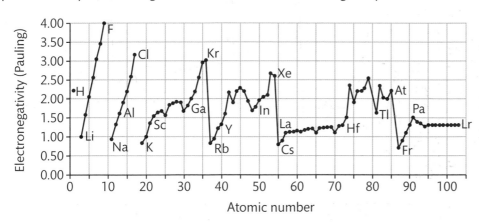

▶ **Figure 1.20**: Periodic trends in electronegativity

Trends in the periodic table are usually identified across periods or down groups. However, there are often similarities between elements that are diagonal to each other. For example, beryllium and aluminium have identical electronegativity. There is an increase across the period from group 2 to group 3, but then as electronegativity decreases down a group, this increase is balanced out. Other diagonal pairs also have similar electronegativity, for example, lithium and magnesium. These similarities mean they form similar bonds and may show similar chemistry.

⏸ PAUSE POINT   Explain how electron affinity affects the reactivity of atoms.

> **Hint**   You will need to use the terms nuclear charge, shells and shielding.
>
> **Extend**   Use this information to explain why potassium is a very reactive metal.

First ionisation energy and reasons for trends

First ionisation energy is the minimum energy needed for one mole of the outermost electrons to be removed from one mole of atoms in a gaseous state. One mole of positively charged ions is formed. First ionisation energies of the elements in a period show **periodicity**. There is an overall trend of first ionisation energy increasing across the period.

This trend is shown in the graph in Figure 1.21.

It takes more energy to remove an electron as you go across the period. This is because the number of protons increase across the period so the positive charge on the nucleus increases. This means that the force of attraction pulling on the outer electron increases. However, you can see there is not a steady increase in first ionisation energy. There is a pattern in the dips and increases for each period.

Across period 2, it dips at group 3 and group 6 elements. You can see the same pattern in period 3. Period 4 is a little different because it also contains the transition elements.

Key terms

First ionisation energy – the energy needed for one mole of electrons to be removed from one mole of gaseous atom. For example, the equation shows one mole of potassium atoms losing one electron to become one mole of positive ion:

$K(g) \rightarrow K^+(g) + e^-$.

Periodicity – the repeating pattern seen by the elements in the periodic table.

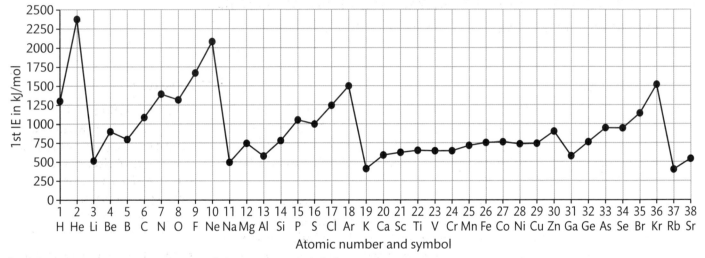

▶ **Figure 1.21:** First ionisation energies of elements in periods 1–4

For periods 2 and 3, there is a pattern which suggests that electrons removed from the third energy level are arranged in different sub-levels. Across period 2, the first electrons are removed from the 2s sub-level. The value for beryllium is higher than for lithium because beryllium has one more proton in its nucleus. There is a decrease for boron where the electron is taken from the 2p sub-level. This is a higher energy level than the 2s sub-level and so the electron is easier to remove – that is, the 2s sub-shell shields the 2p sub-shell, making it easier to remove an electron.

Carbon and nitrogen show the expected increase because the electron removed from each element is in the same 2p sub-level. These electrons occupy the orbital on their own; all are unpaired. There is a second dip in the first ionisation energy at oxygen. Here the electron removed is also in the 2p sub-level, but it is paired with another electron in that level. The electrostatic repulsion between the two electrons in the orbital means that it is easier to remove this electron. The first ionisation energy increases then for fluorine and neon because they have increasing positive charge. A similar pattern is seen for period 3 where the electrons are removed from the 3s and 3p sub-levels.

In period 4, a similar pattern is seen for elements in groups 1 to 7 and 0. The transition elements' first ionisation energy does increase across the period, but only a little. Their outer electron always comes from a 4s sub-shell because this has a higher energy than a 4d sub-shell in the transition elements. As you go across the group, the number of protons increases, but so does the number of 3d electrons. These 3d electrons provide shielding and so cancel out the effect of the extra proton, so ionisation energy only increases slightly across the d block.

As you go down a group, the first ionisation energy generally falls. You can also see this on the graph. Sodium has 8 more protons than lithium, so it may be expected that this increase in positive charge would increase the first ionisation energy. However, the outer electron in sodium is further away from the nucleus and has more shielding.

In lithium, the outer electron is attracted to 3 positive protons and is shielded by 2 negative electrons, so there is an overall attractive charge of +1. In sodium, the outer electron is attracted to 11 positive protons and is shielded by 10 negative electrons so the overall attractive charge is also +1. The outer electron in sodium is further away from the nucleus and this lowers the effect of the +1 charge, and so the first ionisation energy is lowered. This trend continues down the group and can also be observed in other groups such as groups 2 and 7.

Assessment practice 1.7

Evaluate the factors that affect first ionisation energies of elements in a period and in a group.

Electron affinity

Electron affinity can be simply defined as an atom's ability to gain an electron and become a negative ion. It is the change in energy (kJ mol^{-1}) of a neutral gaseous atom when an electron is added to the atom to form a negative ion. First electron affinity is when a –1 ion is formed. First electron affinities are negative. Table 1.4 gives the first electron affinities for group 7 elements.

The negative sign shows that energy is released. The amount of energy released generally decreases as you go down group 7. Fluorine is an exception. Electron affinity indicates how strong the attraction is between the nucleus of an atom and the incoming electron. If this attraction is strong, more energy is released. Just as in first ionisation energy, number of protons (or nuclear charge), distance from nucleus and shielding all have an effect on electron affinity. As you go down the group, the nuclear charge increases. However, there is also extra shielding from electrons as further shells are compressed. The further down the group you go, the further the outer shell is from the positive pull of the nucleus and so the attraction becomes weaker. This means that less energy is released when the ion is formed.

Fluorine is a very small atom and this is why it does not follow this pattern. When fluorine gains an electron to become fluoride, this new electron is added to a region that is already full of electrons and so there is repulsion from these.

Group 6's electron affinities follow a similar pattern to that in group 7 with it decreasing as you go down the group. Oxygen does not follow this pattern. It has a lower electron affinity than sulfur for exactly the same reason as fluorine having a lower electron affinity than chlorine.

Overall, group 6 elements have lower electron affinities than group 7 elements in the same period. This is because they have one less proton but the same amount of shielding. (Remember that elements in a period have the same number of electron shells as each other.)

Group 6 elements will also have a second electron affinity where the negative ion gains a second electron forming a charge of –2. It is the change in energy (kJ mol^{-1}) of

> **Key term**

> **Electron affinity** – the change in energy when one mole of a gaseous atom gains one mole of electrons to form a mole of negative ion. For example, for oxygen:
> $O(g) + e^- \rightarrow O^-(g)$.

▶ **Table 1.4:** First electron affinities

Element	First electron affinity kJ mol^{-1}
Fluorine	–328
Chlorine	–349
Bromine	–324
Iodine	–295

a mole of gaseous –1 ions when an electron is added to the ion to form a –2 ion. The two negative charges will repel, so this change in energy will be positive as energy will be needed to force an electron into the negative ion. Group 7 elements can also have a second electron affinity.

The first electron affinity for oxygen is –142 kJ mol[-1]. The second electron affinity for oxygen is +844 kJ mol[-1]. The high energy is needed to overcome the repulsion between the negative electron and negative ion.

$$O(g) + e^- \rightarrow O^-(g)$$
$$O^-(g) + e^- \rightarrow O^{2-}(g)$$

Type of bonding in the element

The electronegativity of elements can be used to predict the type of bonding in a compound. Bonding is a spectrum from ionic to covalent bonding with most compounds sitting somewhere between the two. It is rare to have a wholly ionic or wholly covalent compound. In a hydrogen molecule, both hydrogen atoms have the same electronegativity. This means that they form a covalent bond that is not polar. When hydrogen bonds with fluorine to make hydrogen fluoride, a polar covalent molecule is formed. This is because fluorine has a high electronegativity and so attracts the bonding pair. This gives the fluorine atom a positive charge and the hydrogen atom a negative charge.

You cannot directly measure electronegativity of an element. The chemist Linus Pauling produced a scale that gives a relative value for the elements and this allows you to predict how ionic a covalent bond will be.

Electronegativities for elements in periods 1 to 3

H 2.1

Li 1.0   Be 1.5   B 2.0   C 2.5   N 3.0   O 3.5   F 4.0

Na 0.9   Mg 1.2   Al 1.5   S 1.8   P 2.1   S 2.5   Cl 3.0

If the difference between the electronegativities of the elements forming the bonds is low, then the covalent bond will be less polar than when the difference between the electronegativities is high. As the difference increases, the covalent bond will become more polar. If the difference is very large, then the bond becomes ionic.

Ionic bonds can also show polarity. The extent of the polarisation will depend on whether:

▸ either ion is highly charged
▸ the cation is relatively small
▸ the anion is relatively large.

A small cation that is highly charged will tend to draw electrons towards it. A large anion that is highly charged will have an electron cloud that is easily distorted. This means that some of the negative charge is shared with the cation. This gives the ionic bond some covalent characteristics.

Period 2	Li		Be		B		C	N_2	O_2	F_2	Ne
Period 3	Na		Mg		Al		Si	P_4	S_8	Cl_2	Ar
Structure	giant metallic						giant covalent	simple molecular			
Forces	strong forces between positive ions and negative delocalised electrons						strong forces between atoms	weak intermolecular forces between molecules			
Bonding	metallic bonding						covalent	covalent bonding within molecules intermolecular bonding between molecules			

Trends: melting point and boiling point

The elements in the periodic table also show periodicity for melting and boiling points. Melting and boiling points depend on the strength of the forces between the atoms in an element. Going down group 1, the melting and boiling points decrease. This means that the forces of attraction get weaker. The melting and boiling points increase as you go down group 7. This means that the forces of attraction get stronger.

When an element melts, energy is used to overcome some of the attractive forces holding the atoms or molecules of the element together. When an element boils, most of the rest of the attractive forces are broken. The stronger the forces between the atoms, the higher the melting and boiling point will be. The melting and boiling points peak in the middle of period 2 and 3. The lowest values are found in group 0.

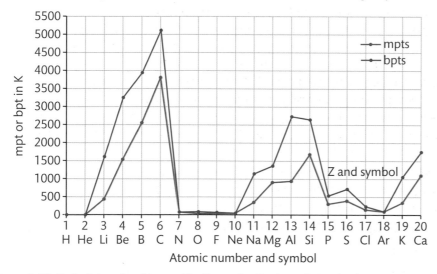

▶ **Figure 1.22:** Periodicity of melting and boiling points for Periods 1–3 (and start of Period 4)

Period 2

As you go across groups 1 to 3, metals have increasing nuclear charge because they have increasing number of protons and increasing number of delocalised electrons and so have stronger metallic bonding. This means the melting and boiling points increase as you go across the metals in the period. Carbon has giant covalent bonding forming a giant lattice structure with each atom bonding to 4 other carbon atoms. So its melting and boiling points are very high because it has strong covalent bonds that need a large amount of energy to break. The non-metals in groups 5 to 7 have small separate molecules and so have low melting points. There are only weak van der Waals forces that need to be overcome.

Period 3 follows the simple trend of period 2 with a few small exceptions. Sulfur, in group 6, has a higher melting and boiling point than the rest of the non-metals. This is because of the different size of the molecules of each of these elements. Phosphorus exists as P_4, sulfur as S_8, chlorine as Cl_2 molecules and argon as Ar atoms. The strength of the van der Waals forces increases as the size of the molecule increases. Therefore, because sulfur has the biggest molecule, it has the strongest van der Waals forces and so the highest melting and boiling points.

Assessment practice 1.8

Evaluate how type of bonding, intermolecular forces and molecule size affects the melting point in elements in period 3 and groups 2 and 6.

Physical properties of metals: electrical conductivity, thermal conductivity, malleability, ductility

Key terms

Malleable – can be hammered into shape without breaking.

Ductile – can be hammered thin or stretched into wires without breaking.

Metallic bonding allows for electrical conductivity through a solid or liquid metal. The delocalised electrons carry the electric charge. Copper is an excellent conductor of electricity. In fact, it has the best conductivity of any metal except for silver and so is used for electrical cables and wires.

The delocalised electrons in metals also absorb heat energy which gives them kinetic energy. This energy is then transferred through the metal by these electrons. Metals are good thermal conductors. This makes many metals such as aluminium and copper useful for saucepans, heat sinks in computers, and radiators.

The structure of metals also explains why they can be **malleable** or **ductile**. The atoms in the layers are able to roll over each other. They can move to new positions without breaking the metallic bonds. Aluminium is very malleable which, along with its high thermal conductivity, makes it useful for aluminium foil.

Research

Research the uses and applications of metals based on their properties. Can you find more examples of how each of the properties listed above make the metal useful?

Chemical properties of elements

The reactions between oxygen and metals are very important. How easily they react and the product they make can influence how a metal is used. For example, iron reacts very easily with oxygen and forms rust, so it is often painted to protect it from oxygen in the air. Table 1.5 shows how elements react with oxygen.

▶ **Table 1.5:** Products and reactivity of all period 2 and 3 elements with oxygen

Group	Element	Reactions with oxygen	Equations
1	Lithium	Rapid, burns with red flame. Metal oxide produced that forms a **alkaline solution** when dissolved in water.	$4Li\,(s) + O_2(g) \rightarrow 2Li_2O(s)$
	Sodium	Very vigorous, burns with orange flame. Metal oxide produced that form basic solution when dissolved in water.	$4Na\,(s) + O_2(g) \rightarrow 2Na_2O(s)$ $2Na\,(s) + O_2(g) \rightarrow Na_2O_2(s)$
2	Beryllium and magnesium	Needs heat to react as do group 1 elements. Very vigorous reactions.	$2Be\,(s) + O_2(g) \rightarrow 2BeO(s)$
3	Aluminium	Vigorous at first. Rapidly forms a water insoluble coating of Al_2O_3. This layer prevents the aluminium below from corroding and so makes aluminium an extremely useful material.	$4Al\,(s) + 3O_2(g) \rightarrow 2Al_2O_3(s)$ It is **amphoteric**.
4	Carbon	Forms slightly acidic oxides. Shows reaction with heat.	$C(s) + O_2(g) \rightarrow CO_2(g)$ $2C(g) + O_2(g) \rightarrow 2CO\,(g)$ – this is incomplete combustion
	Silicon	No reaction	$Si(s) + O_2(g) \rightarrow SiO_2(s)$ – weak acidic
5	Nitrogen	Forms a range of oxides with different **oxidation** states. A high temperature is needed for these reactions to take place.	It can produce NO, and NO_2 and N_2O_5.
	Phosphorus	Burns vigorously with a white flame.	P_4O_6 if limited oxygen, P_4O_{10} if excess oxygen.
6	Oxygen	In ozone layer. O_2 and O_3 are **allotropes**.	$O(g) + O_2(g) \rightarrow O_3(g)$
	Sulfur	Two oxides form. Burns slowly with a blue flame.	$S(g) + O_2(g) \rightarrow SO_2(g)$ $2SO_2(g) + \frac{1}{2}O_2(g) \rightarrow 2SO_3(g)$
7	Most halogens react	Unstable oxides form.	Not usually formed by direct reaction.
0	Neon Argon	No reaction.	

As you move across periods 2 and 3, the general pattern for the oxides formed from left to right are 'ionic bonding to giant covalent structure to small covalent molecules'. The products change in nature from solids to gases and from alkaline to amphoteric to acid. This is due to the changes in bonding across the period.

> **Key terms**
>
> **Alkaline solution** – a solution with a pH above 7.
>
> **Oxidation** – loss of electrons from an atom/ion.
>
> **Allotropes** – two or more different physical forms that an element can exist in, e.g. graphite and diamond are allotropes of carbon.
>
> **Amphoteric** – substance that can act as both an acid and a base.

Products and reactivity of metals with oxygen, water, dilute hydrochloric acid and dilute sulfuric acid

Reactions with oxygen

Group 1 metals react rapidly with oxygen. Lithium, sodium and potassium are stored under oil to prevent contact with air due to this. The more reactive group 1 metals are usually stored in sealed glass tubes to ensure that no air or oxygen is present. The reactions of lithium and sodium with oxygen are shown in Table 1.5. Their oxides contain the simple ion O^{2-}. Sodium and potassium can also form the peroxide M_2O_2 containing the molecular ion O_2^{2-}. Potassium, rubidium and caesium ignite in air to form the super-oxides KO_2, RbO_2 and C_5O_2. These contain the molecular ion O_2^{2-}.

These more complicated ions are unstable near a small positive ion. The covalent bond between the two negative oxygen ions in O_2^{2-} is weak. The electrons in the peroxide ion will be attracted to a positive ion but the positive ion can polarise the negative ion.

Lithium only has a +1 charge, but it is a small ion and so it has a high charge density. This causes the peroxide ion to break into an oxide and an oxygen atom. The super-oxide ions are even more unstable and are only stable in the presence of the larger, non-polarising ions at the bottom of group 1.

Group 2 metals tend to burn in oxygen or air to form metal oxides. This is the general equation.

$$2M + O_2 \rightarrow 2MO$$

Beryllium tends to form a coating of beryllium oxide. This makes it resistant to further oxidation.

Strontium and barium will also form peroxides. This is the general equation.

$$M + O_2 \rightarrow MO_2$$

Group 3 metals react with oxygen and this is the general equation for the reaction.

$$4M + 3O_2 \rightarrow 2M_2O_3$$

Thallium will also react to produce Tl_2O. Aluminium, like beryllium, also forms an outer coating of aluminium oxide. This means that it behaves as an unreactive metal.

The group 4 metals lead and tin can also produce oxides with the formula MO and MO_2.

When d block metals react with oxygen, the oxides are often brittle. Iron oxide is rust. Some d block metals become resistant to corrosion because they quickly form an unreactive outer oxide layer that prevents any more of the metal from reacting.

Titanium oxide is an example of this. The d block metals can form a range of oxides. Transition metals are much less reactive than group 1 and 2 metals in general.

Reactions with water

Group 1 metals are called alkali metals because when they react with water they produce a basic solution. Here is the general equation.

$$2M(s) + 2H_2O(l) \rightarrow 2M^+(aq) + 2OH^-(aq) + H_2(g)$$

They react violently with water. The reaction becomes more violent as you go down the group.

Group 2 metals also produce hydroxides in the reaction with water. Here is the general equation.

$$M(s) + 2H_2O(l) \rightarrow M(OH)_2(aq) + H_2(g)$$

Magnesium only reacts with steam, while the metals below magnesium will react increasingly easily with water. Beryllium does not react with water.

Group 3 metals are not very reactive with water. Aluminium does not appear to react at all due to its outer aluminium oxide layer. Group 4, 5 and 6 metals do not react with water. Transition metals react slowly with water and some do not react at all.

Reactions with dilute acids

Metals above copper in the reactivity series can react with dilute acids to form metal salts (an ionic compound formed from a neutralisation reaction) and hydrogen. For example, magnesium reacts with dilute hydrochloric acid to give magnesium chloride and hydrogen:

$$Mg + 2HCl \rightarrow MgCl_2 + H_2$$

It reacts with dilute sulfuric acid to give magnesium sulfate and hydrogen:

$$Mg + H_2SO_4 \rightarrow MgSO_4 + H_2$$

Sodium reacts with hydrochloric acid to form sodium chloride and hydrogen:

$$2Na + 2HCl \rightarrow 2NaCl + H_2$$

This is a very violent reaction and is too dangerous to carry out in a school/college laboratory.

The reactions of calcium, strontium and barium with sulfuric acid are a little more complicated. This is because the sulfates of these metals are insoluble. They form a protective layer that prevents more of the metal reacting.

❚❚ PAUSE POINT   List the different types of metal reaction.

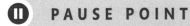   Think about the chemical substances they react with and what products they form.

   Describe how the reactions change as you go down group 1 metals.

Positions of metals in the reactivity series in relation to position in the periodic table

The reactivity series (see Figure 1.22) is a list of metals in order of how reactive they are with oxygen, acids and water. The higher a metal is in the series, the more reactive it is. This is because it has a higher tendency to lose an electron and form a complete outer shell. The more reactive a metal is, the more difficult it is to extract from its ore and the more likely it is to be found in a compound.

The most reactive metals are in group 1, as reactivity decreases across the period. It also increases down the group, which means that francium is the most reactive metal and is at the top of the reactivity series. Most reactivity series do not list francium, as it is so radioactive and unstable that it is rarely seen uncombined. Most reactivity series have potassium at the top and gold and platinum at the bottom as they are so unreactive. It can be useful to place carbon and hydrogen. Carbon is present as it can extract/displace some metals from their compounds and ores, which explains why some metals cannot be extracted from their ores by reaction with carbon. Hydrogen is present as metals below hydrogen will not react with dilute acids or water to displace hydrogen. The order of reactivity is group 1, group 2, group 3, group 4, transition metals. The key reason is that, going across the period, the nuclear charge increases, therefore it is harder to lose an electron and react.

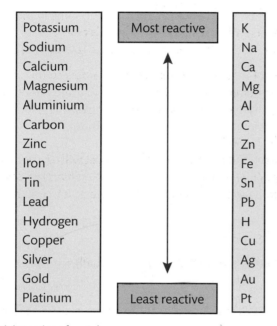

▶ **Figure 1.23:** Reactivity series of metals

Oxidation and reduction

An atom becomes an ion when it loses or gains an electron or electrons. The term **redox** refers to the transfer of electrons that occurs during chemical reactions. When atoms of an element lose electrons, it is called oxidation. For example, $Mg \rightarrow Mg^{2+} + 2e^-$ (this is a **half equation**).

When electrons are gained, it is called **reduction**. For example, $\frac{1}{2}O_2 + 2e^- \rightarrow O^{2-}$.

The two half equations together show the reaction between magnesium and oxygen:

$$Mg + \frac{1}{2}O_2 \rightarrow MgO$$

This process of electron transfer allows the reaction, oxidation and reduction to occur simultaneously.

In reactions that make ionic compounds, it is easy to see where electrons are lost or gained. You cannot write half equations for reactions where covalent compounds are formed, for example, in the formation of water. This is where **oxidation states** come in. Oxidation states of atoms in a molecule of an element are always zero. So in O_2 both oxygen atoms have an oxidation state of 0. In water the oxygen is more electronegative than hydrogen, so it has more power to attract electron density. So oxygen can be given the oxidation state of –2 as if it has taken two electrons in the covalent bond.

Key terms

Redox – the transfer of electrons during chemical reactions.

Reduction – when an atom/ion gains electrons. The phrase OIL RIG will help you remember the difference between oxidation and reduction: **O**xidation **I**s **L**oss (of electrons), **R**eduction **I**s **G**ain (of electrons).

Half equation – an equation that shows the loss or gain of electrons during a reaction.

Oxidation state – the number assigned to an element in a chemical compound. It is a positive or negative number depending on how many electrons the element has lost or gained. (Also called oxidation number.)

Each hydrogen can be said to have 'lost' an electron so will have an oxidation state of +1. Compounds and molecules always have an overall oxidation state of 0, as the oxidation numbers of all the elements in the compound will add up to 0.

So oxidation occurs when the oxidation state of an atom increases and reduction occurs when the oxidation state of an atom decreases. In the formation of water, hydrogen has a change in oxidation state from 0 to +1 and oxygen changes from 0 to –2.

Not all reactions are **redox reactions**. If the oxidation states do not change, then the reaction is not a redox reaction. For example, the reaction between hydrochloric acid and sodium hydroxide:

$$HCl + NaOH \rightarrow NaCl + H_2O$$
oxidation state   +1 –1   +1 –2 +1   +1 –1 2(1) –2

In this case, the oxidation states for each atom are the same in the reactants as in the products, so this is not a redox reaction.

Step-by-step: Assigning oxidation states
`7 Steps`

1 The oxidation state of an atom in an element is always zero. For example, in sodium, Na, it is 0 and in O_2, oxygen, it is 0.

▼

2 The oxidation state in an element or its ion is always its charge.

▼

3 The oxidation state of fluorine in a compound is always –1 as it is the most electronegative element.

▼

4 The oxidation state of oxygen is nearly always –2 (except in peroxides and FO, where it is –1 and +1).

▼

5 The oxidation state of chlorine in a compound is usually -1 unless bonded with F or O.

▼

6 The oxidation state of hydrogen is +1 unless bonded to a metal when it is –1. Group 1 metals are +1, group 2 metals are +2, aluminium is +3.

▼

7 The sum of oxidation states in a compound is always 0. In polyatomic ions, the sum of the oxidation state of each element in the formula is the overall charge.

Variable oxidation states of transition metal ions

Transition metals have variable oxidation states due to their highest energy electrons being in the d sub-shell. This is a defining property of transition elements. When a transition metal loses electrons to form a positive ion, the 4s electrons are lost first, followed by the 3d electrons. The maximum oxidation state increases as you go along the period until manganese, which has a maximum oxidation state of +7 where 2 electrons are lost from the 4s, and 5 from the 3d orbitals. There is no simple rule to predict possible oxidation states, so you may want to learn some of the common states for the commonly used elements.

Scandium and zinc only have one oxidation state when in a compound. The others in the first period have two or more. For example, iron has possible oxidation states of +2 or +3 and these are written as Fe (II) and Fe (III).

Transition metals and their compounds have a large range of uses. Most of these uses are because of their variable oxidation state. Many are used as **catalysts**. For example, iron is used in the Haber process and platinum is used in catalytic converters in cars. Some transition metal compounds are also used as catalysts. For example, vanadium (V) oxide is used in the process for making sulfur dioxide (contact process). In the decomposition of hydrogen peroxide, manganese (IV) oxide is used as a catalyst. It is oxidised by the hydrogen peroxide to form manganese (VII) oxide. This then decomposes back to manganese (IV) oxide and oxygen. This means that the catalyst is ready to be reused.

Transition metals also have the metal properties discussed previously. They are good electrical and thermal conductors and they are malleable and ductile. This makes them useful when a conducting material is needed, as well as in structural materials. They have greater strength and are less likely to corrode than group 1 metals, and so are often added to alloys to improve the properties of the material.

> **Key term**
>
> **Catalysts** – substances that increase the rate of a chemical reaction but are unchanged at the end of the reaction.

> **Theory into practice**
>
> Transition metals are extremely important in the chemical industry. They are used as a catalyst in a range of manufacturing processes.
> 1 Research one use of a transition metal as a catalyst.
> 2 Describe how the transition metal is used in the process.
> 3 Explain how the transition metal acts as a catalyst to increase productivity.

⏸ PAUSE POINT
Work out the oxidation state of chlorine in the following compound HCl, $HClO$, $NaClO_2$, $KClO_3$, ClO_2, Cl_2O_7.

Hint Think about the number of electrons lost or gained. Follow the rules above.

Extend In the following equations, which elements have been reduced, oxidised, neither oxidised nor reduced?

$$2KBr + Cl_2 \rightarrow 2KCl + Br_2$$

$$3Cu + 8HNO_3 \rightarrow 3Cu(NO_3)_2 + 2NO + 4H_2O$$

Displacement reactions of metals/halogens

A metal will displace a less reactive metal in a metal salt solution. For example, when iron is added to blue copper sulfate solution, the solution will lose its colour as iron sulfate is formed. You will also see a pink-brown metal forming as copper is displaced out of the solution. Here is the equation for the reaction.

$$Fe(s) + CuSO_4(aq) \rightarrow FeSO_4(aq) + Cu(s)$$
oxidation state   0   +2 +6 4 (−2)   +2 +6 4 (−2)   0

The iron has been oxidised and the copper has been reduced. Iron is more reactive than copper as can be seen in the reactivity series. You can predict which metals will displace which from their salts by using the reactivity series (see Table 1.6).

▶ **Table 1.6**

	Magnesium	Zinc	Iron	Copper
Magnesium sulfate	No reaction	No reaction	No reaction	No reaction
Zinc sulfate	Displacement	No reaction	No reaction	No reaction
Iron sulfate	Displacement	Displacement	No reaction	No reaction
Copper sulfate	Displacement	Displacement	Displacement	No reaction

Halogens are **oxidising agents** which means they withdraw electrons from another atom or ion. The oxidising power of a halogen decreases as you go down group 7. If chlorine reacts with potassium bromide, then the bromine will be displaced and potassium chloride will form.

$$CL_2 + 2KBr \rightarrow 2KCl + B_2$$

This is because chlorine is a stronger oxidising agent than bromine and so withdraws an electron from the bromide ion (see Table 1.7).

▶ **Table 1.7**

	Chlorine	Bromine	Iodine
Potassium chloride	No reaction	No reaction	No reaction
Potassium bromide	Displacement	No reaction	No reaction
Potassium iodide	Displacement	Displacement	No reaction

Uses and applications of substances produced within this unit

Knowing the chemical and physical properties of elements and the compounds is important to chemists when they are researching for a substance for a specific industrial application. The substances in this unit have a range of applications. Some are given below, but it would be useful for you to research more applications whenever you discuss or investigate a chemical substance.

▶ Metal and non-metal oxides have a range of applications. For example, magnesium oxide is used as a starter material for industrial processes such as producing magnesium alloys or fibreglass.
▶ Metal salts are used to make the colours in fireworks.
▶ Sodium chloride is used for many different manufacturing processes such as making glass, paper and rubber, as well as being used in water softening systems.
▶ Sulfates are used in detergents.
▶ Copper sulfate is used in water treatment to kill algae.

Discussion

The uses listed here are not exhaustive. Research further uses of the products in this unit. Can you link the uses to their physical and chemical properties? Discuss this with your group.

⏸ PAUSE POINT How do the physical properties of elements change across the periods?

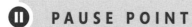 Think about reasons for trends in ionisation energy across periods 2–4 and down groups 1, 2 and 7.

 Research how physical properties and ionisation energies may affect reactions in industry. You may want to pick one specific reaction.

Further reading and resources

www.rsc.org The website of the Royal Society of Chemists.
www.sciencebuddies.org A website giving hands-on science projects.
www.virtlab.com A series of hands-on experiments and demonstrations in chemistry.

Getting started

Biology is the study of living organisms. Cells are found in all living organisms. They are the fundamental unit of structure and function in all living organisms. From single prokaryotic cells, to the millions of cells that make up animals and plants, cells are vital to life. It is essential that you understand the structure and function of cells in order for you to understand the fundamental concept of biology. See if you can list parts of a plant and animal cell. When you have completed this unit, you should be able to add more to your list.

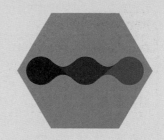

B   Structure and function of cells and tissues

B1 Cell structure and function

In this section you will learn about cell theory, microscopy, and the ultrastructure, and function of animal, plant and **prokaryotic cells**. You will also use micrographs to identify cell organelles and carry out **magnification** calculations.

Key terms

Prokaryotic cell – a cell with no true nucleus or nuclear membrane.

Magnification – the number of times larger the image appears compared to the actual size of the object being viewed.

Cell theory

Cell theory is the concept that cells are the fundamental unit of structure, function and organisation in all living organisms. Cell theory states that both plant and animal tissue are composed of cells and that cells are the basic unit of life. It also states that cells can only develop from existing cells. In 1655, the English scientist Robert Hooke used an early light microscope to observe the structure of finely sliced cork. He made observations and described what he saw as 'cells'. This was the start of the development of cell theory. Developments in microscopy meant cells could be observed in detail for the first time. Figure 1.24 shows the timeline for cell theory development.

1839:
Universal cell theory
Matthias Schleiden suggested that all plant material is composed of cells. Jan Purkyne observed that animal tissue is composed of cells and the structure is similar to plant tissue. The scientist credited for the Universal Cell Theory is Theodor Schwann, a German physiologist. He proposed that all living things are composed of cells.

1860:
Spontaneous generation disproved
Louis Pasteur demonstrated that bacteria will only grow in sterile nutrient broth after it has been exposed to air. This disproved the theory of spontaneous generation of cells.

1665:
Robert Hooke first described cells.

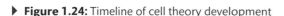

1674–1683:
The first living cell was observed
Anton van Leeuwenhoek was the first person to observe bacteria and protoctista from pond water samples, after developing powerful glass lenses.

1831:
Nucleus observed
Robert Brown, an English botanist, was the first to observe and describe the nucleus in a plant cell.

1852:
Evidence for the origin of new cells
Robert Remak observed cell division in animal cells. His findings were not accepted at the time, but in 1855 Rudolf Virchow published the findings as his own to show new cells form from existing ones.

▶ **Figure 1.24:** Timeline of cell theory development

Organelle – specialised structures found within a living cell.

Resolution – the ability to distinguish between objects that are close together.

Nucleus – an organelle found inside a cell which contains genetic information.

Mitochondria – an organelle where aerobic respiration takes place.

Chloroplast – a plant organelle where the stages of photosynthesis take place, found in plant cells, photosynthetic bacteria and algae.

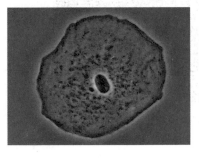

▶ Human cheek cells seen under a light microscope

Microscopy

Before microscopes were invented, people knew nothing about cells, sperm and bacteria or any other micro-organisms. Microscopes have given us the power to see these sorts of things in microscopic detail. With high-power microscopes, it is possible to observe cell **organelles**.

A microscope is an instrument that is used to magnify objects that are too small to see with the naked eye. Using microscopes to see distinct cells that make up multi-cellular organisms allows us to observe how their structure relates to their function. When hospitals receive tissue samples, the cytology department need to determine if these samples are healthy or diseased. Histopathologists will analyse the samples using microscopy. It is important that they are able to recognise what they see and record observations accurately.

Light microscopy

Light microscopes were first developed in the 16th century and continue to be improved and developed. Light microscopes use visible light and magnifying lenses to observe small objects. There are limitations to using light microscopes because they have a lower magnification and **resolution** than other, more advanced, microscopes. The maximum magnification of a light microscope is × 1500 and the maximum resolution is 200 nm. However, light microscopes do allow us to observe sub-cellular structures, known as organelles. For example, a light microscope can magnify a cell **nucleus**, **mitochondria**, and **chloroplasts** in plant cells. The image on the left shows a human cheek cell observed down a light microscope. You can clearly see the nucleus in the photograph.

Electron microscopy

Electron microscopes were first developed in the 20th century. They use a beam of electrons in a vacuum with a wavelength of less than 1 nm to visualise the specimen. They allow much more detail of cell ultrastructure to be observed and produce images called electron micrographs, with a magnification of up to × 500 000 and higher resolution, as great as 0.1 nm. Samples are stained using methylene blue for light microscopes. Radioactive salts for electron microscopy can be stained with this stain but there are others that can be used also.

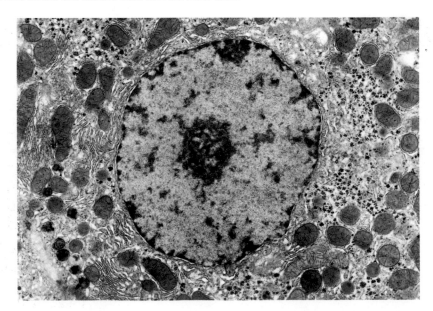

▶ Electron micrograph of animal cell

Calculating magnification

We can use the equation below to work out magnification.

$$\text{Magnification } (M) = \frac{\text{size of image } (I)}{\text{actual size } (A)}$$

The size of the image refers to the length of the image when you measure it with a ruler. Ensure that you always measure in millimetres and convert the actual size to the same units that you have measured in. You will usually be given the magnification or the actual size in the exam question. You will therefore have one unknown and you can rearrange the equation to work out the unknown answer. Always include units in your answer and place your answer on the given line in the exam question. Finally, make sure you show your working out, including the equation above.

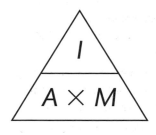

▶ **Figure 1.25:** Use this triangle to help you to rearrange the magnification equation.

Worked Example

Calculate the magnification of the image. Use the equation above to work out the magnification.
Remember to convert all units to make them the same.
1000 nanometres (nm) = 1 micrometre (µm)
1000 micrometres (µm) = 1 mm
1000 mm = 1 m

1   Use your ruler to measure the size of the image in mm. The line measures 50 mm.

2   The image states that the actual size is 50 µm. You need to convert this to mm so they are both in the same units.
$$\frac{50}{1000} = 0.05 \text{ mm}$$

3   Magnification = 50/0.05

4   50/0.05 = 1000

5   Magnification = × 1000

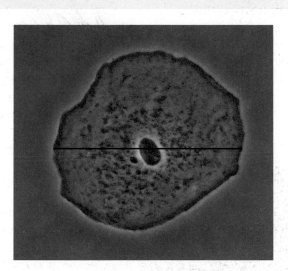

▶ Calculating magnification

Assessment practice 1.9

Work out the magnification for the diagram.
The actual size of the cell shown in the image is 200 µm.

1   Use your ruler to measure the size of the cell shown in the image in mm.

2   The actual size of the cell is 200 µm. You need to convert this to mm so they are both in the same units.

3   Put both figures into the magnification equation and work out the magnification.

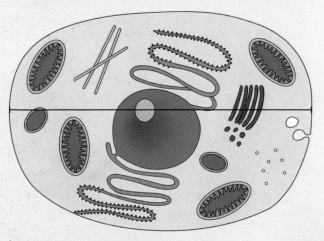

▶ Calculating the magnification of a cell

Work out the magnification of the nucleolus in the image.

1 Use your ruler to measure the size of the nucleolus in mm (size of image).

2 The actual size of the nucleolus is stated on the picture.
You need to convert this to mm so they are both in the same units.

3 Put both figures into the magnification equation and work out the magnification.

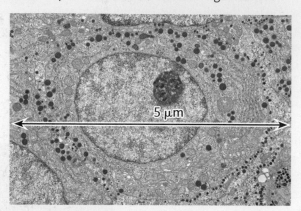

5 μm

▶ Calculating the magnification of a nucleolus

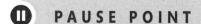

Ⅱ	**PAUSE POINT**	Can you explain the concept of cell theory?
	Hint	Close the book and see if you can produce a time line with important dates in relation to the development of cell theory.
	Extend	Think about the differences between the light microscope and the electron microscope.

Ultrastructure and function of organelles in cells

A cell is the basic unit of life. You will need to be able to recognise different types of cells when using a microscope, by observing the differences in their ultrastructure.

There are two types of cell.
▶ **Prokaryotic** cells are single-celled organisms. They are simple structures and do not have a nucleus or any membrane-bound organelles.
▶ **Eukaryotic** cells make up multi-cellular organisms such as plants and animals. They are complex cells with a nucleus and membrane-bound organelles.

> **Key term**
>
> **Eukaryotic** – an organism that contains the genetic information as linear chromosomes within the nucleus of the cells and numerous specialised organelles.

Eukaryotic cells

Eukaryotic cells are approximately 10–100 μm and the ultrastructure can be seen using an electron microscope. Chemical reactions occur in the cytoplasm of a cell. The cell surface membrane or plasma membrane separates the cell cytoplasm from the external environment. Inside the cell cytoplasm there are a number of different structures called organelles. There are a number of organelles that are common in both plant and animal cells. You will study the structure of plant cells later in this unit.

Animal cell ultrastructure

Figure 1.26 shows the ultrastructure of an animal cell.

plasma membrane – regulates the transport of materials in and out of the cell.

mitochondrion – this is the site of aerobic respiration.

centrioles – take part in cell division, they form spindle fibres that move chromosomes during cell division.

nucleus – contains genetic informationa and controls/regulates metabolic cell activity.

nucleolus – dense spherical structure inside the nucleus that produces ribosomes and RNA.

Golgi apparatus – here newly made proteins are modified and then packaged into vesicles.

rough endoplasmic reticulum (ER) – has ribosomes attached; it synthesises and transports proteins.

vesicle – these transport materials around the cell or out of the cell.

cytoplasm – where metabolic reactions take place.

ribosomes – responsible for protein synthesis when attached to ER.

lysosomes – they are vesicles that contain hydrolytic enzymes. They break down waste material inside the cell.

smooth endoplasmic reticulum (ER) – flattened cavities surrounded by a thin membrane which do not have anything attached. They synthesise carbohydrates and lipids.

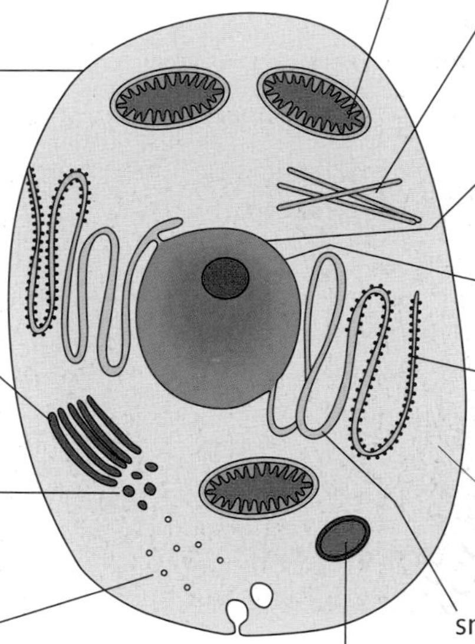

▶ **Figure 1.26:** Ultrastructure of an animal cell with organelles labelled

Table 1.8 describes the structure and function of organelles in an animal cell.

▶ **Table 1.8:** Structure and function of animal cell components

Organelle	Description of structure	Function
Plasma membrane	Composed of a phospholipid bilayer, with proteins embedded in the layer.	The membrane is selectively permeable and regulates the transport of materials into and out of the cell. Separates cell contents from the outside environment.
Cytoplasm	Cytoplasm is a thick, gelatinous, semi-transparent fluid.	The cytoplasm maintains cell shape and stores chemicals needed by the cell for metasolic reactions.
Nucleus	The nucleus is the largest organelle and is surrounded by a nuclear envelope. The envelope has nuclear pores which allow the movement of molecules through it. The nucleus contains chromatin.	The nucleus controls/regulates cellular activity and houses genetic material called chromatin, DNA and proteins, from which comes the instruction for making proteins.
Nucleolus	Dense spherical structure in the middle of the nucleus.	The nucleolus makes RNA and ribosomes.
Rough endoplasmic reticulum (ER)	Network of membrane bound flattened sacs called cisternae studded with ribosomes.	Protein synthesis takes place on the ribosomes and the newly synthesised proteins are transported to the Golgi apparatus.
Smooth endoplasmic reticulum (ER)	Network of membrane bound flattened sacs called cisternae. No ribosomes.	Responsible for synthesis and transport of lipids and carbohydrates.
Golgi apparatus	A stack of membrane bound flattened sacs.	Newly made proteins are received here from the rough ER. The Golgi apparatus modifies them and then packages the proteins into vesicles to be transported to where they are needed.

Organelle	Structure	Function
Vesicles	Small spherical membrane bound sacs with fluid inside.	Transport vesicles are used to transport materials inside the cell and secretory vesicles transport proteins that are to be released from the cell, to the cell surface membrane.
Lysosomes	Small spherical membrane bound sacs containing hydrolytic enzymes.	They break down waste material including old organelles.
Ribosomes	Tiny organelles attached to rough ER or free floating in the cell. They consist of two sub-units and they are not surrounded by a membrane.	Protein synthesis occurs at the ribosomes.
Mitochondria	They have two membranes. The inner membrane is highly folded to form cristae. The central part is called the matrix. They can be seen as long in shape or spherical depending on which angle the cell is cut at.	They are the site of the final stages of cellular respiration.
Centrioles	They are small tubes of protein fibres.	They form spindle fibres during cell division.

Function of animal cells

One of the key functions of a cell is to synthesise proteins for use inside the cell, to lead to cell multiplication and for secretion out of the cell, for example, insulin. Proteins are synthesised on ribosomes attached to rough endoplasmic reticulum. The newly synthesised proteins are transported through the cisternae of the rough ER and packaged into vesicles. They are transported to the Golgi apparatus, where vesicles fuse with the surface of the Golgi apparatus and the proteins enter. It is here that the newly synthesised proteins are modified and then packaged into vesicles. Secretory vesicles will transport proteins that are to be released from the cell to the cell surface membrane. They will fuse with the membrane and release the protein by **exocytosis**.

> **Key term**
>
> **Exocytosis** – process of vesicles fusing with plasma membrane and secreting contents.

Plant cell ultrastructure

Plant cells have all the cellular components that are listed in the animal cell except centrioles (see Table 1.8). However, plant cells have additional structures and centrioles because their main function is to produce carbohydrates during photosynthesis. (See Table 1.9.)

▶ **Table 1.9:** Structure and function of plant cell components

Plant cell structure	Structure	Function
Cell wall	Made of cellulose forming a sieve-like network.	Protects and supports each cell and the whole plant.
Chloroplast	Has a double membrane and is filled with a fluid called stroma. The inner membrane is a continuous network of flattened sacs called thylakoids. A stack of thylakoids is called a granum (grana is plural). Grana contain chlorophyll pigments.	Site of photosynthesis. Light energy is trapped by the chlorophyll and used to produce carbohydrate molecules from water and carbon dioxide.
Vacuole	Membrane-bound sac in cytoplasm that contains cell sap.	Maintains turgor to ensure a rigid framework in the cell.
Tonoplast	The partially permeable membrane of the vacuole.	Selectively permeable to allow small molecules to pass through.
Amyloplast	A double membrane-bound sac containing starch granules.	Responsible for the synthesis and storage of starch granules.
Plasmodesmata	Microscopic channels which cross the cell walls of plant cells.	Enable transport and communication between individual plant cells.
Pits	Pores in the cell walls of the xylem.	Allow water to enter and leave xylem vessels.

Figures 1.27 and 1.28 show the ultrastructure of a plant cell.

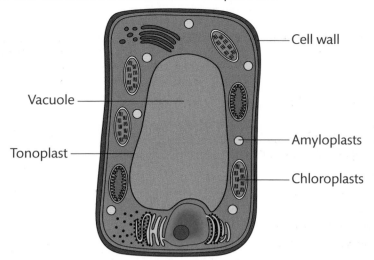

▶ **Figure 1.27:** Ultrastructure of a plant cell

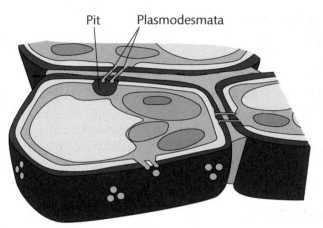

▶ **Figure 1.28:** Ultrastructure of a plant cell with plasmodesmata and pits, adapted from *Biology*, 7 ed. (Raven, P., 2005) Figure 5.5 p.84 McGraw-Hill Education

 **PAUSE POINT** Can you list all the organelles present in a eukaryotic cell?

Hint Think about the differences and similarities between a plant and animal cell.

Extend Find different images on the Internet of plant and animal cells, and identify the organelles.

Prokaryotic cell (bacteria) ultrastructure

Prokaryotes are single-celled micro-organisms that are much smaller than eukaryotic cells. They are generally 1–5 μm in diameter. They are simple in structure, with no **membrane-bound organelles** and fewer organelles. Their **DNA** is not contained in a nucleus. Table 1.10 lists the parts of a prokaryotic cell, and their structures and functions.

> **Key terms**
>
> **Membrane-bound organelles** – organelles surrounded by a phospholipid membrane. For example, lysosomes and Golgi apparatus.
>
> **DNA** – deoxyribonucleic acid, the hereditary material in cells.

Organelle	Structure	Function
Cell wall	Prokaryotic cells are surrounded by a cell wall made of peptidoglycan.	Protects and supports each cell.
Capsule	Slippery layer outside the cell wall of some species of bacteria.	Protects the cell and prevents dessication.
Ribosomes	Smaller than ribosomes found in eukaryotic cells. They consist of two sub-units and they are not surrounded by a membrane.	Protein synthesis occurs at the ribosomes.
Nucleoid	The nucleoid (meaning nucleus-like) is the irregularly shaped region that holds nuclear material without a nuclear membrane and where the genetic material is localised. The DNA forms one circular chromosome.	The nucleoid is the region where generic information can be found and controls cellular activity.
Plasmid	Small loops of DNA.	Plasmids carry genes that may benefit the survival of the organism.

Figure 1.29 shows the ultrastructure of a prokaryotic cell with flagellum.

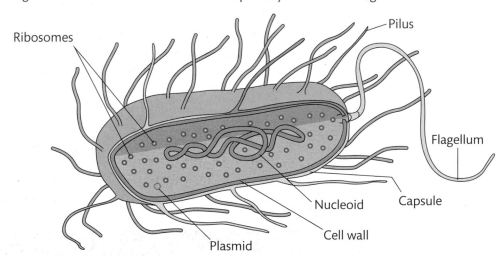

▶ **Figure 1.29:** Ultrastructure of a prokaryotic cell

Ribosome size is determined by their ability to form sediment in a solution. Eukaryotic ribosomes are determined as 80S, whereas prokaryotic cell ribosomes are smaller and are 70S.

Function of bacterial cells

Bacterial cells produce and secrete toxins that have an effect on other organisms. DNA is free in the cytoplasm of a prokaryotic cell in the area called the nucleoid. A section of DNA containing a genetic code for a metabolite unwinds and hydrogen bonds break. RNA nucleotides line up (**complementary base pairing**). Messenger **RNA** is formed. This process is known an transcription. The next process is the production of the bacterial protein. This is called translation and it occurs at the ribosomes. Transcription and translation can occur simultaneously because the genetic material is free in the nucleoid surrounded by ribosomes. The newly made protein/toxin is moved to the surface membrane ready to be secreted to cause infection. Note that many bacteria are beneficial to humans and to eukaryotes.

> **Key terms**
>
> **Complementary base pairing** – the way in which nitrogenous bases in DNA pair with each other. Adenine (A) always bonds with Thymine (T) (or Uracil (U) in mRNA) and Guanine always bonds with Cytosine.
>
> **RNA** – ribonucleic acid, a molecule with long chains of nucleotides.

> **Link**
>
> Go to *Unit 11: Genetics and Genetic Engineering Learning aim A* to find more information about the DNA base pairing rule and about transcription and translation.

Classifying bacteria as Gram positive or Gram negative

It is important that microbiologists can correctly identify bacteria that cause infections to enable them to decide the most effective treatment.

Gram stain

Hans Christian Gram, a Danish microbiologist, developed a staining technique to distinguish between two groups of bacteria:

▶ Gram positive
▶ Gram negative.

Both types of bacteria have different cell wall structures and respond differently to antibiotics. Penicillin stops the synthesis of the cell wall on growing Gram-positive bacteria, but it does not have the same effect on Gram-negative bacteria. Gram-negative bacteria have a thinner cell wall and two lipid membranes.

During the staining technique, two stains are added to the bacterial smear: crystal violet and safranin. If you see a purple stain when observing the smear under a microscope, it means that Gram-positive bacteria are present. If the smear has retained the pink safranin stain, this indicates that Gram-negative bacteria are present. This is because their thinner cell walls and lipid membranes allow ethanol (applied during the method) to wash off all the crystal violet purple stain and to then retain the pink safranin stain.

 PAUSE POINT   Do you know the functions of both animal and bacterial cells?

    Hint   Think about the products made by each cell.

    Extend   Can you think what might go wrong if these processes were interrupted?

Assessment practice 1.11

Produce revision cards on all the organelles you would find in an animal cell, plant cell and bacterial cell.
- On one side, write the name of the organelle.
- On the other side, write its function.

Assessment practice 1.12

Copy and complete the following table to show **three** ways in which prokaryotic and eukaryotic organisms **differ** in the **structure** of their cells.

Prokaryotic	Eukaryotic
1	....
2	....
3..	....

Copy and complete the table below, then put a tick or a cross in each box to indicate whether the feature is present or absent.

Feature	Cell type		
	Plant	Animal	Bacteria
Mitochondria			
Chloroplast			
Cellulose cell wall			
Nucleus			
Ribosome			

B2 Cell specialisation

Many organisms are multi-cellular, meaning that they are made from billions of cells. It is important that cells within these organisms become specialised for different roles with particular functions. Multi-cellular organisms in higher animals and higher plants are organised as follows:

- ▶ specialised cells
- ▶ tissues
- ▶ organs
- ▶ organ systems
- ▶ organism.

Cell specialisation: structure and function

Palisade mesophyll cell

Palisade mesophyll cells found in leaves are rectangular box-shaped cells that contain chloroplasts (see Figure 1.30). The chloroplasts are able to absorb a large amount of light for photosynthesis. They also move around in the cytoplasm in order to maximise the amount of light absorbed. These cells are closely packed together and form a continuous layer in the leaf. Palisade cells are surrounded by a plasma membrane and a cell wall made of cellulose. This helps to protect the cell and keep it rigid. They also have a large vacuole to maintain **turgor** pressure (the plasma membrane pushes against the cell wall of the plant to maintain its rigid structure).

Key term

Turgor – rigidity of plant cells due to pressure of cell contents on the cell wall.

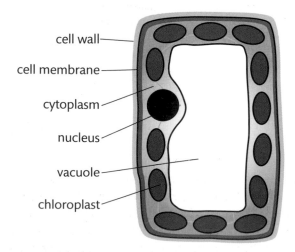

cell wall
cell membrane
cytoplasm
nucleus
vacuole
chloroplast

▶ **Figure 1.30:** Basic structure of a palisade mesophyll cell found in leaves

Root hair cell

These cells are found at a plant's roots, near the growing tip (see Figure 1.31). They have long hair-like extensions called root hairs. The root hairs increase the surface area of the cell to maximise the movement of water and minerals from the soil into the plant root. The cells have thin cellulose walls and a vacuole containing cell sap with a low **water potential**. This encourages the movement of water into the cell.

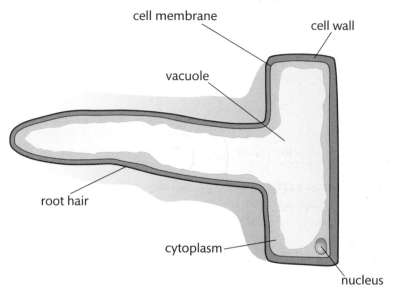

▶ **Figure 1.31:** Basic structure of a root hair cell in plants

Sperm cells

Sperm cells are male **gametes** in animals (see Figure 1.32). They have a tail-like structure called a undulipodium so they can move. They also contain many mitochondria to supply the energy needed for this movement. In human sperm, the mid-piece of the tail is 7 µm long and the end is approximately 40 µm in length. The sperm head is 3 µm wide and 4 µm long. It is made up of an acrosome, which contains digestive enzymes. These enzymes are released when the sperm meets the egg, to digest the protective layer and allow the sperm to penetrate. The sperm's function is to deliver genetic information to the egg cell or ovum (female gamete). This is fertilisation.

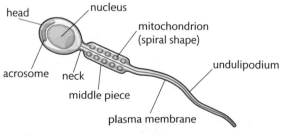

▶ **Figure 1.32:** Sperm cell

Egg cells

Egg cells, or ova, are the female gametes in aminals (see Figure 1.33). An egg cell is one of the largest cells in the human body, and is approximately 0.12 mm in diameter. It contains a nucleus, which houses the genetic material. The zona pellucida is the outer protective layer/membrane of the egg. Attached to this is the corona radiata, which consists of two or three layers. Its function is to supply proteins needed by the fertilised egg cell.

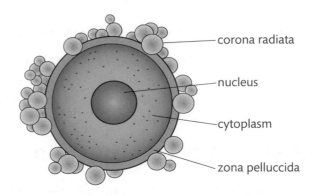

▶ **Figure 1.33:** Basic structure of an ovum

Red blood cells

Red blood cells or erythrocytes are a biconcave shape (where both sides concave inwards, see Figure 1.34). This increases the surface area to volume ratio of an erythrocyte. They are flexible so that they can squeeze through narrow blood capillaries. Their function is to transport oxygen around the body. In mammals, erythrocytes do not have a nucleus or other organelles. This increases space for the **haemoglobin** molecules inside the cell that carry oxygen.

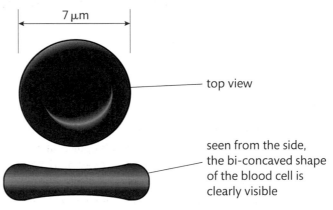

▶ **Figure 1.34:** Erythrocyte

White blood cells

Neutrophils are a type of white blood cell and they play an important role in the immune system (see Figure 1.35). They have multi-lobed nuclei, which enables them to squeeze though small gaps when travelling to the site of infection. The cytoplasm holds lysosomes that contain enzymes that are used to digest **pathogens** that are ingested by the neurophil.

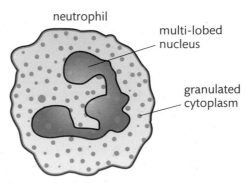

▶ **Figure 1.35:** Basic structure of a neutrophil

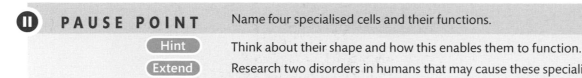

PAUSE POINT   Name four specialised cells and their functions.

> Hint   Think about their shape and how this enables them to function.
>
> Extend   Research two disorders in humans that may cause these specialised cells to change in structure and therefore not be able to carry out their function efficiently.

B3 Tissue structure and function

A collection of differentiated cells that perform a specific function is called a tissue. There are four main tissue types in animals:

- epithelium
- muscle
- connective (supports, connects or separates different types of tissues and organs in the body)
- nervous.

Epithelial tissue

Epithelial tissues are found lining organs and surfaces. Epithelial tissues can be divided into different types:

- squamous epithelial tissue
- columnar epithelial tissue
- endothelium tissue.

Squamous epithelial tissue

Simple squamous epithelial tissue is a lining tissue and is one cell thick (see Figure 1.36(a)). It is made from flattened specialised squamous epithelial cells. These cells form a thin, smooth, flat layer. This makes them ideal when rapid diffusion is necessary. They line various structures. An example is the alveoli in the lungs, which provide a short diffusion pathway to allow rapid diffusion of oxygen into the blood and carbon dioxide into the lungs.

Epithelium cells can be damaged by smoking. Smoking irritates and causes inflammation and scarring in the epithelium tissue of the lungs. The alveoli walls become thicker due to scarring and produce more mucus. The damage to the air sacs causes emphysema and the lungs lose their natural elasticity. This causes:

- breathlessness
- persistent coughing
- phlegm.

These symptoms are all associated with Chronic Obstructive Pulmonary Disorder (COPD).

Ciliated columnar epithelial tissue

Ciliated columnar epithelium tissue is made up of column-shaped **ciliated cells** with hair-like structures called cilia covering the exposed cell surface (see Figure 1.34(b)). Ciliated epithelium line the trachea in the respiratory system in order to protect the lungs from infection. They do this by sweeping any pathogens away from the lungs. Goblet cells are column shaped and are also present in the respiratory tract. They secrete mucus to help trap any unwanted particles that are present in the air that you breathe. This protects your lungs because it prevents bacteria reaching the alveoli.

> **Key term**
>
> **Ciliated cells** – cells with tiny hair-like structures.

(a)

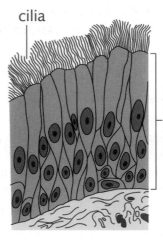

(b) cilia

pseudo-stratified epithelial layer

▶ **Figure 1.36:** Epithelial tissue

Endothelial tissue

Endothelial tissue consists of a layer of flattened cells, one layer thick. It is found lining the heart, blood vessels and lymphatic vessels (vessels that make up the lymphatic system). The cells provide a short diffusion pathway for the movement of various substances, such as:

▶ products of digestion into blood capillaries

▶ blood plasma and tissue fluid in and out of blood capillaries.

There are a number of risk factors that can cause damage to the endothelium. Carbon monoxide and high blood pressure can damage the inner lining of the arteries. White blood cells repair the damage and encourage the growth of smooth muscle and the deposition of fatty substances such as cholesterol under the endothelium lining of arteries, not on the surface. This process of deposition is called atherosclerosis. These deposits, called atheromas, may build up enough to break through the inner endothelial lining of the artery, eventually forming plaque in the **lumen** of the **artery**. This reduces the size of the lumen and restricts blood flow.

Key terms

Lumen – the space inside a structure.

Artery – blood vessel that carries blood away from the heart.

Assessment practice 1.14

Copy and complete the table below, and compare two types of epithelium: squamous and ciliated. For each type, state one function and one specific location in the human body where it can be found.

Type of epithelium	Function of tissue	Specific location in the human body
Squamous		
Ciliated		

Muscle tissue

Muscles are composed of cells that are elongated and form fibres. Muscle cells contain protein filaments called actin and myosin that enable muscles to contract and cause movement.

There are three types of muscle tissue:

▶ **Skeletal** muscle is found attached to bones. You can control its contraction and relaxation, and it sometimes contracts in response to reflexes.

▶ **Cardiac** muscle is found only in the heart. It contracts at a steady rate to make the heartbeat. It is not under voluntary control.

▶ **Smooth** muscle is found in the walls of hollow organs, such as the stomach and bladder. It is also not under voluntary control.

Skeletal muscle fibre

Muscle tissue needs to be able contract (shorten in length) in order to move bones. In a muscle, cells join up to make muscle fibres. These are long strands of cells sharing nuclei and cytoplasm, which is known as the sarcoplasm. Inside the muscle cell cytoplasm are many mitochondria, specialised endoplasmic reticulum known as sarcoplasmic reticulum and a number of microfibrils. Each muscle fibre is surrounded by a cell surface membrane called the **sarcolemma**.

Skeletal muscle shows a stripy/banding appearance under a microscope. Skeletal muscle is made up of thousands of muscle fibres. Each muscle fibre is made up of myofibrils.

> **Key terms**
>
> **Sarcolemma** – cell membrane of a striated muscle cell.
>
> **Myofibril** – basic rod-shaped unit of muscle cell.

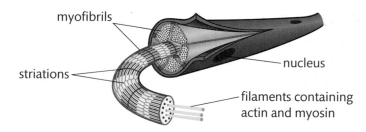

▶ **Figure 1.37:** Structure of a muscle fibre

Myofibril fibres are made from proteins called myofilaments, which enable contraction to take place because of the contractile nature of the proteins in the filament. They appear as different coloured bands: A-band and I-band (you can remember which is which as A-bands are d**A**rk and I-bands are l**I**ght).

Sarcomere

The span from one z-line to the next in Figure 1.38 is known as the sarcomere. When the muscle is relaxed, this is approximately 2.5 μm in length. This length reduces when the muscle contracts because the I-band and H-zone lengths are reduced. The A-band does not change in length during contraction.

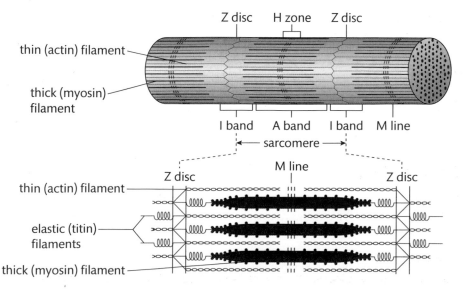

▶ **Figure 1.38:** Muscle fibre microscopic structure

There are two protein filaments found in muscle cells. This filament made of actin and thick filaments made of myosin. During muscle contraction, the thin actin filaments move and overlap the thick myosin filaments. The sarcomere shortens, decreasing the size of the overall muscle.

There are two types of muscle fibres: slow twitch and fast twitch. These fibres influence how muscles respond during physical activity. Human muscles contain a mixture of both.

Slow twitch muscle fibres

Slow twitch muscles are more effective at using oxygen to generate energy in the form of **ATP**, for continuous and extended muscle contractions over a long time. These fibres help marathon runners and endurance cyclists to continue for hours. Slow twitch fibres have:

▶ less sarcoplasmic reticulum

▶ more mitochondria for sustained contraction

▶ more myoglobin

▶ a dense capillary network.

These fibres release ATP slowly by **aerobic respiration**.

Fast twitch muscle fibres

Fast twitch muscle fibres can be divided into two different kinds.

▶ Fast twitch oxidative muscle fibres are similar in structure to slow twitch muscle fibres. They contain many mitochondria, myoglobin and blood capillaries, but they are able to **hydrolyse** ATP much more quickly and therefore contract quickly. They are relatively resistant to fatigue.

▶ Fast twitch glycolytic muscle fibres have relatively less myoglobin, few mitochondria and few capillaries. They contain a large concentration of **glycogen** that provides fuel for **anaerobic respiration**. They contract rapidly but also fatigue quickly.

> **Key terms**
>
> **ATP** – adenosine triphosphate, an enzyme that transports chemical energy within cells for metabolism.
>
> **Aerobic respiration** – respiration with oxygen.
>
> **Hydrolyse** – a chemical reaction involving breaking down a compound with water.
>
> **Glycogen** – many glucose molecules bonded together and stored in the liver and muscles.
>
> **Anaerobic respiration** – respiration without oxygen.
>
> **Dendrons** – extension of a nerve cell.

Nervous tissue

The central nervous system (CNS) consists of the brain and spinal cord. It is made up of billions of non-myelinated nerve cells and longer, myelinated axons (axons with myeline sheath) and **dendrons** that carry nerve impulses. Nervous tissue is made of nerve cells called neurons.

Neurons

Neurons are cells that receive and facilitate nerve impulses, or action potentials, across their membrane and pass them onto the next neuron. They consist of a large cell body called a soma with small projections called dendrites and an axon. The end of the axon is called the axon terminal. It is separated from the dendrite of the following neuron by a small gap called a synapse.

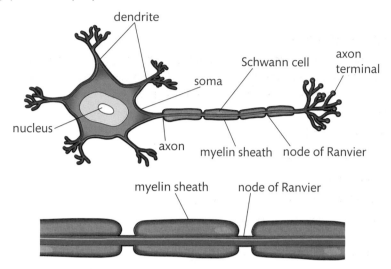

▶ **Figure 1.39:** Myelinated neuron

Information travels along neurons in the form of electrical signals called nerve impulses. A nerve impulse is known as an action potential. Action potentials arise from a change in the ion balance in the nerve cell which spreads rapidly from one end of the neuron to the other. Neurons are bundled together to form nerves and nerves form a network all around the body. When the action potential travels to the axon terminal, neurotransmitters (chemicals) are released across the synapse and bind to the post-synaptic receptors, continuing the nerve impulse in the next neuron.

Sensory neurons receive information from receptors, for example, ears, and take this information to the CNS. The brain processes the information, then motor neurons take the information from the brain to the effector, for example, muscle.

Resting potential and action potential

Resting potential is the term given to a neuron that is not transmitting an action potential and is at rest. However, the neuron is actively transporting sodium and potassium ions across its membrane to maintain a negative potential in the interior of the cell compared to the outside. The membrane is more permeable to potassium ions, so for every three sodium ions actively transported across the membrane, only two potassium ions are actively transported. This process requires energy in the form of ATP. The cell membrane is described as polarised. At rest, the gated sodium ion channels in the membrane are closed and it is a sodium/potassium pump that is used to transport the sodium and potassium ions across the membrane creating a potential difference of −60 **mV**.

When a nerve impulse is stimulated by a receptor cell or another neuron, an action potential is generated. The neuron is always ready to conduct an impulse. The axon membrane is polarised, which means that the fluid on the inside is negatively charged with respect to the outside. An action potential is the electrical potential which results from the process of ions moving across the neuron cell membrane when the correct channels are open in response to a stimulus, causing the inside of the neuron to be more positive than the outside.

Link

Go to *Unit 9: Human Regulation and Reproduction Learning aim A* to find more information about the nervous system.

Key term

mV – millivolts, a small voltage/potential across a cell membrane.

Depolarisation

At rest, there are more positive ions outside the neuron. When an action potential is generated, there is a quick change in the permeability of the axon membrane that spreads down the whole neuron as a wave of depolarisation (see Figure 1.40). The voltage across the membrane changes. A small number of voltage gated sodium channels detect this change, and open to allow a few sodium ions to diffuse into the axon. The membrane depolarises and becomes less negative than the outside, with a potential difference of -50 mV (threshold). If the stimulus is large enough and this threshold is reached, then the rest of the sodium gated channels open to allow rapid diffusion of sodium ions into the axon, making the inside positively charged in comparison to the outside, with a potential difference of +40 mV across the membrane. Sodium ion channels close and potassium channels open. Potassium ions therefore diffuse out of the cell. This makes the inside of the axon negatively charged again. This is called repolarisation and it restores resting potential.

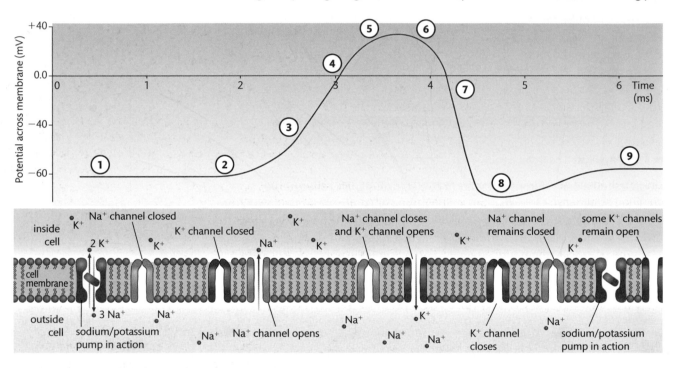

▶ **Figure 1.40:** Depolarisation

Myelinated neurons and saltatory conduction

Some neurons have an axon covered with a fatty sheath called myelin (see Figure 1.39). Myelin is made from specialised cells called Schwann cells that wrap themselves around the axon when they develop in an embryo. The Schwann cells are thick and form a lipid insulating layer around the neuron called the myelin sheath. This insulates the axon and makes the action potential travel faster.

Saltatory conduction happens only in myelinated nerves and it greatly increases the speed of the action potential. The myelin sheath insulating the axon means that ion exchange can only occur at the **nodes of Ranvier** (see Figure 1.39) that are in between the Schwann cells, where the axon membrane is exposed and not covered with Schwann cells.

Saltatory conduction is the process of the signal jumping (saltatory comes from the Latin *saltare*, meaning 'to dance'). When the action potential reaches a node of Ranvier, sodium ions diffuse into the axon membrane. They displace the potassium ions down the axon because they are both positively charged, and like charges repel. The movement of the potassium to the node further down the axon makes the next node more positive and depolarises it until the threshold is reached. The impulse quickly jumps from node to node, making the action potential quicker. Only a small part of the axon is being used, so less ATP is needed and fewer ions are being exchanged.

The speed of an action potential in humans

The speed at which a nerve impulse travels in humans is 1 to 3 m/s in unmyelinated fibres and 3 to 120 m/s in myelinated fibres. The speed of travel (conduction) depends on:

▶ axon diameter – the larger the axon, the faster the conduction

▶ myelination of neuron – the nerve impulse travels faster if the neuron is myelinated

▶ number of synapses involved – the fewer synapses there are to cross, the faster the communication.

Synapses

When the nerve impulse reaches the end of the neuron, it must cross a gap called a synapse (see Figure 1.41) to get to the next neuron or the effector cell. A nerve impulse crosses the synapse in the form of a chemical transmitter called a neurotransmitter. Neurotransmitters diffuse across the synapse and initiate an action potential in the neuron at the other side. The presynaptic neuron ends in a swelling called the synaptic bulb and it contains many mitochondria as ATP is needed. The neurotransmitters are stored in temporary vesicles in the synaptic bulb that can fuse with the surface to release the neurotransmitters into the synapse. They also contain voltage-gated calcium ion channels.

There are hundreds of neurotransmitters. The most common neurotransmitter is acetylcholine and synapses that have this as their transmitter are called cholinergic synapses. Acetylcholine molecules are released by exocytosis and they diffuse across the cleft. The acetylcholine molecules bind to the receptor sites on the sodium ion channels in the postsynaptic neuron to generate a new action potential.

Step by step: Chemical transmission across the synapse   8 Steps

1 The action potential arrives at the synaptic bulb.

2 Calcium channels open in the presynaptic membrane. Calcium ions diffuse into the neuron membrane down a concentration gradient.

3 As the calcium concentration increases, the synaptic vesicles containing neurotransmitters move towards the presynaptic membrane.

4 The vesicles fuse with the membrane and release the chemicals into the synaptic cleft.

5 Neurotransmitters diffuse across the synaptic cleft. This is known as synaptic delay because it is slower than an electrical signal travels.

6 The neurotransmitter binds to the postsynaptic cell membrane receptor sites on the sodium channels.

7 Some neurotransmitters open sodium channels in the membrane, causing sodium ions to pass in. This creates a excitatory postsynaptic potential (EPSP) and makes the membrane receptive to the signals coming in. If this reaches the threshold, the action potential is generated.

8 The neurotransmitter will excite the cell and, once it has acted on the membrane, enzymes act on the neurotransmitter to break them down.

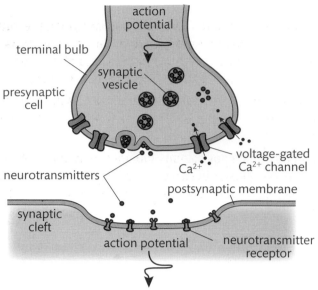

▶ **Figure 1.41:** A synapse

An electroencephalograph (EEG) is a test that looks at the activity of the brain cells. When the brain is working, nerve impulses travel from one cell to another. These produce electrical signals that can be picked up by detectors attached to a person's skull. The detectors send the signals that they detect to a recorder. The recorder produces a graphical trace which can be interpreted to see whether there is any abnormal activity, such as that which may suggest that an epileptic seizure has occurred. An electrocardiogram (ECG) detects electrical signals in the heart. This can show whether the heart is working properly.

Problems that can occur

Parkinson's disease is a **genetic** disease that affects the nervous system. Parkinson's sufferers are not able to produce the naturally occurring chemical dopamine, a neurotransmitter that helps smooth and normal movements. Without this, people show symptoms of:

▶ slow movement

▶ speech problems

▶ tremors when moving

▶ poor balance.

The drug, L-dopa, replaces the dopamine that is lost in people with Parkinson's disease.

Serotonin is another of the body's naturally occurring neurotransmitters. It is normally active in the brain and can cause problems if it is not produced. Some forms of depression are caused by a reduced amount of serotonin in the brain.

Ⅱ PAUSE POINT   Synapses are an integral part of the nervous system. Outline the role of the synapse in the nervous system.

Hint   Draw a synapse and add commentary to show the function of the synapse on the nervous system.

Extend   Research beta blockers and Prozac™ and discuss their effect on the synapse.

Further reading and resources

Boyle, M. and Senior, K. (2008). *Human Biology* (third edition). Collins Educational (ISBN 9780007267514).

Kennedy, P. and Sochacki, F. (2008). *OCR Biology AS*. Oxford: Heinemann Educational (ISBN 9780435691806).

Getting started

Modern communications involve technology: phone, email, radio and TV, social media. In scientific work, modern technology is used to make and record observations as well as to analyse and share them: spectroscopy, endoscopy, data-logging, satellite imaging and so on.

All these communications and measurement technologies depend on ways in which waves behave. To get the best out of them, you need to understand waves.

What do you already know about waves? How many kinds of wave can you think of and picture? Write a list of the main terms you know for describing waves. When you have completed this unit, see if you can add to that list.

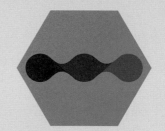

C Waves in communication

C1 Working with waves

Waves generally start with a disturbance – for example, wind blowing across the surface of the sea, or a stone being thrown into a pond. The energy imparted by that disturbance causes a regular repeating motion, backwards and forwards or up and down, which is called an **oscillation**.

Oscillations, period and amplitude

Examples of oscillations are the pendulum of a clock, a child on a swing, a weight bouncing on a spring. Often, as in the suspension of a car, our aim is to try to damp down the oscillations as soon as possible. These oscillations are not themselves waves because they do not travel anywhere, but like waves they do have a **frequency** and **periodic time**, which describe the rate at which the oscillation repeats itself.

In an oscillation, something is displaced from its rest position, but it also has a tendency to bounce back. In a physical oscillation, like the examples above, the **displacement** is a distance moved by something from rest. But, for example, in an electrical oscillation, the displacement would be a change of voltage or of current going regularly up and down in value. In either case, we measure the size of an oscillation by its **amplitude**.

How can an oscillating system sometimes produce a wave motion?

Wave motion

Waves transfer energy from one place to another, but without causing any net movement of material.

The energy transfer depends on the way an initial oscillating system is connected to its surroundings. If that connection can carry some energy from that first oscillation and transfer it to a similar system next to it, then that system will also start oscillating. However, that second oscillation will not be quite in time with the first one.

> **Key terms**
>
> **Oscillation** – a regularly repeating motion about a central value.
>
> **Frequency** – $f = \dfrac{1}{T}$ – i.e. the number of whole cycles occurring in one second. (Symbol: f; SI unit: Hertz, Hz.)
>
> **Period** (or 'periodic time') – the time taken for one whole cycle of an oscillation, i.e. before the motion starts to repeat itself. (Symbol: T; SI unit: s.)
>
> **Displacement** – how far the quantity that is in oscillation has moved from its mean (rest) value at any given time. (Symbol and unit: various according to what the quantity is that is oscillating.)
>
> **Amplitude** – the maximum value of displacement in the oscillation cycle – always measured from the mean (rest) position.

Have you experienced a heavy person sitting down next to you on a springy sofa or bed? As they sit down at one point, you, some distance away, may find yourself going up. If it is a very springy seat, then you might both find yourselves bouncing a couple of times before the oscillation dies away. If you watch children jumping on a bouncy castle, or two gymnasts sharing a trampoline, you might see something similar happening. This is the start of a wave.

When a wave transfers the energy of an oscillation, it takes time. So a short distance away, though a similar oscillation happens, it is delayed in time. If you travel with the wave for one whole **wavelength**, you will find another place where the oscillation does once again occur exactly in time with the first one. In fact, it is now delayed by one whole cycle – that is, the time delay is equal to the oscillation period. Two such points along the wave are said to be 'in phase' with one another. (We will explore the idea of phase a bit more in the section below about graphs of waves.)

Wave speed

A wave travels one wavelength during its periodic time. So that means you can calculate its speed, v, as wavelength, λ, divided by periodic time, T. However, instead of the periodic time, frequency is more commonly used, f, where $f = \frac{1}{T}$. Frequency is measured in cycles per second or Hertz (Hz). It is often easier to use frequency, because periodic times are usually tiny fractions of a second. The faster the oscillations, the larger is the value of the frequency. Using frequency, you can rewrite the equation for the speed of a wave as:

$$v = f\lambda$$

Graphical representation of wave features

One example of an oscillating system is a piston in a motor car engine. The piston is connected to a rotating crankshaft, and that is what (via the transmission system) drives the vehicle's wheels in circular motion (see Figure 1.42). One complete oscillation of the piston corresponds to one whole turn of the crankshaft. They both have the same periodic time (and frequency).

So the mathematics of oscillation and of circular motion are closely connected. Figure 1.43 shows you how. This is why the graphs you draw of oscillations and waves are typically sine waves. The sine is a mathematical function of the angle through which you can imagine a crankshaft turning to drive the motion.

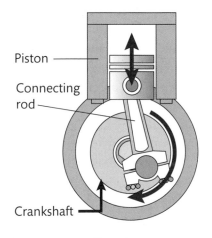

Piston

Connecting rod

Crankshaft

▶ **Figure 1.42:** Piston in a motor car engine.

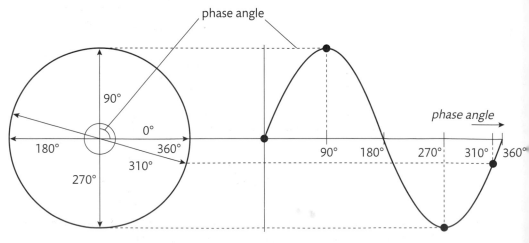

▶ **Figure 1.43:** How a rotating vector generates a sine wave

You can use this idea of the angle generating a cycle of oscillation when you compare two wave motions that are not in phase with one another. The **phase difference** is usually given as an angle, where 360° (or 2π radians) equates to a whole cycle – a shift equivalent to one wavelength in distance or one period in time.

▶ **Table 1.11:** Phase angles in degrees and radians, and also compared to wavelengths

Wave cycle position	Start	$\frac{1}{4}$ of a cycle	$\frac{1}{2}$ of a cycle	$\frac{3}{4}$ of a cycle	1 whole cycle	1.5 cycles	2 whole cycles
Phase/°	0	90	180	270	360	540	720
Phase/rad	0	$\frac{\pi}{2}$	π	$\frac{3\pi}{2}$	2π	3π	4π
Number of wavelengths	0	$\frac{\lambda}{4}$	$\frac{\lambda}{2}$	$\frac{3\lambda}{4}$	λ	$\frac{3\lambda}{2}$	2λ

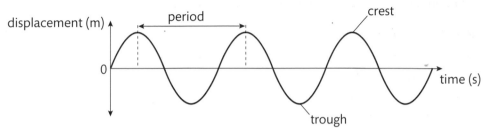

▶ **Figure 1.44:** Graph showing how the displacement varies over time at one fixed position in space as a wave travels past

> **Key term**
>
> **Phase difference** – the difference in phase angle between two waves of the same frequency and wavelength, where 360° (2π radians) represents a single whole cycle of the waveform.

Look at the two graphs in Figures 1.44 and 1.45. They represent the same wave motion, but there is too much going on in it for everything to be captured in one still picture. So one graph concentrates on the changes happening at a fixed point in space, while the other is for a fixed point in time and shows how the wave extends through space.

The quantity represented on the vertical axis of both these wave graphs is the displacement of whatever is oscillating from its rest value. So for different kinds of wave, this will be a different physical property.

▶ For water ripples it is the water level.

▶ For sound waves it is microscopic movements of molecules linked to pressure variations.

▶ For light, radio and other electromagnetic waves it is an oscillating electric field.

Waves create a pattern of displacements in space as well as in time. So, taking a snapshot in time, you can picture a wave using a graph of displacement against position along the direction in which the wave travels (propagates). See Figure 1.45.

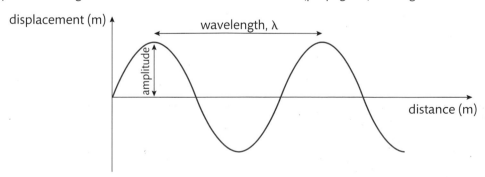

▶ **Figure 1.45:** The same wave, but showing how displacement varies with distance along the wave in the direction of its travel: a snapshot at one fixed point in time

On this graph, one whole cycle along the horizontal distance axis marks out a wavelength. Note that moving forward along the wave in distance is equivalent to moving backwards in time so far as the phase of the oscillations at that point is concerned.

PAUSE POINT

Close the book and try using graphs to explain some of the key wave terms: displacement, period, phase, frequency, wavelength, amplitude. Draw a rotating vector and wave diagram using a pair of compasses and some graph paper.

Hint　　On the time axis mark off the wave period, allowing space on the axis for two whole cycles. Mark phase angles along the same axis. Use measurements of displacement taken from the generating circle to help you plot out the points.

Extend　　What happens to the shape of the wave if you double the frequency?

What other type of graph can you draw in order to show the wavelength?

Types of wave motion: transverse and longitudinal

When the displacement occurs in the same direction that the wave travels, for example, in a sound wave, it is a longitudinal wave. By contrast, in a transverse wave the displacement is at right angles to the direction of propagation of the wave, for example, water ripples and electromagnetic waves (see Figure 1.47).

Transverse waves are easy to picture because they look like the sine wave graphs you normally draw, with displacement on the vertical axis and the distance travelled by the wave plotted horizontally. (See Figures 1.45 and 1.47.)

However, in a longitudinal wave, the different displacements of particles along the direction in which the wave is propagating, lead to a series of compressions (where particles are packed closer together) and rarefactions (where they are further apart). You can create a longitudinal wave in a spring by making it oscillate along its length (see Figure 1.46). A soft 'slinky' spring is best for seeing clearly the compressions and rarefactions that travel down its length.

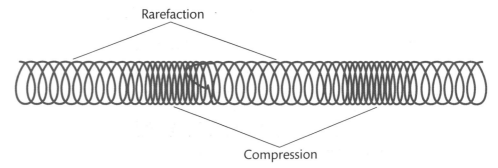

▶ **Figure 1.46:** Compressions and rarefactions travelling down a spring

Because of this, sound and other longitudinal waves are sometimes described as pressure waves: oscillations in pressure travelling through a solid or fluid medium (see Figure 1.48).

Earthquakes and other seismic events below the earth's surface generate two types of shock wave: a longitudinal 'pressure' wave and a transverse 'shaking' wave. They travel at different speeds and so will each arrive at different times, making earthquakes quite complex events to study.

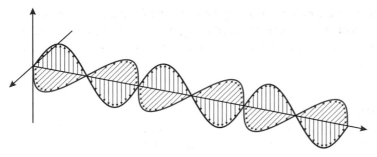

▶ **Figure 1.47:** Electromagnetic waves are transverse in three dimensions: so some can be 'polarised' in one plane and some in another plane at 90° to the first one

<table>
<tr><td>

Theory into practice

Because electromagnetic waves are transverse, they can be polarised. The direction of displacement plus the direction of propagation define a plane of polarisation. Passing electromagnetic radiation through a polarising filter will remove all the components that have oscillations in the plane at 90° to that of the filter. Polarising sunglasses work because when sunlight is passed through a polarising filter, half of sunlight's intensity is removed. As light that has been reflected tends to be mostly polarised in one plane, polarising sunglasses are great for cutting out glare from reflective surfaces.

Explain why there cannot be an equivalent to this for sound waves. What is different about them?

</td></tr>
</table>

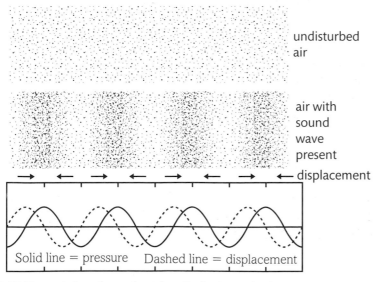

undisturbed air

air with sound wave present

displacement

Solid line = pressure   Dashed line = displacement

▶ **Figure 1.48:** Transmission of sound as a longitudinal wave in air

Diffraction gratings

Diffraction

Diffraction is a key characteristic of all waves. It means the tendency of a wave to spread out in all directions, transferring energy to its surroundings as it does so.

When a wave is moving straight forward, for example, in a beam of light, you can picture moving wave-fronts. Each is at right-angles to the direction of propagation (i.e. the direction of travel and energy transfer).

These wave-fronts will also be straight lines. In three dimensions they form a flat plane. (See Figure 1.49.)

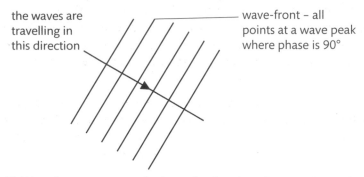

the waves are travelling in this direction

wave-front – all points at a wave peak where phase is 90°

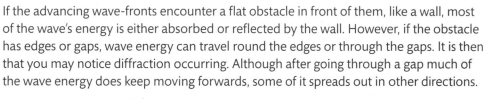

▶ **Figure 1.49:** Wave-fronts are perpendicular to the direction of a ray, and show the points where the wave oscillations have the same phase

▶ Ripple tank photo showing diffraction through a gap

If the advancing wave-fronts encounter a flat obstacle in front of them, like a wall, most of the wave's energy is either absorbed or reflected by the wall. However, if the obstacle has edges or gaps, wave energy can travel round the edges or through the gaps. It is then that you may notice diffraction occurring. Although after going through a gap much of the wave energy does keep moving forwards, some of it spreads out in other directions.

A good way to see this effect is to use water ripples generated in a specially designed glass-bottomed tray (a 'ripple tank'). By shining a light downwards through the water in the tray onto a horizontal white screen or sheet of paper, the moving ripples can be clearly seen as bright lines. If obstacles are introduced with edges, or with gaps about whose size is a few times the wavelength of the ripples, you can observe diffracted ripples with curved wave-fronts, even though the original ripples had straight-line wave-fronts.

Every point along a wave-front has oscillations and energy. So each point on the wave-front can act as a secondary source of circular ripples spreading out in all directions. In this way, all the little secondary ripples add together to make a straight (or plane) wave-front keep moving forward as a straight line, until it meets an obstacle.

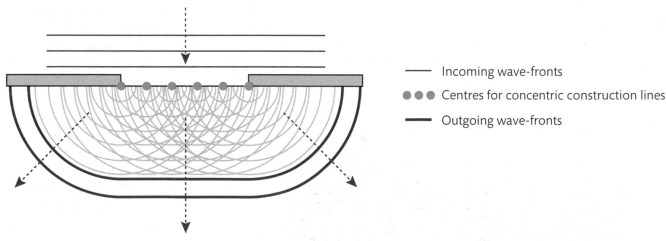

—— Incoming wave-fronts

●●● Centres for concentric construction lines

—— Outgoing wave-fronts

▶ **Figure 1.50:** Huygens' construction correctly predicts diffraction through a gap

The Dutch mathematician and scientist, Christiaan Huygens, developed a geometrical construction to predict the shape of waves in water (see Figure 1.50). In 1678, he was the first to apply this wave-front propagation principle to light and he showed that wave theory could explain all the behaviour of lenses and mirrors. However, it was Thomas Young's experiments on diffraction and interference of light in 1801 that finally convinced the whole scientific community to use a wave motion theory for light.

Gratings

A diffraction grating is a flat plane object. It has a series of regular lines formed on it that block parts of an advancing wave-front. For microwaves, these lines could be a series of regularly spaced metal bars or wires. For light you would usually use a piece of glass with a series of very fine and regularly spaced scratches on its surface. You can get a similar effect for X-rays by using the regularly spaced rows of atoms or molecules in a crystal.

When a wave-front meets a diffraction grating, some of the wave's energy continues propagating forward through the gaps between the grating lines. This is **transmission**. Some more of the wave's energy may be absorbed in the grating itself, but the remainder of the energy is scattered backwards as a **reflection**.

> **Key terms**
>
> **Transmission** – wave energy passing through an object, e.g. a diffraction grating, and mostly continuing forward in the original direction, though some energy will be diffracted through angles of less than 90°.
>
> **Reflection** – wave energy that bounces off a surface and has its direction of travel altered by more than 180°.

Think just about the forwards (transmission) direction.

The grating creates a set of regularly spaced secondary sources, where each gap in the grating allows energy through. Each gap generates a set of circular wave-fronts spreading out from that location. The spacing, d, between the grating lines is very important. If it is close in size to the wavelength, λ, of the incident radiation (the incoming waves) then the grating will produce an **interference pattern** of regularly spaced bright and dark lines (fringes). These correspond to strong intensity created at certain angles after diffraction and no wave energy at all at certain other intermediate angles.

The interference pattern is due to **superposition** of the waves from the separate, regularly spaced sources, coming through the gaps in the diffraction grating. Wherever the **path difference** between waves from adjacent sources works out to be a whole number, n, of wavelengths, then the displacements due to waves from all the separate sources will be in phase. They add together constructively to give a bright spot of high intensity of radiation at that point. The same effect can be clearly demonstrated with water ripples (see below).

> **Key terms**
>
> **Interference pattern** – a stationary pattern that can result from the superposition of waves travelling in different directions, provided they are **coherent**.
>
> **Coherent** – literally means 'sticking together' and is used to describe waves whose superposition gives a visible interference pattern. To be coherent, waves must share the same frequency and same wavelength and have a constant phase difference.
>
> **Superposition** – the adding together of wave displacements that occurs when waves from two or more separate sources overlap at any given location in space. The displacements simply add mathematically.
>
> **Path difference** – the difference in length between two (straight line) rays, e.g. one from a particular grating gap to a given point in space and the ray from the next-door grating gap to the same point.

▶ Superposition of waves from two separate sources demonstrated in a ripple tank

Assessment practice 1.15

The graph shows two waves of the same frequency passing through a single point in space.

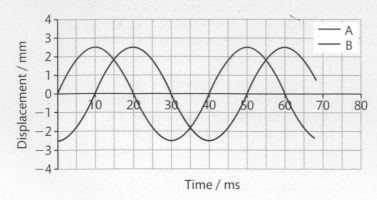

a) What is the phase difference between wave A and wave B?

b) Sketch a graph of the resultant wave motion at that point in space and determine:
   i) its amplitude
   ii) its phase angle relative to wave A.

Where the path difference between waves from adjacent gaps in the grating works out to be half a wavelength – or any odd number of half wavelengths – the interference between the waves will be destructive. That means that the wave displacement due to the wave energy coming from one grating gap is completely out of phase with that coming from the next-door gap. They have a phase difference of 180° (π radians) so that whenever one has a positive displacement, its neighbour has an equal but negative displacement. This means they cancel one another out and the resulting wave intensity is zero – a dark spot.

Bright positions of high intensity radiation do not only occur in the straight ahead 'transmission beam' direction. They also occur at any angle, ϑ, to the transmission beam direction that satisfies the equation:

$$n\lambda = d \sin\vartheta$$

where n = order of diffraction, λ is wavelength, d is grating spacing and ϑ is diffraction angle.

The bright interference line at the angle where $n = 1$, i.e. the where path difference is one wavelength, is called first order diffraction. Where $n = 2$, it is second order diffraction and so on.

If the radiation consists of a mixture of waves of different wavelengths, for example, white light, then the transmission beam ($n = 0$, so $\vartheta = 0$) has all the wavelengths in one place and so is still 'white'. However, for the first and higher orders of diffraction the angle, ϑ, will vary with the wavelength, λ, and as a result you will see a separated (or 'dispersed') spectrum of different coloured lines. The separation of the lines is greater for the second order diffraction compared with the first, but the intensity of the lines lessens as the angle ϑ increases.

It is easy to imagine and to demonstrate diffraction by a grating in transmission mode, as described above, for water ripples and for radio waves or microwaves. Diffraction of light can also be done with a grating in transmission mode. However, for light, and even more for X-rays and γ-rays, it is more common to use reflection mode.

▶ Coloured spectra produced by diffraction of white light by a grating

Gratings in reflection mode

In reflection mode, instead of looking at what comes through a grating, you look at the part of the wave energy that is bounced back off the grating surface. Once again, because the grating lines are regularly spaced, an interference pattern is produced. The geometry and equations are just the same as for transmission, but on the opposite side of the grating. That is because you are still looking at a regularly spaced set of secondary sources of waves, but now it is the reflected wave-front from the grating surface between each line that counts.

The advantage of reflection mode is that you do not have to worry about the transparency of the grating. So long as the wave energy hitting the grating 'lines' gets lost, it does not matter whether it has been absorbed or transmitted. It is the reflectivity of the grating's surface that counts for giving a strong, measurable signal.

Coherent light sources

Wave theory is used to describe light (and all the other kinds of electromagnetic radiation) because it gives a good explanation of basic light properties you observe, notably diffraction and refraction, but also reflection.

However, when light is emitted from or absorbed by matter, you can only explain what happens by thinking of light as being composed of tiny particles or 'packets of energy', which are called **photons**.

So, when thinking about the coherence of light, you have to combine ideas from wave theory with the idea of individual photon particles – what is called 'wave-particle duality'. To try to visualise this, you can use the concept of 'wave packets' or 'wave trains' (see Figure 1.51). These are snippets of wave motion of a given frequency that start up from zero with a growing amplitude and then die away again.

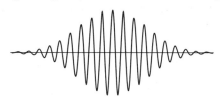

▶ **Figure 1.51:** The displacement-time and displacement-distance graphs for a photon wave packet might look like this – a very short burst of waves or 'wave train'

More evidence about the nature of light and about coherence comes from interference, for example, the patterns produced by diffraction gratings. You can only demonstrate interference between beams of light that were produced from the same source but which have subsequently been split up, for example, by:

▶ mirrors
▶ passing through a double slit (Young's experiment)
▶ the lines of a diffraction grating.

However, if the path difference between the beams gets larger than a certain size – called the coherence length – then interference fringes fade and disappear. That can happen if the spacing of a double slit is gradually increased. Similarly, if the phase difference is increased beyond a value equivalent to a time known as the coherence time, by adjusting the position of a beam-splitting mirror, again the interference pattern disappears.

So, for light and other electromagnetic waves, the conditions for coherence go beyond the simple frequency and phase requirements that apply for other waves like water ripples or sound.

Key terms

Photon – a **quantum** of electromagnetic radiation. Photons have zero mass and zero charge, but a definite energy value linked to their frequency.

Quantum – the smallest unit that can exist independently. A quantum has clearly defined values of energy, mass, charge and other physical quantities.

No detector is fast enough to directly record the frequencies and phase relationship of two light waves. So you cannot directly measure whether or not they are coherent. Instead, you have to base your understanding of light wave/particle packets (i.e. photons) on the results of interference experiments.

Photons produced from separate sources are emitted randomly at different instants, and it seems they do not generally overlap in time enough to give interference. So they must be described by bursts of waves of very short duration. Measuring the energies of individual photons, emitted at different moments from the same source, gives a small spread of energies (bandwidth) and hence also of frequencies – another obstacle to coherence.

It is because they have a long coherence length that all the photons produced in the same pulse from a laser are able to pack so closely and to travel together in such a tightly directed beam.

But if you direct light beams from two different LED sources, or even from two separate lasers, at the same screen, you never see interference patterns occurring. They are not coherent.

PAUSE POINT

What *is* light really? Is it a wave motion, a stream of particles – or can it be both at the same time? Try out the idea of wave trains to see if you can explain the conditions under which interference will or will not occur.

Hint

Sketch two wave trains. Show them overlapping by different amounts along the horizontal axis. Think about how superposition will work and what the resultant wave form will look like. Then explain what coherence length means.

Extend

a) Why do displacement-time and displacement-distance graphs have a similar shape?

b) A source like an LED emits photons with very slightly differing frequencies. What effect will that have on the coherence time?

Emission spectra

The **quantum theory** of light and other electromagnetic radiations is based on the experimental observation that there is a simple relationship between the frequency, f, of the radiation and the energy, E, carried by each photon:

$$E = hf$$

where h is the Planck constant, $-6.626\,070 \times 10^{-34}$ Js. That constant of proportionality between energy and frequency has been very precisely measured and experiments indicate it is universal.

> **Key term**
>
> **Quantum theory** – combines ideas from wave motion and particle mechanics theories to create a new 'wave mechanics'. At the sub-atomic level all the particles – protons, neutrons, electrons, photons, etc. – also behave like waves (e.g. they can be diffracted). When they are bound into an atom or molecule, these particles behave like stationary waves with a fixed wavelength and energy.

If a chemical element or compound is vaporised by heating in a flame, or if you pass an electric current at high voltage through a gas, you typically see light emitted of a characteristic colour, according to the chemical nature of the material you are testing.

When you look at the spectrum of that light, by splitting it up using a prism or a diffraction grating, what you see is a number of bright, coloured lines at definite frequencies. This is an emission spectrum. Each line in the spectrum matches to photons all emitted with very nearly the same frequency – and therefore they also each have virtually the same energy.

You can explain these lines by thinking about the possible **energy levels** for electrons in the atom or molecule concerned. When a gas or vapour is cool, most of the atoms/ molecules will be in the **ground state** where their orbital electrons are in the lowest energy state possible. Just as water naturally flows downhill, things in general tend to gravitate towards the lowest energy state possible.

However, in a hot or electrically excited gas or vapour, many of the outer electrons get knocked into higher energy levels. Then, as they begin to drop back down to lower levels and eventually back to the ground state, they have to lose energy.

De-excitation occurs one electron at a time, at randomly unpredictable instants in time. Electrons in highly excited states (i.e. those in higher energy levels) may make the journey back down to the stable ground state in two or three jumps. Rather like the balls in an arcade game of bagatelle, they may spend some time resting in one of the intermediate energy levels. Each transition from a higher to a lower energy level means that a specific amount of energy (the difference between the two energy levels) has to be lost by the electron.

That extra energy is emitted as a photon of light. Because the energy levels are fixed, the energy differences between them are always the same and are typical of the particular atom or molecule in which the electron is bound. The gas samples that are investigated, for example, by passing an electric current through them in a discharge tube, contain vast numbers of atoms. So there are always many atomic electrons making these de-excitation transitions, and you see a spectrum of lines (light frequencies) corresponding to all the transition paths they can take. Even for hydrogen, the simplest atom of all with just one electron, there are several energy levels and even more possible electronic transitions between them – so there are lots of lines in the hydrogen spectrum.

Key terms

Energy level – one of the fixed, allowed values of energy for an electron that is bound in an atom, or for a proton or neutron that is bound in a nucleus.

Ground state – the lowest energy state possible for a given bound particle.

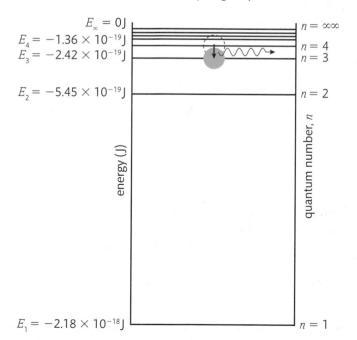

▶ **Figure 1.52:** Electron de-excitation in a hydrogen atom – a photon of a specific frequency is emitted

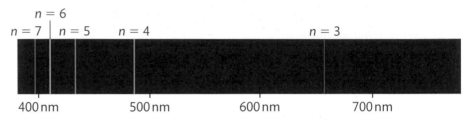

▶ Hydrogen visible emission line spectrum (Balmer series)

Worked Example

What is the frequency and wavelength of the photon emitted in Figure 1.52?

The electron drops from level $n = 4$ down to level $n = 3$.

The energy difference between the levels (the energy lost by the electron) must equal the energy of the emitted photon.

From the diagram, the energy level difference, ΔE, is from -1.36×10^{-19} J down to -2.42×10^{-19} J.

$$\Delta E = E_3 - E_4 = -2.42 \times 10^{-19} - -1.36 \times 10^{-19} = -1.06 \times 10^{-19} \text{ J}$$

The fact that this energy value is negative indicates that energy is given out here.

But ΔE = the photon energy = hf

So $f = \dfrac{\Delta E}{h}$

$$f = \frac{1.06 \times 10^{-19} \text{ J}}{6.63 \times 10^{-34} \text{ Js}} = 1.60 \times 10^{14} \text{ Hz}$$

$c = f\lambda$ where c is speed of light, 3×10^8 m s^{-1}

$\lambda = \dfrac{c}{f}$

$$\lambda = \frac{3.00 \times 10^8 \text{ m s}^{-1}}{1.60 \times 10^{14} \text{ Hz}} = 1.875 \times 10^{-6} \text{ m} = 1875 \text{ nm, so this spectral line will appear in the infra-red.}$$

Ⅱ PAUSE POINT Test your understanding of spectra by investigating further the hydrogen spectrum. In what parts of the electromagnetic spectrum would you find lines corresponding to the highest energy electron transitions?

Hint Reading off data from Figure 1.52, use the method shown in the worked example to also calculate frequencies for the transitions $n = 3$ to $n = 2$ and from $n = 2$ to $n = 1$.

Extend The hydrogen spectrum contains several series of lines (for example, Lyman, Balmer, Paschen, Brackett) in different parts of the spectrum. Is it the upper or the lower energy level of an electron transition that determines to which series it belongs? Why?

Stationary waves resonance

In a **stationary wave** (or standing wave), energy is stored rather than transferred to other locations. Oscillations of different amplitudes occur along the length of the wave in a pattern that does not change over time. Points of minimum (ideally zero) amplitude are called **nodes** and occur at every half-wavelength along the wave's extent. Intermediate between the nodes are **antinodes** – points of maximum amplitude. You can think of a stationary wave as being made from two coherent travelling waves that pass through each other in opposite directions (see Figure 1.53).

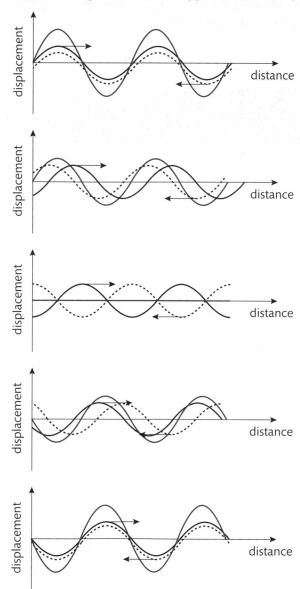

> **Figure 1.53:** When two coherent waves pass through each other, the resultant wave (shown by a red line) is a stationary wave pattern with nodes and antinodes

Stationary wave patterns most often occur in resonators, where the wave motion is confined in a fixed space. The resonator has boundaries that prevent the wave progressing further and reflect its energy back. So you can picture a travelling wave starting down in one direction, hitting the boundary and being reflected back along its own path. The result of superposing these two waves of more or less equal amplitude, but travelling in opposite directions, is a stationary wave pattern. You can demonstrate this in a laboratory by causing vibrations in a stretched string.

Key terms

Stationary waves (or standing waves) – wave motions that store energy rather than transferring energy to other locations.

Nodes – points along a stationary wave where the displacement amplitude is at a minimum (ideally zero).

Antinodes – points of maximum amplitude that occur halfway between each pair of nodes.

Resonance – the storing of energy in an oscillation or a stationary wave, the energy coming from an external source of appropriately matched frequency.

Forcing frequency (or driving frequency) – the frequency of wave energy from an external source that is coupled to a resonator. Efficient energy transfer into the resonator only occurs when this is close to one of the natural frequencies.

Natural frequency – a resonator has a series of natural frequencies (or 'modes' or 'harmonics'), each of which corresponds to an exact number of half wavelengths fitting within its boundaries.

The resonator will also have a mechanism for interacting with and absorbing travelling wave energy from outside itself. Small amounts of energy collected over a period of time can be stored up in the stationary wave and build up a much larger amplitude oscillation. This effect is **resonance**. It happens when the wave energy coming in from outside has a **forcing frequency** equal or very close to a **natural frequency** of the resonator.

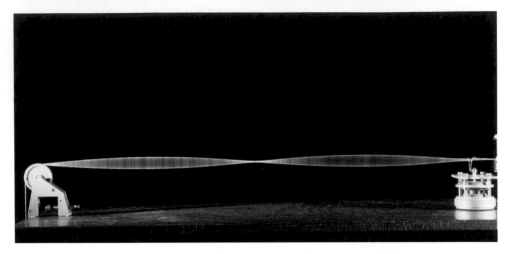

▶ Stationary waves in a stretched string can be demonstrated with a vibration generator, a pulley and weights to adjust the tension

A resonator can be set into stationary wave motion by a sudden impact that transfers a large amount of energy in an instant, which is then stored in the wave. This is the case for a gong or bell. Alternatively, resonators can be excited continuously by an external vibration, which is the case for most musical instruments other than percussion instruments.

Musical instruments

Both stringed and wind instruments depend on resonance to produce their musical notes.

In a stretched string, the oscillations are transverse, and the speed, v, at which waves travel down its length, L, depend on the string tension, T, and on the string's mass, m, per unit length, μ $(= \frac{m}{L})$. The wave speed can be calculated using the formula:

$$v = \sqrt{\frac{T}{\mu}}$$

The fixed ends of the string are nodes because they are points of no vibration. So the harmonic with the lowest frequency (the fundamental harmonic) has just one antinode between those nodes and a wavelength equal to twice the length of the string (i.e. $\frac{\lambda_1}{2} = L$). Using the wavelength and the wave speed, you can determine the frequency, f_1, of this fundamental harmonic.

But there are lots of other harmonics with higher frequencies that will fit neatly onto the same stretched string, with first one and then more extra nodes appearing between those at either end. Any number of half wavelengths can be fitted into the string's length, giving harmonics that make the string resonate at frequencies: f_1, $2f_1$, $3f_1$, $4f_1$, etc. The smooth, rich tone of a violin and other stringed instruments is due to the fact that bowing the string excites not just the fundamental harmonic frequency, but also smaller amounts of vibration at all these higher harmonics.

Tubes can also act as resonators, and are the basis of all wind instruments.

An open-ended cylinder (pipe) will naturally have maximum oscillations (i.e. oscillation antinodes) at both its open ends, and one or more nodes in between. This gives a set of harmonic frequencies similar to those for a stretched string. However, you cannot control the speed of the waves, which is just fixed at the speed of sound in air, so to tune a pipe you need to alter its length.

Another useful way to look at sound vibrations in pipes is to think about the wave pressure rather than the oscillation of the molecules of air. Pressure is always 90° out of phase with molecule displacements, so wherever there is an oscillation antinode it corresponds to a pressure node – and vice versa. So open-ended pipes have a pressure node at each end.

However, if you open a hole in the side of a pipe, you can reduce the pressure there to near atmospheric and so force a pressure node (oscillation antinode) to occur there. The woodwind family of instruments operates in just that way – opening and closing holes in the side of the pipe to alter the stationary wave pattern produced inside. The recorder, the flute, the piccolo and many organ pipes all operate as open-ended pipes, with the sound vibrations being excited by blowing air over a sharp edge.

A closed-ended cylinder instrument actually has one end closed but the other open to the air to let the sound out. That makes for a displacement node at the closed end with an antinode at the open end and gives lower musical notes: the fundamental harmonic will have just a quarter wavelength fitting in the pipe.

1st Harmonic

▶ **Figure 1.54:** Fundamental harmonic of an open-ended pipe

1st Harmonic   3rd Harmonic
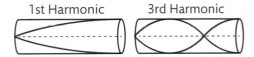

▶ **Figure 1.55:** Harmonics of a pipe with its left-hand end closed

---

⏸ PAUSE POINT   Brass instruments are closed-ended cylinders, in which different harmonic notes can be blown by the players' altering the pressure and vibration of their lips on the mouthpiece. Which harmonics are possible in a closed-ended instrument?

Hint   Sketch the first few harmonics that will fit in a pipe, giving a node at one end and antinode at the other end.

Extend   Why do brass instruments sound harsher than stringed or woodwind instruments? Think about how the wavelengths and frequencies of the brass instruments' harmonics compare with that of the fundamental harmonic. Which ratios of harmonic are missing?

▶ Making music using stationary waves resonance

Other applications of stationary waves

Radio and TV antennas have a reflector element that bounces the incoming waves back and creates a stationary wave pattern. The detector is placed at an antinode position for the particular wavelength of radiation the aerial has been designed to pick up.

In microwave ovens, stationary wave patterns, caused by reflections from the metal sides of the oven, cause hot and cold spots corresponding to antinodes and nodes, hence the need for a turntable.

Bound electrons in atoms and molecules behave like stationary waves, bouncing around in the space they are restricted to by the attraction of the nuclear positive charge. The discrete energy levels that electrons can occupy each correspond to a stationary wave pattern. Wave patterns with higher numbers of nodes correspond to higher energy levels.

Assessment practice 1.16

The free section of a stretched guitar string is 70 cm long and it produces a fundamental harmonic with a frequency of 450 Hz.

a) What is the wavelength of the fundamental vibration in the string?

b) What is the speed of the waves travelling up and down this string?

c) If the mass per unit length of this string is 0.001 kg/m, what must be the tension in the string?

d) Explain why the waves travelling up and down the string produce a stationary wave pattern at that frequency but not at lower or slightly higher frequencies.

e) What is the next higher frequency at which a stationary wave could be formed in this string?

C2 Waves in communication

Fibre optics have become a vital backbone for modern communication systems. In this section you will learn how and why they work, and about their importance in scientific investigations.

The principles of fibre optics

To understand fibre optics, you need first to know about refractive index and total internal reflection.

The laws of refraction

Light (or electromagnetic radiation of other frequencies) travels best through a vacuum. Its rapidly oscillating electric field generates an oscillating magnetic field, and the changing magnetic field in turn generates another nearby oscillating electric field. And so the wave progresses rapidly through space.

When the waves have to travel through matter, their progress is impeded by the electronic charges in the atoms and molecules. Metals, which are full of freely moving electrons, just stop the wave oscillation completely. So the light wave energy is reflected back, just as a sound wave is reflected back at the fixed end of a vibrating string. Metals therefore look shiny and make good mirrors. Many other materials absorb some or all of the light and so look coloured or even black.

In transparent materials, like water, glass and many plastics, the waves are not stopped or absorbed, but they are slowed down. The ratio of the speed of light in vacuum, c, to its speed in the material medium, v, is called the **refractive index**, n, of the medium. That is:

$$n = \frac{c}{v}$$

Key term

Refractive index – of a transparent medium is the ratio of the speed of light in vacuum to its speed in the medium.

Discussion

'Refraction' of a marching column: Try acting out this 'thought experiment' with some colleagues.

You are marching or walking holding hands with your friends in a line. On a good hard surface you can all keep the same speed, so the line stays straight and moves directly forwards.

Then suppose the person on your left finds herself reaching the edge of the paved area and stepping into long grass. As she slows down she pulls back and twists the line. Then you enter the grass too, and you also slow down.

In which direction is the marching line turned?

If there were several rows of marchers, what would happen to the distance between the marching lines as they crossed onto the grass?

Figure 1.56 shows some wave-fronts of light waves entering a smooth-sided glass block at an angle:

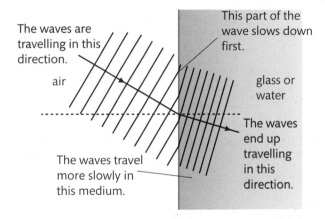

▶ **Figure 1.56:** Light wave-fronts refracted as they enter a glass block

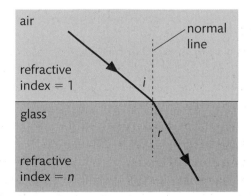

▶ **Figure 1.57:** Defining refraction

Normal line – a line at right angles to the surface of a transparent medium (e.g. glass or water) that passes through the point where a ray enters or exits that medium. The direction of rays is always described by measuring the angle between the ray and the normal line.

Incidence – the direction of the incoming ray.

Refraction – means bending of the direction of travel, so it describes the direction of an outgoing ray after bending.

In a similar way to the thought experiment with a marching column of people, the light wave-fronts are turned as they slow down in the glass. The wave travels more directly into the glass block – that is, closer to an imaginary line drawn at right-angles to the surface of the block, which is called the **normal line**.

If, instead of wave-fronts, you just draw light rays, the diagram looks like Figure 1.57.

You can label and measure two important angles: the angle of **incidence**, i, and the angle of **refraction**, r.

When you do the mathematics of this geometry, it turns out that:

$$n = \frac{c}{v} = \frac{\sin i}{\sin r}$$

where n is the refractive index of the glass, v is the speed of light in the glass, and c represents the speed of the light just before it entered the glass, which is actually its speed in air but that is almost the same as its speed in vacuum, hence the use of the letter c. So this equation describes how the speed change on entering the glass is what determines the change in direction of the ray – that is, how much it is refracted (bent) from its incident angle.

Research

Carry out your own practical investigation of refraction, for example, using a light ray box and a glass block with parallel sides. Mark the path of the rays as they enter and leave the block and construct the ray inside the block by joining up the points of entry and exit. Measure the angles of the rays. (Remember: always draw a 'normal' line perpendicular to the side of the glass block, and measure the angles the rays make with that.)

What happens to the equation for refraction when you apply it to a ray leaving the glass block? (Hint: The angles i and r seem to change places.) Can you explain why that is?

Experiment with using larger angles of incidence for the incoming light ray. What happens to the intensity of the outgoing ray? Can you spot where the light has gone?

Total internal reflection – calculation of critical angles

When the light ray comes to the other side of the glass block and tries to leave, back out into the air, or ideally into a vacuum, the light wave will speed up. That makes the wave-front turn back towards the direction in which it was travelling before it entered the glass.

However, speeding up is yet another change, and it is not that easy, particularly if the direction change is substantial. So some of the wave energy gets reflected back inside the glass. This is **internal reflection**. The larger the angles involved, the more light is reflected and the less energy gets through in the refracted beam to escape from the glass into the air.

Key term

Internal reflection – when a wave that is already in an optically dense medium (e.g. glass) hits the boundary with a less dense medium (e.g. air or water) and energy is reflected back into the denser medium.

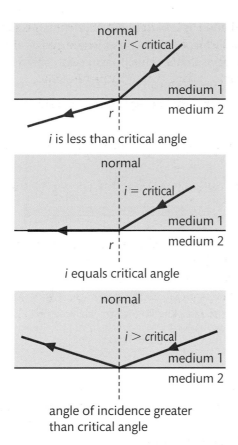

i is less than critical angle

i equals critical angle

angle of incidence greater
than critical angle

▶ **Figure 1.58:** Moving from refraction to total internal reflection

When the angle in the glass between the ray and the normal line is increased to a **critical angle**, C, the refracted ray is bent so far it would need to run at 90°, that is, along the surface of the glass-air interface. That is not possible, so from that incident angle upwards all the wave energy is reflected internally and there is no refracted beam at all. This effect is called **total internal reflection** (see Figure 1.58). Mathematically, since the sine of 90° is 1, the equation for refraction becomes:

$$\frac{1}{n} = \frac{v}{c} = \sin C \qquad \text{i.e. } \sin C = \frac{1}{n}$$

Assessment practice 1.17

The sparkle of diamonds is due to light being very effectively trapped in them by total internal reflection. The critical angle for a diamond-air interface is only 24.4°. Calculate:

a) the refractive index of diamond

b) the speed of light in diamond.

Optical fibres

Optical fibres are very long thin cylinders of glass or, sometimes, plastic. Light is fed into the cut end of the fibre, so when it hits the sides of the fibre, it almost always does so at angles greater than the critical angle. That means all the rays of light get totally internally reflected and keep bouncing down the length of the fibre. No wave energy gets lost through the walls of the fibre, although as glass is not perfectly transparent, some is gradually absorbed. When the light waves arrive at the far end of the fibre, up to a few kilometres away, their intensity is still large enough to measure as a signal. If the joints between them are carefully made, optical fibres joined end to end can pass their light signals on to the next stage in a fibre network, again with only a small loss in intensity.

This makes light in optical fibres a much more efficient way of transmitting signals than sending electrical pulses down copper cables. Copper cables suffer from quite large losses due to electrical resistance, meaning that after a few hundred metres most of the signal has been attenuated away and amplifiers are needed to boost it up again.

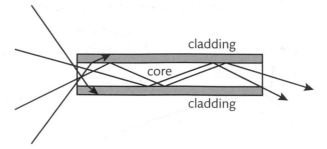

▶ **Figure 1.59:** Light paths through a multimode optical fibre: total internal reflections off the lower density glass cladding material

Theory into practice

The dense glass core of a communications fibre is not surrounded by air, but by a cladding material (see Figure 1.59). The cladding material is normally another type of glass with a lower refractive index than the core. Light in the fibre is totally internally reflected at the core/cladding interface.

Explain why the critical angle is larger than it would be for a bare glass fibre. (Hint: How much speed change is there for light between cladding and core glass? Compare that with air and glass.)

You might think glass fibres would be far too brittle to use. However, the brittle nature of glass is due to microscopic scratches that are easily introduced onto the outer surface. So, when first drawn, glass fibres (i.e. core plus cladding) are scratch free, strong, flexible and durable. Each glass fibre is given a plastic protective sheathing which keeps it scratch free. Sketch and label a diagram of a communications optical fibre.

Using fibres to carry signals

Installing and jointing optical fibres is a job for specially trained technicians using the right specialist tools and equipment. The jointing can be temporary, using a proprietary connector, or permanent. Permanent joints can be achieved by compression or by a welding process using heat.

Reliability of signal transmission depends on the quality of joints between fibres, both along a cable run and also those to the optical transmitter (an LED or laser) at one end and the optical receiver at the other. These are more or less the only places where substantial losses of signal intensity occur, that is, unless the cable is damaged.

Otherwise the attenuation of signal along a fibre is gradual, and a readable output of 1% or more of the input intensity can still be obtained after a kilometre or more. The optical signal can then be boosted in line by an optical amplifier. These devices are an offshoot of solid state laser technologies, and they increase the intensity of the light beam without needing to convert it back into an electrical signal.

Applications of fibre optics in medicine

Fibre optics are widely used in endoscopes (see Figure 1.60). Endoscopes are optical instruments with long tubes that can be inserted into a body organ through an opening such as the throat, nose, ear canals or anus. These allow a trained medical practitioner to see inside a body organ, for example, the upper oesophagus and stomach or the colon and intestines, without undertaking surgery. Endoscopes are also used during keyhole surgery to guide the use of surgical instruments with remote handling, which are often incorporated into the same tube system.

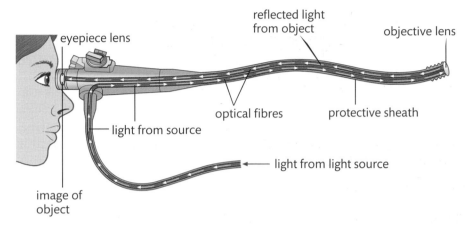

▶ **Figure 1.60:** Medical endoscope used to reduce the need for invasive surgery

Light is piped in from a source outside the body using a small bundle of optical fibres. The image from inside the body is then focused by an objective lens, like that in a camera. To pipe the image back out to be displayed for the medical practitioner, some shorter rigid tubular endoscopes use a series of relay lenses. In longer flexible 'fibrescopes', the image is also conveyed by a second fibre bundle back to an eyepiece or a camera lens.

Each fibre in the bundle is as thin as a human hair and consists of:

▶ a core

▶ cladding

▶ a protective plastic buffer coating.

The image transmitted is pixelated (formed of coloured dots), since each fibre only transmits one pixel of coloured light. So the resolution of the image depends on the number of fibres in the bundle.

Both the fibre bundles are protected inside the endoscope's long, flexible insertion tube. This tube also carries control lines that can move the bending section at the distal end. That is where the fibres terminate and interface with the lenses, one to distribute the incoming light and the other to collect and focus the image. Sometimes there may be additional control lines to independently move and operate a surgical tool.

Applications of fibre optics in communication

Analogue and digital signals

The information transferred by each fibre of an endoscope is an **analogue signal**. The colour and intensity of the light from a single point on the image formed by the lens inside the patient's body is directly relayed to reproduce an equivalent image point at the eyepiece or camera outside the body. Just one dot in the picture (i.e. one pixel of information) is delivered by each fibre.

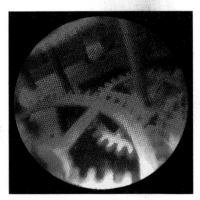

▶ The image from a low-resolution medical endoscope. Each fibre produces just one dot of colour.

Other examples of analogue signals are:

▶ the electrical signals made by a microphone, which mimic the shape and intensity of the sound waves they are detecting
▶ the position of the pointer on a pressure dial gauge
▶ the waveform displayed on a cathode ray oscilloscope, which copies and shows the variation of an AC voltage with time.

An optical fibre could convey much more information than just one pixel of coloured light. The light travelling down the fibre is a very high frequency wave. However, provided that you operate at significantly below that frequency, you can chop the light signal into on/off bursts and use it to transmit **digital signals**. Digital signals are numbers in binary code, that is, a series of ones and zeros.

Because of the very high frequencies being used, a huge amount of digital data can be transmitted in a short time. So, instead of just one pixel, a whole television picture can be digitised as a string of numbers that represent the brightness and colour of every pixel in it, and that can arrive in your home travelling along a single digital fibre cable. And not just one TV picture, but a whole host of other channels all available at the same time. Current data networks operate at up to order of Gigabit per second rates, but higher frequencies are possible with future technological development.

Key terms

Analogue signal – a signal with strength proportional to the quantity it is representing.

Digital signal – conveys in binary code a number that represents the size of the measured quantity.

Ⅱ PAUSE POINT

Up to about what frequency could you potentially operate an optical switch and send readable signals down a fibre using near infra-red radiation?

Hint

Use the formula $v = f\lambda$ to calculate the frequency of near infra-red light, wavelength 830 nm. (The speed of light is approximately 3×10^8 ms^{-1}.) Assume that you should allow at least 100 cycles on followed by 100 cycles off to represent a single bit of data.

Extend

What would currently limit the bit rates at which data can be transmitted?

Digitising information not only makes it possible to send more data faster than using analogue transmission. It also makes the transmission much more reliable and interference free. After a kilometre of travel down an optical fibre, the colour (i.e. frequency) of the light will be maintained, but its intensity may have dropped to only about 1% of the original. Nevertheless, if the signal is chopped into digital bursts, you can still reliably read whether it is on or off. That signal can be amplified and no information will have been lost.

Converting a signal from analogue to digital is carried out electronically using an analogue to digital (A to D) converter. At regular intervals this device samples (i.e. measures the size of) the analogue signal. Each sample value is measured and converted into a whole number of units, according to what sensitivity has been. So if the sensitivity is 0.1 volt, then values of 0, 0.1, 0.2, 0.3, 0.5 volt etc can be recorded, but no values between these. The sample value is then output as a digital signal which is a binary code number.

Binary code uses base 2 rather than base 10, and so represents any number as a series of ones and zeros. Electrical or optical signals represent this by switching on for a 'one' and off for a 'zero'. The switching is done at a predetermined speed called the clock speed. The clock speed determines how much data can be transmitted in a given time. The sampling rate needs to be set at least an order of magnitude (i.e. 10 times) higher than the frequency of any waveform you want to represent. For example, for sound signals, the sampling rate needs to be set at 10 times the frequency of the highest pitch harmonic within the audible range you want to reproduce. (The reason that voices sound odd on the telephone is because only a limited range of frequencies is reproduced by the digital sampling of telephone systems. This saves data space but loses quality.)

Step by step: Analogue to digital conversion   6 Steps

1 Select a transducer, a device that produces an analogue electrical voltage signal proportional to the quantity you want to measure, for example, a pressure sensor, a thermocouple or thermistor for temperature, a microphone for sound.

▼

2 Connect the output of the transducer to the input of an A to D converter, using a screened cable. To avoid picking up electrical interference the screening must be well earthed.

▼

3 Set up the A to D converter to sample the analogue signal. This is equivalent to taking measurements to plot results out on a voltage-time graph.

▼

4 Select an appropriate sampling rate, which is your sensitivity on the time axis.

▼

5 Select an appropriate sensitivity for the conversion of the voltage signal into a number. The smallest difference you will be able to convey with the digitally converted data is one unit.

▼

6 Connect the A to D converter output to a switch/transmitter to send the digital information: either electronically down copper cables; wirelessly using Bluetooth, WiFi or similar protocols; or by optical signals along a fibre network. Optical fibres are free from electromagnetic interference and virtually impossible to hack into.

Case study

Fibre optic broadband networks

Broadband is used as a relative term to indicate the speed and carrying capacity of a data channel. In connection with the Internet it has been used to market the improvement from earlier telephone dial-up connections, which were very limited and slow. Fibre optic broadband has been progressively replacing copper cable connections with consequent gains in data speed. Speeds of up to 100 Mbit per second to the desktop are now not uncommon. Speeds in the network backbone have, naturally, to be higher.

The name 'broadband' derives from the idea of 'bandwidth'. That is the practice of dividing the signal space up into a number of separate frequency bands. Each band can then be used to carry a separate channel of data, all at the same time. In the case of optical fibres, this is achieved by 'wave division multiplexing' (WDM). In WDM, light of several different colours (frequencies) is sent down the same fibre at the same time. Each different frequency of light carries another channel of information.

Multimode fibre is the standard fibre cable used for sending optical signals over short to medium distances – for example, connections to instruments, jumpers in cabinets, small local area networks. (It is also used elsewhere where high light power is needed, for example, in endoscopes.) The optical cores are around 100 µm (microns) in diameter, which is wide enough to allow a variety of separate beams to bounce along the inside of the fibre, striking the outer surface of the core with various different angles. If this is used for distances longer than a few hundred metres, it results in some of the beams travelling a significantly longer path and thus arriving late, which then degrades the quality of the signal. The 'ones' and 'zeros' of digital signals will thus no longer have sharp beginnings and ends.

Single mode fibre (see Figure 1.61) has an even narrower core (8 μm to 10 μm), which is less than ten wavelengths of the infra-red light that is used in them. This means there is just no space for different beams travelling at different angles down the core. Instead, the light wave moves as a single wave-front straight down the centre of the fibre, and all the signal energy reaches the far end of the fibre at the same instant. Millions of kilometres of this high-quality cable is laid every year to build the fibre optic networks for telephone, cable TV and broadband Internet communications.

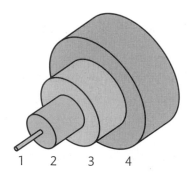

▶ **Figure 1.61:** Structure of a typical single mode optical fibre cable: 1. Core 8 μm diameter, 2. Cladding 125 μm dia., 3. Buffer 250 μm dia., 4. Jacket 400 μm dia.

Check your knowledge

1 What distances are involved in the local area network (LAN) that serves your laboratory?

2 What broadband data speeds do you need?

3 Would standard multimode fibre be adequate for the LAN cables? For which data connections are you still using copper cables?

Assessment practice 1.18

Describe how a medical endoscope transmits an image from inside the body.

How do the signals in the fibres of an endoscope differ from those transmitted through a fibre optic broadband network?

C3 Use of electromagnetic waves in communication

Speed of electromagnetic waves in a vacuum

Light, and all forms of electromagnetic radiation, travel at the same speed through a vacuum: $2.997\,925 \times 10^8$ m s^{-1}. This is a physical constant value that is usually denoted by the letter c. The fact that light always travels at this huge speed, and that nothing has ever been observed travelling faster than that, is not a 'law'. It is an experimentally observed fact, and it is the basis for Einstein's theories of relativity. Scientists are always on the lookout for any exceptions, but so far none have ever been seen.

Inverse square law for intensity of a wave

Waves transfer energy, and energy is a quantity that is always conserved. Wave-fronts propagating out from a point or a spherical source will themselves be spherical. As each wave-front increases in radius it also increases in area. The formula for the surface area of a sphere of radius r is $4\pi r^2$. The energy in the moving wave-front is distributed over that expanding area, and so its intensity decreases accordingly:

$$I = \frac{k}{r^2}$$

where I is intensity of wave, k is a constant and r is distance from source.

Regions of the electromagnetic spectrum

Although all types of electromagnetic radiation have the same nature and can be described by the same equations – Maxwell's equations – you experience them quite differently.

▶ Your eyes can only detect a very small range of frequencies. These are visible light.

▶ You can sense frequencies just a little lower than that of red light as radiant heat warming you. These are infra-red radiation (IR).

▶ There are frequencies just above your visible range that can be seen by bees and some other animals, which help plants grow and which cause sunburn. These are ultra-violet light (UV), because the frequencies are above those of violet light.

The remaining types of radiation are named according to how they are produced. At the highest frequencies the frequency ranges for X-rays and for γ-rays (gamma rays) overlap somewhat. X-rays are produced by high-energy atomic electron transitions and are just a higher energy version of light and UV radiation. On the other hand, γ-rays come from nuclear disintegrations and from collisions between high-energy sub-atomic particles.

For every frequency of radiation, f, you can calculate a corresponding wavelength in vacuum, λ, using the speed of light, c in the wave equation, $c = f\lambda$.

Table 1.13 shows the main types of electromagnetic radiation, how each is produced and some of the things they are used for.

▶ **Table 1.12:** SI unit prefixes

Prefix	Symbol	× 10 to the power	Factor
exa	E	18	1 000 000 000 000 000 000
peta	P	15	1 000 000 000 000 000
tera	T	12	1 000 000 000 000
giga	G	9	1 000 000 000
mega	M	6	1 000 000
kilo	k	3	1000
(none)	(none)	0	1
milli	m	-3	0.001
micro	μ	-6	0.000 001
nano	n	-9	0.000 000 001
pico	p	-12	0.000 000 000 001
femto	f	-15	0.000 000 000 000 001
atto	a	-18	0.000 000 000 000 000 001

▶ **Table 1.13:** Frequencies, sources and applications

The e/m spectrum	Frequency range	Wavelengths	Produced by		Used for		
Radio: Long wave Medium wave Short wave (HF) VHF UHF	30 to 300 kHz 300 kHz to 3 MHz 3 to 30 MHz 30 to 300 MHz 300 MHz to 3 GHz	10 to 1 km 1 km to 100 m 100 to 10 m 10 to 1 m 1 m to 100 mm	Electronic oscillators coupled to broadcast antennas	Astronomical radio sources for example, neutron stars	Radio and TV broadcasting Mobile phones (UHF) Plasma heating for fusion reactors Industrial ovens		
Microwaves SHF EHF	3 to 30 GHz 30 to 300 GHz	100 to 10 mm 10 to 1 mm	Klystron or magnetron tubes, or solid state diodes	Thermal emission Cold → Very hot	Domestic ovens RADAR Satellite and terrestrial communications links		Spectroscopy
Infra-red (IR): Far Mid-range Near	300 GHz to 30 THz 30 to 120 THz 120 to 400 THz	1 mm to 10 µm 10 to 2.5 µm 2500 to 740 nm	Light emitting diode (LED) or laser		Night vision cameras Optical fibre comms. Movement detectors Remote controls	Photosynthesis in plants	Spectroscopy
Visible light	400 to 800 THz	740 to 370 nm	Emission by outer electron transitions		Illumination Imaging Signalling	Photosynthesis in plants	
Ultra-violet (UV)	800 THz to 30 PHz (10^{15} Hz)	370 to 10 nm			Insect vision Tanning		
X-rays	30 PHz to 30 EHz (10^{18} Hz)	0.01 to 10 nm	Inner electron excitation and decay		Medical imaging		
γ-rays	generally > 30 EHz	generally < 0.01 nm	Nuclear reactions and particle decays		Radiation sterilisation Medical tracing		

▶ **Table 1.14:** Frequencies, sources and applications

Application	Power and mode of transmission	Frequency band	How it is used and regulated
Satellite communications	High power signals over very long distances; concentrated by dish antennae.	1 to 40 GHz (microwaves)	Satellite transponders receive incoming upload signals, amplify them and retransmit them as a download signal on a different frequency band. *For more info search 'satellite frequency bands' on the European Space Agency website www.esa.int*
Mobile phones	High power networked system, range several km.	800 MHz to 2.6 GHz (UHF radio to microwave borderline)	5 or 10 MHz bands allocated to different operators. 2G, 3G and 4G cellular networks offering increasing speeds for data. Higher frequencies have greater data capacity but travel less distance through air and penetrate into buildings less well.
Bluetooth®	Low power device to device links, range up to about 10 m.	2.4 to 2.4835 GHz – the Industrial, Scientific, Medical (ISM) unlicensed band – borderline between UHF radio and microwave frequencies	Early Bluetooth devices interfered with Wi-Fi devices because both would use the same channel for an extended period of time. Modern Bluetooth uses frequency-hopping – i.e. broadcasting in short bursts on a number of different frequency channels across the band. This reduces the amount of data lost, and in most cases both Bluetooth and Wi-Fi can maintain service. *For more info search for 'Bluetooth and Wi-Fi' at IntelligentHospitalToday.com*
Wi-Fi	Medium power networked system, range ~10 to 100 m.		
Infrared	Low power device to device links, range only a few metres.	IR wavelength 870 nm or 930 to 950 nm (frequency about 320 THz)	Used for remote controls and for data transfer between computers, phones, etc. The longer wavelength band is better because it does not suffer from 'sunlight blinding'. Atmospheric moisture blocks that range in sunlight.

ⅠⅠ PAUSE POINT   What types of material are transparent or partially transparent to radio waves or to microwaves? How do you know that? What benefits or problems might be caused by these waves being diffracted or refracted?

> **Hint**   What happens to phone and radio reception inside buildings? What kinds of structure completely block reception inside? In other buildings, what could be the explanation for dead spots and good reception points?

> **Extend**   Look at some rooftop UHF TV antennas. Can you identify the reflector? Sketch how the reflector forms a stationary wave pattern along the antenna. Mark where the nodes occur and where the detector dipole is placed.

Worked Example

Calculate how many times smaller the intensity of light falling on a surface 6.0 metres away from a light bulb will be compared with the intensity at just 0.5 metres away.

Answer

$$I = k/r^2$$

So

$$I_1 / I_2 = (r_2/r_1)^2 = (0.5/6.0)^2 = (1/12)^2 = 1/144$$

The intensity will be 144 times smaller at 6.0 m compared with just 0.5 m away from the light bulb.

Assessment practice 1.19

Use the inverse square law to calculate approximately how many times smaller a mobile phone signal will be 2 km from the transmitter compared with reception 500 m from the transmitter mast.

What other factors can affect signal strength?

Assessment practice 1.20

Infra-red and Bluetooth® are used for device-to-device signal transmission. What are the advantages and disadvantages of each? Compare and contrast these two with mobile phone communications.

Further reading and resources

www.esa.int: European Space Agency. This site gives lots of useful information about satellites and the way they are used for communications. The page on 'satellite frequency bands' shows shows how the SHF band is subdivided and what applications use each sub-band.

www.intelligenthospitaltoday.com: Intelligent Hospital Today. If you search this site for 'Wi-Fi and Bluetooth' you should find a very useful article that explains the history of Bluetooth® and how it has been adapted to co-exist with Wi-Fi, which uses the same frequency band and so could potentially cause interference and loss of signals.

THINK ▶▶FUTURE

Sara Logan,

Laboratory Technician in a Consumer Product Company

I started work as a laboratory technician straight from sixth form college. The company I work for makes toiletries such as shampoo and shower gel. They also make cleaning products like bathroom and kitchen cleaners. When I started, I mainly carried out tests on the new products to check they were safe. This included carrying out titrations to ensure the products were the correct pH and that they contained the necessary chemical substances in the right proportions.

This is still a large part of my job but now I have worked here for five years, I am also involved in the development of new products. This means my knowledge of the periodic table and bonding has been invaluable. I know what substances will react and what sort of properties the products will have before I start experimenting. I also understand which chemical substances are safe to react and which ones are too dangerous to use.

We are given a brief that tells us what the product should do, whether it is a tear-free shampoo for babies or a strong cleaner for filthy ovens. We experiment and test using our knowledge of other products. This usually takes many months.

It is important for us to be able to communicate well and work in a team. The whole team work together to research, produce and test a new product. I have to listen carefully to my brief/instructions and ensure I carry them out in order to meet deadlines. I must write a report at the end showing my findings. Sometimes I have to present this report to the client and my boss. I often have to use Excel and PowerPoint to produce my report and presentations.

It is always an exciting day when we realise we have a product that exactly matched the brief.

Focusing your skills

Think about the role of a laboratory technician. Consider the following:
- What types of people will you work with and how will you support them?
- What role will you play in helping them achieve their goals?
- What types of company will you work for? Will you work in research and development or quality assurance, for example?
- What skills do you currently have? What skills do you think may need further development?

Getting ready for assessment

This section has been written to help you to do your best when you take the assessment test. Read through it carefully and ask your tutor if there is anything you are still not sure about.

About the test

The assessment test will last 90 minutes and there are a maximum of 90 marks available. The test is in 3 sections and will ask a range of short answer questions as well as some longer answer questions worth up to 6 marks.
Each section, Chemistry, Biology and Physics, will include:

- short answer questions worth 1–4 marks
- a longer answer question worth 6 marks.

Remember that all the questions are compulsory and you should attempt to answer each one. Consider the question fully and remember to use the key words to describe, explain and analyse. For longer questions you will be required to include a number of explanations to your response; plan your answer and write in detail.

Preparing for the test

To improve your chances on the test, you will need to make sure you have revised all the key assessment outcomes that are likely to appear. The assessment outcomes were introduced to you at the start of this unit.

To help plan your revision, it is very useful to know what type of learner you are. Which of the following sounds like it would be most helpful to you?

Type of learner	Visual	Auditory	Kinaesthetic
What it means	Need to see something or picture it, to learn it	Need to hear something to learn it	Learn better when physical activity is involved – learn by doing
How it can help prepare for the test	• Colour-code information on your notes • Make short flash cards (so you can picture the notes) • Use diagrams, mind-maps and flowcharts • Use post-it notes to leave visible reminders for yourself	• Read information aloud, then repeat it in your own words • Use word games or mnemonics to help • Use different ways of saying things – different stresses or voices for different things • Record short revision notes to listen to on your phone or computer	• Revise your notes while walking – use different locations for different subjects • Try to connect actions with particular parts of a sequence you need to learn • Record your notes and listen to them while doing chores, exercising, etc. and associate the tasks with the learning

Do not start revision too late! Cramming information is very stressful and does not work.

Useful tips

- **Plan a revision timetable** – schedule each topic you need to revise, and try to spend a small time more often on each of them. Coming back to each topic several times will help you to reinforce the key facts in your memory.
- **Take regular breaks** – short bursts of 30–40 minutes' revision are more effective than long hours. Remember that most people's concentration lapses after an hour and they need a break.
- **Allow yourself rest** – do not fill all your time with revision. You could schedule one evening off a week, or book in a 'revision holiday' of a few days.
- **Take care of yourself** – stay healthy, rested and eating properly. This will help you to perform at your best. The less stressed you are, the easier you will find it to learn.

Sitting the test

Listen to, and read carefully, any instructions you are given. Lots of marks are often lost because people do not read questions properly and then do not complete their answers correctly.

Most questions contain command words. Understanding what these words mean will help you understand what the question is asking you to do. Remember the number of marks can relate to the number of answers you may be expected to give. If a question asks for two examples, do not only give one! Similarly, do not offer more information than the question needs: if there are two marks for two examples, do not give four examples.

Planning your time is an important part of succeeding on a test. Work out what you need to answer and then organise your time. You should spend more time on longer questions. Set yourself a timetable for working through the test and then stick to it. Do not spend ages on a short 1 or 2 mark question and then find you only have a few minutes for a longer 4 or 6 mark questions. It is useful when reading through a question to write down notes on a blank page. This way you can write down all the key words and information required and use these to structure an answer.

If you are writing an answer to a longer question, try to plan your answers before you start writing. Have a clear idea of the point your answer is making, and then make sure this point comes across in everything you write, so that it is all focused on answering the question you have been set.

If you finish early, use the time to re-read your answers and make any corrections. This could really help make your answers even better and could make a big difference in your final mark.

Hints and tips for tests

- Revise all the key areas likely to be covered. Draw up a checklist to make sure you do not forget anything!
- Know the time of the test and arrive early and prepared.
- Ensure that you have eaten before the test and that you feel relaxed and fresh.
- Read each question carefully before you answer it to make sure you understand what you have to do.
- Make notes as you read through the question and use these to structure your answer.

- Try answering all the simpler questions first then come back to the harder questions. This should give you more time for the harder questions.
- Remember you cannot lose marks for a wrong answer, but you cannot gain any marks for a blank space!

Q. Ethane and fluoromethane have similar-sized molecules.

▶ **Figure 1.62:** Fluoromethane and ethane

Explain why fluoromethane has a higher boiling point than ethane. (2)

Fluorine has the highest electronegativity of all elements.
This means there is a larger permanent dipole on the fluoromethane.

This is an 'explain' question. The examiner is looking for a fact and a reason or a 'because'. The question is worth 2 marks so 2 points must be made to gain these marks. The answer above would be worth both marks.

Q. Potassium fluoride has many uses in manufacturing, synthesising and refining. It converts chlorocarbons to fluorocarbons. It is used to etch glass and in the making of disinfectants. However, potassium fluoride is not made in the laboratory by reacting potassium with fluorine as this reaction is not safe.

Discuss why it is not safe to react fluorine with an alkali metal in a school laboratory. (6)

> This is a 6-mark levelled question. It is worth 2 pass marks, 2 merit marks and 2 distinction marks. You gain marks for showing understanding rather than there being 1 mark per point. The more detailed and in-depth your discussion, the more likely you are to gain 6 marks. You would be expected to use all your knowledge about fluorine to answer this question. You should consider position in the periodic table, reactivity, bonding.

Question number	Answer	Mark
	Indicative content	(6)
	▶ Reactions between fluorine and alkali metals are vigorous and explosive.	
	▶ Group 1 metals are the most reactive metals/as you go left to right on periodic table metals become less reactive.	
	▶ As you go down group 1 the metals become more reactive.	
	▶ Group 1 metals only have one electron in their outer shell.	
	▶ Reactivity increases in the non-metals as you go from left to right across the period.	
	▶ Reactivity decreases as you go down group 7/ fluorine is the most reactive non-metal.	
	▶ Fluorine has a small radius because positive protons strongly attract negative electrons to it.	
	▶ It has 5 electrons in its $2p$ sub-shell.	
	▶ The p sub-shell can hold 6 electrons so fluorine is close to its ideal electron configuration.	
	▶ Strong attraction by positive nucleus due to small size/one shell shielding.	
	▶ Fluorine is a strong oxidising agent.	
	▶ It forms strong ionic compounds with metals.	
	▶ Fluorine easily gains one electron and alkali metals easily lose an electron.	
0	No rewardable content.	0
Pass level	A simple description of why it is unsafe. Learner will use the periodic table to show that the reactants are very reactive.	1–2
Merit level	Links ideas of electronic structure to reaction.	3–4
Distinction level	Links ideas of electronic structure to reaction in detail. May discuss strength of oxidising agent to reaction.	5–6

Ans 1. It is not safe to react fluorine with an alkali metal because alkali metals and fluorine are both very reactive meaning that there might be an explosion. This is because alkali metals are on the left hand side of the table and the metals get less reactive as they go across the periodic table.

This would be a pass-level answer. The candidate knows that an explosion will occur and knows how to use the periodic table to show this. The answer is quite simple and shows little understanding of why the elements are so reactive.

Ans 2. Group 1 metals are highly reactive as they only need to lose one metal to form an ionic bond. The repulsion from the positive nucleus makes this easy to react. Fluorine is the most reactive non-metal. It is at the top of group 7 and the trend is that reactivity decreases down the group. Fluorine has a small radius it only has 2 shells so it is already small and the protons in nucleus are able to draw the electrons to it strongly making it even smaller. It has outer electron configuration of $2s^2\,2p^5$. This means it has only got to get one electron to get a full outer shell. The small radius and only one shell shielding the nucleus mean the electron on the metal is strongly attracted to the charge on the positive. Fluorine is a very strong oxidising agent. All of these things means that when fluorine reacts with a group 1 metal the reaction will be vigorous and probably explosive so it is not safe to carry out.

This would be a distinction-level answer. The candidate has discussed reactivity of alkali metals in general in relation to their position on the periodic table. They have discussed the electronic structure of fluorine in detail and have started to relate this to reactivity. They have then said why this makes the reaction unsafe. The ideas are mostly quite detailed and are linked.

Practical Scientific Procedures and Techniques 2

Getting to know your unit

Carrying out practical laboratory techniques correctly and accurately is an important part of the work of the laboratory technician. Techniques developed over a century ago are still used in modern analytical chemistry and are the basis for analysis in a range of occupations related to the chemical industry, medicine and pharmaceuticals, education, forensic investigation and many more. Following correct laboratory practice will improve your analytical skills, develop your transferable skills and help you to appreciate the need for attention to detail in an area that affects all our lives.

Assessment

You will be assessed by a series of assignments set by your tutor.

How you will be assessed

This unit will be assessed using a series of internally assessed tasks within assignments set by your tutor. Throughout this unit, you will find activities that will help you work towards your assessment. Completing these activities will not mean that you have achieved a particular grade, but you will have carried out useful research or preparation that will be relevant when it comes to completing your assignments.

In order for you to achieve the tasks in your assignments, it is important to check that you have met all of the Pass grading criteria. You can do this as you work your way through the assignments. Merit criteria require you to analyse and demonstrate skilful application of procedures, while Distinction criteria require you to evaluate your practice.

The assignments set by your tutor will consist of a number of tasks designed to meet the criteria in the table. This is likely to consist of a written report but may also include activities such as:

▶ demonstrating correct and appropriate practical techniques confirmed by Observational Record and/or Witness Statement

▶ presenting findings to your peers and reviewing the procedures and applications of your work during class discussion

▶ analysing, evaluating and reviewing your own performance in a critique that highlights your strengths and weaknesses.

Assessment criteria

This table shows what you must do in order to achieve a **Pass**, **Merit** or **Distinction** grade, and where you can find activities to help you.

Pass	Merit	Distinction
Learning aim **A** Undertake titration and colorimetry to determine the concentration of solutions.		
A.P1 Prepare and standardise solutions for titration and colorimetry. **Assessment practice 2.1**	**A.M1** Demonstrate skilful application of procedures and techniques in titration and colorimetry to accurately determine the concentration of solutions. **Assessment practice 2.1**	**A.D1** Evaluate the accuracy of procedures and techniques used in titration and colorimetry in relation to outcomes and suggest improvements. **Assessment practice 2.1**
A.P12 Investigate the concentration of unknown solutions, using procedures and techniques in titration and colorimetry. **Assessment practice 2.1**		
Learning aim **B** Undertake calorimetry to study cooling curves.		
B.P3 Obtain data using different equipment to construct cooling curves. **Assessment practice 2.2**	**B.M2** Analyse the rate of cooling of substances from your data using cooling curves to draw conclusions. **Assessment practice 2.2**	**B.D2** Evaluate the accuracy of practical work in calorimetry in relation to the analysis of the cooling curve. **Assessment practice 2.2**
B.P4 Determine the rate of cooling of substances using cooling curves. .**Assessment practice 2.2**		
Learning aim **C** Undertake chromatographic techniques to identify components in mixtures.		
C.P5 Use chromatographic techniques to produce chromatograms. **Assessment practice 2.3**	**C.M3** Analyse own chromatograms and relate the factors that affect the separation of mixtures to the quality of results obtained. **Assessment practice 2.3**	**C.D3** Evaluate the chromatographic techniques used in relation to outcomes and suggest improvements. **Assessment practice 2.3**
C.P6 Explain the use of chromatographic techniques to separate mixtures. **Assessment practice 2.3**		
Learning aim **D** Review personal development of scientific skills for laboratory work.		
D.P7 Summarise key personal competencies developed in relation to scientific skills undertaken. **Assessment practice 2.4**	**D.M4** Analyse skills developed and suggest improvements to own practice. **Assessment practice 2.4**	**D.D4** Evaluate scientific skills developed in terms of potential for future progression. **Assessment practice 2.4**

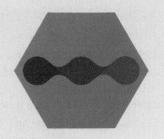

Getting started

Analytical chemistry is an exact branch of chemistry which is performed using a range of laboratory equipment including glassware, digital and mechanical devices. Using a prepared worksheet and examples of the equipment supplied by your tutor, try to memorise the names of the apparatus. Test your answers with a partner.

Undertake titration and colorimetry to determine the concentration of solutions

This section outlines the foundation principles in the use of basic laboratory equipment. The importance of calibrating and testing of measurements in glassware and electronic equipment will be covered with suitable examples to ensure that you have the necessary information to produce valid results from practical activity. You will be guided through titration techniques to determine concentration values of solutions and introduced to colorimetric measurement.

Safety considerations

It is vital to observe essential safety rules and practices while working in a laboratory. You should always think through the safety aspects of practical investigations first. Where appropriate, produce a complete risk assessment and have it checked by your tutor before every series of practical activities is carried out. Remember the following points.

▶ Thoroughly research the topic and be aware of the possible risks.

▶ Familiarise yourself with glassware and instruments.

▶ Learn the hazard symbols which appear on all substance containers.

▶ Identify the risks by using Hazcards (see Figure 2.1) and COSHH information.

▶ Produce a comprehensive risk assessment (RA).

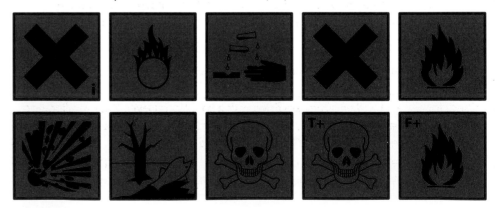

▶ **Figure 2.1:** Which of these common hazard symbols can you identify?

Laboratory equipment and calibration

All analytical laboratories carry out scientific analysis using reliable methods which are linked to the correct use of equipment. This equipment varies slightly between laboratories and is dependent on the specific nature of the analysis undertaken and the specialism required.

Balancing and weighing

Generally, and with few exceptions, most laboratories have access to electronic balances and top-pan balances which can be calibrated regularly with manufacturers' certified weights. The **precision** requirement depends on the circumstance of the laboratory and of the investigations carried out, but it is generally accepted that all laboratories will ensure that balances are available with a 'rough' balance of two decimal places and an 'analytical' balance of up to four decimal places.

▶ Weighing crystals on a typical top pan balance used in all types of chemical laboratories

The laboratory top-pan balance is a fundamental and regularly used piece of precision equipment which deserves to be used and maintained with great respect from learners and technicians.

Depending on the degree of precision, the balance can be influenced by many external factors including; the surface, draughts, **ambient temperature**, vibration, 'warm-up' time, magnetism and static electricity. In general, you need to guard against most of these influences for the more sensitive balances and you should consider re-calibration if you think any of these influences may have affected the balances.

Re-calibrating is quick and easy. Standard weights are used to calibrate balances. By following the manufacturer's table of **tolerance**, simply put a test weight on the balance. If the reading is within tolerance limits, you do not need to take further action. If the reading is outside these limits, then you need to make adjustments.

Using a balance

1   Switch on the balance and allow sufficient time for the device to achieve a thermal equilibrium.

2   Check for correct levelling. Many are fitted with a liquid 'spirit level'. Place a calibration weight on the pan. If the reading is within the manufacturer's tolerance levels, then no adjustment is needed. If it is not, then adjust the reading.

3   Never weigh the material directly onto the balance. Use a suitable container or weighing boat. Weigh the container itself and press 'tare' or 'zero' to eliminate the mass of the container.

4   Remove the container and then fill it with the substance to be weighed carefully and in the middle of the container. If an instruction states 'measure accurately approximately 10 g...' this means as close to the 10 g as possible but not necessarily exact.

5   To maintain the balance, keep it switched on during long activities, keep it clean and regularly check for **drifting** of the measurements by using **calibration** weights. Drifting in measurements shows when a measurement of a constant quantity such as a mass is repeated several times and the measurements drift one way during the experiment. Keep the balance in a clean environment at room temperature.

Key terms

Precision – how close two or more measurements are to each other.

Ambient temperature – the temperature of the surroundings.

Tolerance – the acceptable upper and lower limits for a measurement.

Drifting – variations in the readings of the balance due to internal mechanical wear, for example.

Calibration – to adjust or correct the graduations of a measuring device, when compared to a known value standard.

Volumetric glassware

The most commonly used items of glassware used for volumetric measurement are also the most essential pieces of apparatus for chemical analysis. They include:

▶ teat pipettes (and glass and bulb pipettes)
▶ burettes
▶ filter funnels (plastic versions are also commonly used)
▶ volumetric flasks
▶ conical flasks
▶ beakers.

Table 2.1 summarises glassware that you might see in the laboratory.

▶ **Table 2.1:** Volumetric glassware for standard laboratory analysis

Name	Sizes	Uses	Limitations
Pipette (non-graduated)	5.0 ml to 50 ml	Taking volumes of solutions when accuracy of a single volume is needed, normally to add small volumes dropwise	Can be very fragile when handled and need some practice with pipette fillers to produce accurate volumes
Pipette (graduated)	Graduated between 1 ml and 1/10th ml	Taking volumes of solutions of specific random volumes	Practice is needed with pipette fillers to produce accurate volumes
Burette	25 ml and 50 ml	Measuring accurate volumes with a graduation of 0.1 ml allowing a meniscus to be observed	The tap can become stiff to use so silicon grease is needed for lubrication. The tip can become clogged or fill with air bubbles
Conical (Erlenmeyer) flask	50 ml to 500 ml	Mixing and swirling volumes of solutions with little risk of spillage	No real limitations in its use
Volumetric flask	10 ml to 2000 ml	Measuring, mixing and making up volumes to a given mark with a high degree of accuracy	Glass can expand as a result of chemical reactions producing heat
Beaker	Varying sizes, generally 10 ml to 1000 ml	Measuring non-accurate volumes of solutions, general usage, waste etc with a graduation to +/– 10%	No real limitations as fit for general purpose, but hot alkalis can damage the surface

▶ Standard laboratory pipette using either a bulb or a plunger for filling the graduated tube

Volumetric glassware will have a Certificate of Calibration when it is purchased. However, you must never assume that the measurements you are taking are 'absolutely' accurate in terms of the volume of liquid in your glassware.

The volume occupied by a mass of liquid can vary a lot at different temperatures and so can the volume of the glassware itself. The temperature used to calibrate laboratory glassware has been set at 20 °C, and glass does not expand much with a rise in temperature.

Dilute aqueous solutions can expand by 0.025% per °C. and a 1.000 litre of water (1 kg or 1000 g mass) at 15 °C can now occupy a volume 1.0025 litres at 25 °C. Re-calibration can be carried out to allow for this aspect by weighing the liquid and determining its density at a given temperature using data tables. If the mass and density are established, you can work out the volume using:

$$\text{Volume } (\textbf{\textit{V}}) = \frac{\text{mass } (m)(g)}{\text{density } (d)(g/cm^3)}$$

The density of water is taken to be 0.9982 g ml^{-1} at room temperature and pressure.

▶ **Figure 2.2:** You can use this triangle to rearrange the equation linking volume, mass and density

---

Discussion

Many learners begin the study of chemical analysis believing that digital equipment offers far more precision and accuracy of measurement than laboratory glassware. Carry out some general research into this point with a partner and ask the laboratory technician for examples. Discuss your findings and draw your own conclusions.

Standardisation of solutions using titration

Standard solutions

In **quantitative analysis**, rather than **qualitative analysis**, standard solutions of known concentration are prepared in readiness for titration. You should keep in mind the following guidance when preparing standard solutions.

▶ Make use of the appropriate CLEAPSS Student Safety Sheets. Use goggles and protective gloves where necessary.

▶ Use suitable measuring cylinders for the task. Example: for making a 1000 ml HCl solution of 5 mol dm^{-3}, a 500 ml measuring cylinder for the acid and 1000 ml measuring cylinder for the water is suitable.

▶ Use a graduated pipette of suitable volume for small standard solution quantities. Make the solution in a beaker and transfer to a volumetric flask. Carefully wash out the beaker into the funnel and wash out the funnel before making up to the 'mark' (measurement line on the volumetric flask).

▶ Shake all solutions well.

Key terms

Quantitative analysis – practical experiment producing numerical results (quantities are measurable).

Qualitative analysis – practical experiment producing observational results such as colour, odour, transparency (quantities are not measurable).

a) Procedure by weighing

Worked Example

Here is how to prepare a standard solution of 0.1 mol dm^{-3} sodium hydroxide.

1 From the periodic table, M_r of NaOH = 23 + 16 + 1 = 40

2 For 1 mol dm^{-3} of NaOH you need 40 g of NaOH per 1000 cm^3 distilled water (1000 cm^3 = 1 dm^3).

3 For 0.1 mol dm^{-3} of NaOH you need 40/10 = 4 g of NaOH per 1000 cm^3 distilled water.

To prepare a 0.1 mol dm^{-3} standard solution of NaOH, you will need to dissolve 4.0 g of NaOH in the minimum amount of water that will dissolve the NaOH and carefully transfer it into a 1000 ml volumetric flask. Then add water to make the solution up to 1000 ml.

b) Procedure from known concentration solution

Worked Example

Here is how to prepare a standard solution of 200 cm^3 of 0.1 mol dm^{-3} hydrochloric acid from 1.0 mol dm^{-3} HCl stock solution.

1 Use the formula: $V_s = V_f \times C_f/C_i$

where: V_s – volume of solution needed to be diluted for the required solution
V_f – final volume required, in this case, 200 cm^3
C_f – final concentration required, in this case, 0.1 mol dm^{-3}
C_i – initial concentration of stock solution, in this case, 1.0 mol dm^{-3}.

2 Calculate the volume of solution to be diluted:

$$V_s = \frac{200 \times 0.1}{1.0} = \textbf{20 cm}^3$$

To prepare a 200 cm^3 standard solution of 0.1 mol dm^{-3} HCl from a stock solution of 1.0 mol dm^{-3} HCl, you will need to measure 20 cm^3 of 1.0 mol dm^{-3} HCl stock in a volumetric flask and use distilled water to make up to the 200 cm^3 mark.

Primary and secondary titrimetric standards

We use the word 'standard' in many aspects of general life to mean a level which is accepted, such as standards of education, certain standard of service in a restaurant, etc. In chemical analysis, the word standard refers to a solution of known concentration.

A primary standard is one in which we can have a very high confidence level in its concentration, usually between 99.95 and 99.98%. It must have a known purity and be stable when it is stored for long periods of time. A secondary standard is one in which we do not have such a high level of confidence of concentration value or purity. It is usually compared to the primary standard when trying to determine its concentration. In school or college laboratories, the term primary standard is generally used to identify a pure, 'solid' standard for making a secondary standard of a lesser confidence of concentration. These secondary standards are remade when the need arises.

Ⅱ PAUSE POINT

1. Find the volume of solution to be diluted in order to prepare 500 cm³ of a standard solution of 0.1 mol/dm³ of sulfuric acid (H_2SO_4) from a stock solution of 0.5 mol/dm³.

2. Find the volume of solution to be diluted in order to prepare 500 cm³ of a standard solution of 0.75 mol/dm³ of hydrochloric acid (HCl) from a stock solution of 2 mol/dm³.

Titration

Titration is used in industrial laboratories to find out the concentration of unknown solutions. While the procedure is automated in many laboratories, the basic principles are the same. Figure 2.3 shows the apparatus used for carrying out a titration.

Key term

Titration – the process of determining the concentration of an unknown solution using a solution of known concentration.

Safety tips

You must wear goggles.

Mark the volumetric flask containing solutions with a permanent marker.

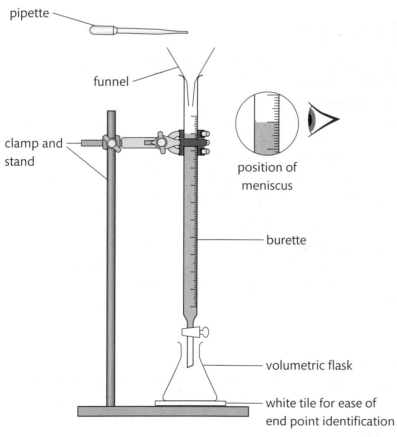

- pipette
- funnel
- clamp and stand
- position of meniscus
- burette
- volumetric flask
- white tile for ease of end point identification

▶ **Figure 2.3:** The apparatus used in a titration

Safety tips

Remove the filling funnel before titrating.

Check the burette tap for blockages before and after.

Take notes at each stage.

Wash your hands afterwards.

Carrying out acid-base titrations

Acid-base titrations are carried out regularly in industry and educational laboratories to determine the unknown concentrations of either acids or alkalis. Although many titrations are now performed using electro-mechanical equipment, the tried and trusted method using laboratory glassware is still commonly used and provides a quick and effective means to produce accurate results.

A rough titration should be performed prior to the actual determination. Fill the burette with the solution. Open the valve and allow the solution to run into the conical flask while swirling it. When the solution changes colour, close the valve. Record the volume at eye level. Perform a calculation to find the concentration value. Now carry out the correct and careful titration using the rough figures as a guide.

Steps in the investigation	Pay particular attention to...	Think about this...
1. Set up the apparatus as shown in Figure 2.3.	Make sure that the burette is absolutely vertical from all directions.	If the burette is at an angle, the measurement of acid or alkalis will be incorrect.
2. Ensure that the tap is open. Rinse the burette with distilled water and then with a small quantity of standard solution.	Fully rinse the burette with the standard solution to be sure that only this is recorded flowing from the burette afterwards.	
3. Close the tap and fill the burette.	Fill the burette above the graduation mark to have sufficient solution to release in order to set the first graduation level.	If you make a mistake, simply add extra standard solution and try again.
4. Release the titre solution slowly until the meniscus is on the first graduation level.	Make sure that the base of the curve sits on the graduation mark.	
5. If air bubbles are present, repeat steps 2 to 4, to ensure that no air bubbles interfere with the readings during titration.	Ensure that no air bubbles interfere with the readings during the titration.	An air bubble could put your eventual results out by a number of cm^3, having an important effect on results.
6. Transfer a known volume of unknown concentration solution to a conical flask.	Operate the pipette carefully and according to good practice.	
7. Add three drops of indicator (e.g. phenolphthalein) to the solution in the conical flask.	Be sure that the colour of the solution is clear enough to enhance the point at which the change in colour first appears.	
8. Slowly titrate – opening the tap and gently swirling the conical flask.	Observe the solution carefully to identify the first signs of colour change.	The swirling motion will ensure that the end point is a true neutralisation mark.
9. At the first indication of colour change, reduce the flow rate. Observe the **end point** as the complete colour change and record the level on the burette.	Your results table should show up to three separate volume readings for the same titration. Once you have two values the same (concordant), use this in your calculation. Other values can be discarded.	

Worked Example

In this experiment, 20.0 cm³ of 1.0 mol dm⁻³ NaOH neutralised 25.5 cm³ of H_2SO_4 whose concentration was believed to be 0.4 mol dm⁻³.

Step 1 Balance the chemical equation:

$$2NaOH(aq) + H_2SO_4(aq) \rightarrow Na_2SO_4(aq) + 2H_2O(l)$$

2 moles of NaOH are needed to neutralise 1 mole of H_2SO_4. The **stoichiometry** is 2:1.

Step 2 Calculate the number of moles of NaOH:

$$20.0 \text{ cm}^3 = \frac{20}{1000} = 2 \times 10^{-2} \text{ dm}^3 = \textbf{0.02 mol}$$

Step 3 Calculate how much H_2SO_4 has been neutralised (note 2:1 ratio)

$\frac{1}{2}$ the amount of NaOH = $\frac{1}{2}$ × 0.02 mol = **0.01 mol**

Step 4 Change the volume of H_2SO_4 to dm³ (1000 cm³ = 1 dm³):

$$25.5 \text{ cm}^3 = 25.5/1000 = \textbf{2.55} \times \textbf{10}^{-2} \textbf{ dm}^3$$

Step 5 The concentration of H_2SO_4 is:

$$\frac{0.01}{2.55} \times 10^{-2} \text{ mol/dm}^{-3} = \textbf{0.392} \text{ mol/dm}^3$$

 PAUSE POINT

Using the procedure shown above, keeping all measured values the same and replacing sulfuric acid for hydrochloric acid (HCl), find out the concentration of HCl by neutralisation with the NaOH.

1. A sample of vinegar (ethanoic acid) was titrated with 0.75 mol/dm³ NaOH. The volume of NaOH needed to neutralise the vinegar was 20.5 cm³. What is the concentration of the vinegar?

2. 19.5 cm³ of 0.1 mol/dm³ NaOH neutralised a solution of 0.5 mol/dm³ nitric acid (HNO_3). What is the concentration of the nitric acid?

Hint   Balance the equation firstly, noting the stoichiometry.

Extend   Does the stoichiometry of the reactants in this example simplify the calculation? How?

pH meter and probes

The acidity of a chemical substance is sometimes defined as its ability to donate a hydrogen ion to another molecule or atom in an ionised state. The pH scale is a measure of how acidic or basic a substance is and ranges from 0 to 14. A neutral pH is 7, a pH less than 7 is acidic, and a pH greater than 7 is basic. The pH is equal to $-\log_{10} c$ (c is the hydrogen **ion** concentration in **moles** per litre) and so each pH unit below 7 is ten times more acidic than the next higher unit. A pH of 3 is ten times more acidic than a pH of 4. A pH of 3 is 0.001 M concentration of hydrogen ions and a pH of 4 is 0.0001 M concentrations of hydrogen ions. Similarly, a pH value of 9 is ten times more alkaline than a pH of 8.

When hydrogen chloride is dissolved in water, it ionises completely and produces H⁺ (aq) and Cl⁻ (aq) ions. The hydrogen ion is basically a proton. Acids are therefore associated with the transfer of the hydrogen ion. The pH meter measures a potential difference between one electrode and another which registers ion activity. The scale converts this activity to a pH reading. In general, the pH meter is a very precise voltmeter.

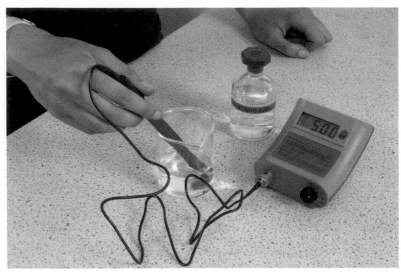

▶ A standard pH meter used in many applications, particularly in determination of the end point in titration

Safety tip

Handle the pH meter with great care to avoid a build-up of static which will affect its measurement. Use Hazcards for help with calibrating buffer solutions.

The method of calibration refers directly to the electrodes which are placed in the solution. Standard solutions of known pH can be bought from chemical supplies as pH calibration buffers.

Calibration and use

▶ Collect **pH calibration buffer** solutions of 4.00, 7.00 and 10.00 if available.

▶ Switch on the machine. Allow time for the electronics to stabilise and set the temperature control to account for the buffer solution depending on the model used.

▶ Remove the protective cap and rinse the electrode with distilled water and immerse in buffer solution 7.00. Repeat for all buffer solutions, rinsing the electrodes with distilled water every time. Adjust where necessary.

▶ Add the electrode to the sample after rinsing with distilled water.

Example Standard Calibration Buffers include ethanoic acid/sodium ethanoate mixture.

The difference between the electronic measure of pH and the litmus paper versions is clearly in the precision to which the measurement is made. This precision is based on the recording of ions which related to the pH of the solution. pH = $-\log$ [concentration of H^+ ions]. The reading on this particular pH meter gives values to 0.01 and so the probable error is quoted as ±0.01.

Key term

pH calibration buffer – an aqueous solution of accurate pH used to set the pH meter levels.

Link

Go to *Unit 3: Science Investigation Skills Learning aim B.*

⏸ PAUSE POINT   After calibrating your pH meter with suitable pH buffer solutions, you decide to test the pH of a strong acid solution and get results of 1.01 and 1.03 from your meter. What could have given rise to the different readings?

Hint   What procedure may not have been followed correctly during calibration?

Extend   Faced with this situation, what should be the best course of action to take in order to obtain accurate results?

Titration and pH/volume graphs

In order to understand and visualise the change of pH as the volume of acid to alkali changes, it is useful to produce a pH/volume graph. The resulting graph is called a **pH curve**.

> **Link**
>
> Go to *Unit 13: Applications of Inorganic Chemistry Learning aim A2.*

From these graphs, a number of important aspects concerning the reaction of an acid with an alkali can be identified.

▶ The pH changes with volume.

▶ The pH changes more sharply when approaching the **equivalence point**.

▶ The pH changes more sharply after equivalence point has been reached.

▶ The pH level may not always be 7.0 at the equivalence point.

> **Key terms**
>
> **pH curve** – a graphical shape describing how pH changes during acid-base titrations.
>
> **Equivalence point** – the point at which the solutions have been mixed in exactly the right proportions relating to the chemical equation (stoichiometry).

Investigation 2.2

Plotting a pH curve

The change in pH of an alkali solution when it is neutralised by adding an acid solution is best displayed using a pH volume graph or 'curve'. By using a calibrated pH meter, it is possible to plot the pH at a series of points related to the volume of acid used. In this way it is easier to visualise both gradual and sharp changes in pH and to identify equivalence points, end points and neutralisation more effectively.

You can use the following procedure if it is decided to add an acid to an alkali.

Steps in the investigation	Pay particular attention to...	Think about this...
1. Produce a 1.0 M standard solution of both hydrochloric acid (HCl) and sodium hydroxide (NaOH).	Check the quantities of acid and alkali accurately. Any errors will be carried through the investigation.	If measurements and quantities are inaccurate at the start of the process, then the results will be invalid.
2. Prepare the apparatus for titration, adhering to safe practices.	Follow correct and appropriate safe working practice at all times.	
3. Transfer 25 cm³ of NaOH to the conical flask.	Ensure that you are now adept at using the pipette filler. It can save you time and help to ensure accuracy in your final results.	
4. Use a pH meter to record the pH value of the NaOH.	Be careful to ensure that the pH meter is correctly calibrated before use.	Try to check the calibration of the meter before each activity, or at least check the calibration report records, if available.
5. Fill the burette with HCl standard and titrate with the NaOH at steps of 5 cm³.	Observe the burette at eye level to eliminate possible errors in the marks observed.	You will need to take notes of readings at every stage and transfer them to the report later.
6. Carefully record the pH of the solution in the conical flask at each 5 cm³ interval in a suitable table.	Make suitable observation of the pH.	
7. Use all of the HCl in the burette for complete results.		
8. Plot a graph of pH on the *y*-axis and volume of HCl on the *x*-axis.	Graph paper is essential if valid conclusions are to be made afterwards.	

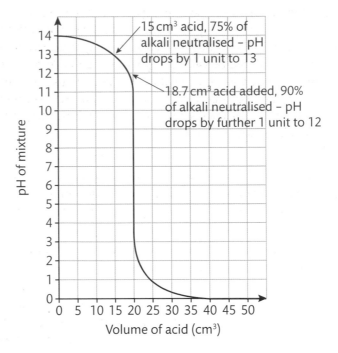

Figure 2.4: Graph of change in pH with change in volume of acid

In Figure 2.4, at ~~22.5ml~~ 18.7cm³ of acid added, about 90% alkali is neutralised so the pH only drops by 1 unit at this point. At ~~24.5ml~~ 19.9cm³ of acid added, about 98% of the alkali is neutralised, so the pH has dropped by 2 units.

⏸ PAUSE POINT   What can you say about the shape of the graph?

Hint   Is the 'end-point' easily identifiable? What happens to the graph when excess acid is added to the alkali?

If the volume of acid is increased, what difference would you see in the graph? If the concentration of the alkali is less, how would this change the graph shape?

Extend   Overlay your graph with a shape expected if you added the alkali slowly to the acid.

Colorimetry

Colorimetry is a technique which measures the intensity of colour. The level of colour in a solution can be used to provide a value of its concentration since the intensity of the colour from a chemical reaction is proportional to the concentration of the substance tested. The colour will be compared to known colours and corresponding concentrations.

Spectroscopy and the spectrometer (spectrophotometer)

This method of chemical analysis is used to determine the purity or concentration of a chemical substance. The principle relies on the ability of the substance under investigation to emit, absorb or scatter **electromagnetic radiation** of differing wavelengths. Specific wavelengths are used in this method. The substance (for example, solution) interacts with the electromagnetic waves to different degrees dependent on the wavelength of the wave, because shorter wavelengths in the

Key term

Electromagnetic radiation – energy released by electrical and magnetic processes ranging from low to high frequency and short to long wavelength. It includes radio, microwaves, infra-red, visible light, ultra-violet, X-rays and gamma waves.

electromagnetic spectrum have greater energies. The molecules of substances are all different in terms of the atoms they are made of and the way in which they are bonded together. Molecules are formed from the bonding of atoms and electron sharing. These can be influenced by electromagnetic radiation. As a result, a chemical substance will absorb electromagnetic radiation at a frequency determined by the energy levels within the atoms. Since no two substances absorb electromagnetic radiation of the same frequency, the spectrum produced will be unique and can be used to identify the substance.

Link

Unit 1: Content area C Waves in communication considers emission spectra.

Research

Solutions appear a certain colour because light of a particular wavelength is absorbed by the molecules while the remaining wavelengths are transmitted to our eyes. Produce a visible range spectrum of colours in a carefully coloured diagram. From research, label the shortest and longest wavelength in the visible spectrum and include all the wavelength boundaries between the colours. These will be measured in nanometres (nm). Identify at least one chemical solution which absorbs a wavelength within each range and label these on your diagram.

Key terms

Electromagnetic spectrum – the range of energies produced by electrical/ magnetic effects.

Nanometres (nm) – measure of wavelength which are 1 000 000 000th of a metre in length (1×10^{-9} m).

Diffraction grating – a set of parallel, closely spaced slits which can separate light out into its specific colours because different wavelengths are diffracted (bent around the openings) at different angles.

Solution – the resulting liquid which has the solute dissolved in a solvent.

How does a spectrophotometer work?

▶ A bright light from two or more lamp types is sent through the device in the wavelength range between 200 nm (**nanometres**) and approximately 800 nm. This covers the near infra-red, visible light and near ultra-violet regions of the electromagnetic spectrum.

▶ The light is put through a **diffraction grating** which acts like an efficient prism and separates out the colours (wavelengths), sending the required wavelength through a narrow slit.

▶ A rotating disc system splits up the beam.

▶ One beam is sent through a sample cell of 1 cm width, the other through a similar reference cell containing the pure **solution** (no **solute** added to the **solvent**).

▶ The device converts the **intensity** of the light through the cells into electric current. The intensity of light through the reference cell is Io and the intensity of light through the sample cell is I.

▶ The ratio of the intensity values (I/I_o) gives the transmittance and the absorbance value can then be found using $A = -\log_{10}$. If I_o is 100% (the intensity of the light through the reference cell) and I is 10% (meaning that 90% of the light through the sample cell has been absorbed) then I/I_o is 100/10, i.e. 10. $\log_{10}$ of 10 = 1.

Key terms

Solute – a substance which is dissolved in another substance and is usually the lesser amount.

Solvent – the liquid in which a solute dissolves.

Intensity – (when related to light) the amount of light energy transmitted. Measured in photons (particles of light energy) per second.

The calibration is set at 0 to 100%. Any value obtained for absorbance is provided as a % and so you should write probable error or accuracy as ±1%. The measured values of % absorbance are directly proportional to the concentration of the chemical at the given wavelength and so provide a linear measurement.

▶ **Table 2.2:** Example data table for use during colorimetry investigation

Test tube	Concentration (mol dm⁻³)	Absorbance (A)
1		
2		
3		
4		
5		
Unknown sample		

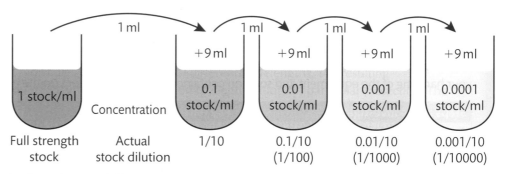

▶ **Figure 2.6:** Serial dilutions showing how an initial 10 ml of known concentration is diluted by a factor of 10 each time

Ⅱ PAUSE POINT   In your own words, describe 'absorbance'.

Hint   Include comments about chemical structure and bonding and wavelengths involved.

Extend   Provide an explanation of how absorbance can determine the concentration or identification of a solution.

The Beer-Lambert law

The Beer-Lambert law provides a mathematical relationship between the absorbance of light and the concentration of a substance. Generally, the more light absorbed by a substance, the greater its concentration.

$$A = elc$$

where:
A – absorbance
e – constant of proportionality or **molar absorptivity** (how well the substance absorbs light and can have the units $cm^{-1} M^{-1}$)
l – the path length (usually the width of the cuvette)
c – concentration of solution (mol dm⁻³).

In basic terms, **absorbance** refers to the amount of light absorbed by the solution, **transmittance** refers to the amount of light which passes through the solution. The two are clearly linked and each can be calculated if the other is known. The relationship between absorbance (A) and transmittance (T) is a logarithmic one:

$$A = -\log T.$$

Worked Example

1 Your colorimeter records a transmittance value of 45%.
What is the absorbance value?
A transmittance value of 45% is the same as 0.45.

Since $A = -\log T = -\log(0.45) = $ **0.347**

2 Your **analyte** sample solution has an absorbance of 0.297, corresponding
to a concentration of 3.0×10^{-4} M from your graph. Your cuvette width is
1 cm and you have obtained your results with the colorimeter set to 480 nm.
What is the molar absorptivity of the analyte (e)?

Using $A = elc$, so that $e = A/lc$.

$0.297/1 \times (3.0 \times 10^{-4}) = $ **990 cm^{-1} M^{-1}**

> **Key term**
>
> **Analyte** – a chemical solution or substance being analysed.

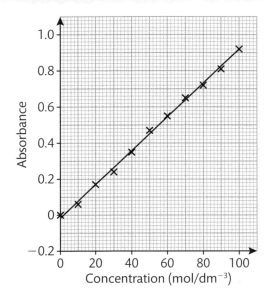

▶ **Figure 2.7:** Graph of absorbance against percentage concentration in a solution

Figure 2.7 shows that, for any value of absorbance, the value of concentration is
proportional since the graph is linear. This is the basis for the Beer-Lambert law.

Assessment practice 2.1

A.P1 A.P2 A.M1 A.D1

Obtain a standard solution of 0.75 M sodium hydroxide and use
to standardise an unknown concentration of hydrochloric acid.
Produce a pH curve of the titration and fully label the important
aspects on the resulting graph.

Calibrate a colorimeter and make a concentration of copper
sulfate. Produce correct serial dilutions to be used and plot a
calibration graph.

Test an unknown concentration of copper sulfate by plotting the
absorbance reading on your graph.

Evaluate the accuracy of your procedures and techniques and
suggest improvements. You will need to consider the probable
error and the precision calibration of your volumetric glassware
for the dilutions, repeat procedures, using the same cuvettes,
re-calibration of colorimeter.

Plan
- What is the task? What am I being asked to do?
- How confident do I feel in my own abilities to
  complete this task? Are there any areas I think I
  may struggle with?

Do
- I know what it is I am doing and what I want to
  achieve.
- I can identify when I have gone wrong and
  adjust my thinking/approach to get myself back
  on course.

Review
- I can explain the results obtained from the task.
- I can apply the activity to other situations.

B Undertake calorimetry to study cooling curves

A calorimeter is simply a container, such as a glass beaker or polystyrene cup, which can be used, with a thermometer, to measure the temperature during a reaction in which heat is exchanged with the immediate outer environment. General uses include:

▶ monitoring endothermic/exothermic reactions

▶ monitoring change of physical phase, for example, freezing (liquid becomes solid), melting (solid becomes liquid)

▶ measuring specific heat capacity.

The process of measurement of the heat transferred is called **calorimetry**.

Thermometers

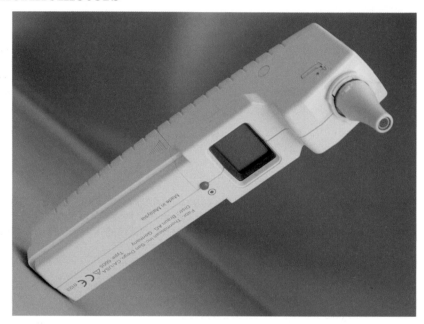

▶ Thermometers currently in use do not always look like the traditional types

The measurement of temperature using a standard liquid expansion thermometer began from the development of the 'centigrade scale', based on divisions from 0 to 100. This is attributed to the work of the Swedish astronomer, Anders Celsius, in 1742. He based the scale on the freezing (0 °C) and boiling point (100 °C) of pure water at a height of sea level and at sea level pressure. The term 'Celsius' is now used.

Thermal energy is defined by the kinetic energy an object may have as a result of the random movement of its particles. All objects possess some thermal energy because all objects have particles within their structure which are moving, however slightly and however low their overall temperature may be. This, of course, is the case unless an object is at the theoretical 'absolute zero' of temperature, −273.15 °C (0°Kelvin). To convert from degrees Kelvin to degrees Celsius, you add 273. To convert from degrees Celsius to degrees Kelvin, you subtract 273. At this temperature the motion of its particles is so negligible that it produces a minimum of heat energy.

To understand what a measure of heat is, you need to appreciate the basic concept of heat in terms of energy and its transfer. Much work was completed on this by Lord Kelvin of Scotland (Sir William Thomson), who produced the Kelvin scale of temperature in 1848.

▶ **Table 2.3:** Types of thermometer and their applications

Thermometer type	Principle of operation	Main applications
Liquid-filled	**Alcohol in glass** – a specified amount of alcohol is coloured and placed in a pressurised glass tube. Alcohol expands with rise in temperature, although it is slow to respond. It only measures to +78°C. The alcohol is not harmful if the glass breaks.	Clinical usage in hospitals, industrial complexes, schools and colleges and for domestic use.
	Mercury in glass – a small amount of mercury is sealed in a glass tube. The metal responds quickly to temperature changes. There are two basic types: a straight tube and another which has a slight bend in the bottom to prevent the liquid from dropping quickly when removed from the mouth. Mercury is poisonous. It is difficult to read the display.	Clinical usage in hospitals. General industrial uses for liquid temperature measurement, schools and colleges, practical scientific investigation.
Electronic	**Thermistor** – a semiconductor component. The electrical resistance of the thermistor decreases in a circuit as the temperature increases. As a result, more current flows. The display may be digital.	Because they are sensitive to temperature changes and link to electrical circuit resistance, they are used in fire alarms and switching circuits for heater systems (range: –250 to 700 °C). Digital thermometers generally used in hospitals for patient care, schools and working environments in conjunction with 'thermometer skin strips'.
	Resistance – a conductive wire in a circuit which increases its resistance to current flow when its temperature rises. Suitable for high temperature changes. Easy to read display. Expensive and can be subject to 'drift'.	Used in industry because it can record temperatures of over 1000 °C.
	Thermocouple – when two wires of different metals are connected together, two junctions are formed which produce a potential difference with temperature. This is a very sensitive device. Short distance measurements can be taken. Difficult to calibrate.	Used in industry to record the temperature of furnaces, ovens, etc. because of its wide range of temperature recording (range: –250 to 1600 °C).
	Rotary – the coiled bimetallic strip principle is used so that when temperatures increase, the strip expands more and touches a calibrated pointer.	Simple construction allows general usage such as in greenhouses or fridge-freezers.
	Infra-red – this type detects various wavelengths of infra-red electromagnetic radiation and is extremely accurate. Can only measure surface temperatures. Affected by radio waves and other waves. Affected by ambient conditions.	Commonly used in hospitals to determine a patient's change in temperature by inserting the device into the ear.

Ⅱ PAUSE POINT Select from Table 2.3 which thermometer type you would use and explain your reasoning for the following situations: a chef testing the temperature when cooking a chicken, a weather monitoring station reporting daily temperatures, a geologist recording the temperature of fresh volcanic lava.

Hint You will need to know the range of temperatures involved for each situation and the material limits from which the thermometers are made.

Extend Identify and explain an alternative thermometer for each situation.

Checking the calibration of a thermometer

The need to check that your thermometers are accurately recording the correct temperature may be more important in some situations than others. It is generally good practice to ensure that you are aware of the possible errors that may develop over time.

Water boils at 100 °C when situated at sea level and a barometric pressure of 760 mm or 1 Bar. If water is under less pressure, such as when at a higher altitude, then the boiling point becomes lower. The boiling point lowers by 1 °C for every 308 m elevation above sea level. This means that when you are checking your thermometer you should take your position above sea level into account.

Discussion

In pairs: find out the height of the Burj Khalifa in Dubai and the height of Mount Everest in the Himalayas. Calculate the temperature at which water will boil at the top of both. Discuss the implications of boiling water in a very deep mine shaft.

The same procedure is used for checking the calibration of all types of thermometer, depending on the temperature scale to be tested. Figure 2.8 shows the steps that can be used to determine the accuracy of calibration for measuring temperatures between 0 °C and 100 °C.

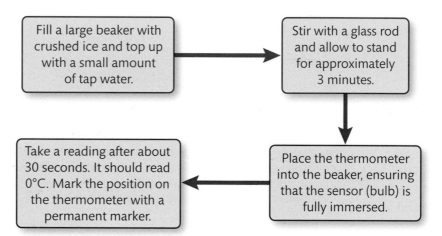

▶ **Figure 2.8:** Checking the calibration of a thermometer

Key term

Boiling point – the temperature at which a substance changes from a liquid to a gas.

Safety tip

If you are calibrating a mercury thermometer and it is broken at any stage, immediately isolate the area and seek help from a technician or other member of staff. Mercury vaporises at room temperature and could be easily inhaled.

You can continue by checking the calibration at **boiling point** (see Figure 2.10).

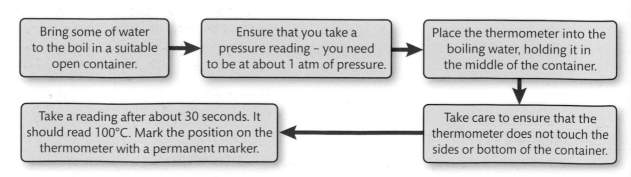

▶ **Figure 2.9:** Checking the calibration of a thermometer at boiling point

Accuracy of thermometers

Many studies have been commissioned to try to determine which type of thermometer provides the most accurate results (gives a reading that is closest to the actual value). This is very important in many branches of science and related subject areas such as industrial development, scientific research and medicine.

As an example, studies for measurement of body temperature in patients have been carried out over many years. New technological advances in thermometers have resulted in a variation in the way that this aspect is measured as a routine part of patient care in hospitals. When digital thermometers were compared to 'mercury-in-glass' types, mercury-in-glass types tended to produce more accurate and reliable readings while the digital thermometers were providing significantly under-recorded temperatures. This was a result of random electronic variations (data based on a clinical study from the University of Stirling). At present, one of the most common and reliable thermometers used to find body temperature is based on the infra-red detection principle. These are simply placed inside a patient's ear for a short time and the result is recorded. These thermometers are therefore termed 'tympanic'. They have the added benefit of being easily read. This is an important factor because many thermometer scales depend on a subjective determination of the value by the person reading the scale.

Case study

Consignment of thermometers

Robert, an experienced laboratory researcher, has received a new consignment of 100 liquid-in-glass thermometers which were ordered two months previously and are well overdue. The consignment consists of mercury thermometers, calibrated for temperatures between 0 °C and 110 °C.

As a senior technician for a pharmaceutical research centre based in the southern Scottish Highlands (approximately 1000 m above sea level), Robert is currently supervising Tim, a new appointee who has recently completed an initial induction in laboratory practices and procedures for the company and has been testing a range of scientific equipment to check their calibration. Tim has been given the task of calibrating the new thermometers before the company fully accepts the order and puts them to use.

Tim opens the consignment in preparation for testing and proceeds in the following manner.

1 He tests all the thermometers.

2 In his results for 0 °C calibration, he uses the same ice cubes throughout, even though they have all melted by the time he has begun testing the last 30 thermometers.

3 He uses freshly boiled water each time to test the 100 °C calibration.

4 Tim decides to save some time and test the thermometers in batches of five. He does not support them in the container.

5 He notices that all the thermometers are reading a little over 97 °C and advises his supervisor, who considers the results carefully.

1 Was it necessary for Tim to test all 100 thermometers?

2 Why should he have replaced the ice cubes before they had melted?

3 What effect could not replacing the melted ice cubes have on the readings of some thermometers?

4 What will happen to temperature readings if some thermometers touch the sides of the container?

5 Why did Tim record a temperature of a little over 97 °C?

Cooling curves

We have all experienced our hot cups of tea, coffee or soup going cold. Hot liquids placed into containers will begin to lose their heat to the surroundings. The rate at which the heat is lost will depend on a number of factors:

- the material of the container
- the starting temperature of the liquid and its surroundings
- the molecular properties of the liquid.

Heat energy will be transferred from the hot liquid by **conduction**, **convection**, **evaporation** and **radiation** to its surroundings until it is at thermal equilibrium with its surroundings.

Producing a cooling curve for water

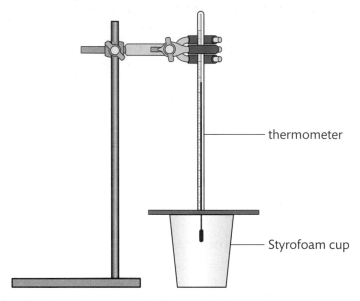

thermometer

Styrofoam cup

▶ **Figure 2.10:** What do you think is the significance of using metal calorimeters rather than other materials in calorimeters?

Key terms

Conduction – the transfer of heat energy in a solid where there exists a difference in temperature.

Convection – the transfer of heat by circulating currents from a region of high density to a region of less density in a gas or liquid.

Evaporation – the change of state of liquid particles to gas near the uppermost surface of a liquid, resulting in a drop in temperature of the remaining liquid.

Radiation – the transfer of energy, such as heat, from a source to its surroundings.

Safety tip

Caution: hot water! Take great care when using boiling water at all times and when handling hot containers. Use tongs or allow things to cool down before moving them.

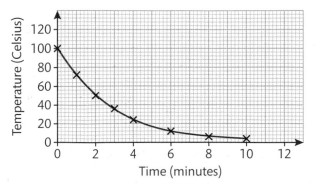

▶ **Figure 2.11:** Cooling curve for freshly boiled water cooling for 10 minutes

1 Set up the apparatus as shown in Figure 2.10.
2 Pour freshly boiled tap water into a 100 ml beaker to the mark.
3 Transfer this water to the calorimeter.
4 Record the temperature of the water and start the stop clock.
5 Record the temperature at 60-second intervals for approximately 10 minutes.
6 Plot the results on a graph of temperature (y-axis) against time (x-axis).
7 Draw a curve of best fit through your plots and draw a tangent to the curve at each point.
8 Calculate the 'rate of cooling' at each point from the gradients of the tangents.

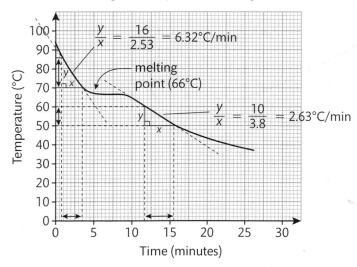

▶ **Figure 2.12:** Cooling curve for stearic acid

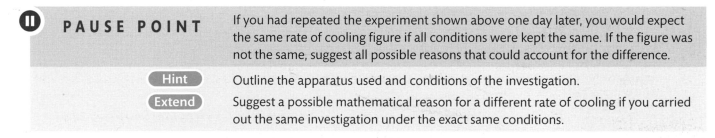

PAUSE POINT   If you had repeated the experiment shown above one day later, you would expect the same rate of cooling figure if all conditions were kept the same. If the figure was not the same, suggest all possible reasons that could account for the difference.

Hint   Outline the apparatus used and conditions of the investigation.

Extend   Suggest a possible mathematical reason for a different rate of cooling if you carried out the same investigation under the exact same conditions.

Determination of melting point

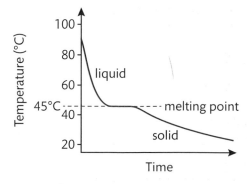

▶ **Figure 2.13:** Where would the freezing point be labelled?

In Figure 2.13, Salol is a liquid at temperatures above 45 °C. It is a solid below this temperature.

The graph also shows a flat section during which time is taken for the Salol to change its physical state from a liquid to a solid. This is because as it changes from a liquid to a solid, energy is removed from the substance with no change in its temperature. If the reverse process was to happen, that is, it was to be heated from solid state to liquid, then there would be a flat section in the heating curve at exactly the same temperature. This shows that energy is taken in while the Salol changes state from a solid to a liquid (its melting point). The explanation for this is that the additional increase in kinetic energy of the particles in the solid allows them to move more freely. The regular arrangement of particles now becomes more random.

Cooling curves like this provide a useful visual and mathematical method to represent the energy changes of substances over a given time, identifying the point at which the substance both changes from a liquid to a solid and changes from a solid to a liquid.

Investigation 2.4

A cooling curve for stearic acid

Stearic acid has been used for thousands of years in a variety of ways. These include adding to cosmetics, food and in hygiene products such as soap and shampoo. The substance is a 'fatty acid' easily obtainable from vegetable and animal fats and is used to provide a hardening agent to many products including candles.

For each step in this investigation, it is important to record the temperature of the stearic acid at precise timings. By careful plotting of results onto your graph, the characteristic changes of state will be easily identified.

Steps in the investigation	Pay particular attention to...	Think about this...
1. Obtain a suitable quantity of stearic acid and place into a boiling tube.		Stearic acid is a fatty acid ($C_{18}H_{36}O_2$) found in many food products such as butter and margarine.
2. Place the boiling tube into a beaker of freshly boiled water.	Check the thermometer reading. It may not fully read 100 °C.	
3. Place a thermometer or temperature probe connected to a digital data logger into the boiling tube (in preparation to record the drop in temperature as the acid cools).	Care must be taken at this stage to avoid the possibility of spills of hot water.	
4. As the stearic acid heats up, it will begin to melt.		
5. Record the highest temperature of the now liquid stearic acid.	Look closely at the stearic acid. It should be fully liquified with no hint of solid material.	
6. Remove the boiling tube from the beaker of hot water and allow to cool.		Note that if the ambient temperature is warm, the boiling tube should be placed into a beaker of cold water.
7. Record the temperature over a period of time at set intervals as the acid cools and for a time after it has become solid.	Take measurements at exact intervals to produce a detailed and accurate record of the cooling of the stearic acid.	
8. Produce a graph of your results (temperature y-axis, time x-axis).		
9. Analyse your results according to the stages shown in Figure 2.14.	Attempt to identify a flattening of the graph representing the changing of state.	

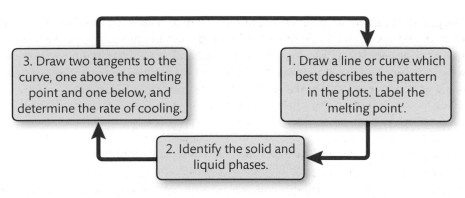

▶ **Figure 2.14:** Stages in analysing your results

The temperature at which melting occurs for a pure form of a material is unique to the material (if the pressure is kept constant) and can be used to identify the material or the purity of it.

You should be aware that the liquid and solid phases are at **thermal equilibrium** at this melting point. As the material cools from a liquid to a solid, the first sign of solidification is crystallisation where **latent heat** is lost to the surroundings and there is a brief halt in cooling of the material. This is shown on the curve by the flattening of the overall cooling curve. After this point is reached, further cooling steadily continues.

Key terms

Thermal equilibrium – point at which there is no temperature change due to heat energy being used to break molecular forces at phase change.

Latent heat – the heat energy taken in or given out when a substance changes state.

Discussion

If your graph is plotted accurately, you will notice that the line marking the melting point is perfectly horizontal. This means that there is one melting point temperature marking the liquid/solid phase change. If your graph showed an uppermost and lowermost point of the melting point temperature (that is, a slight slope), what would this tell you about the purity of the stearic acid?

Theory into practice

Solder is a metal alloy mixture of lead (Pb) and tin (Sn). It is a vital material needed in the electronics and manufacturing industry to ensure that electrical contacts are made between specific components and the circuit deck in devices such as TVs, laptops, video games, display systems, control mechanisms, etc.

As a junior member of the technical team in 'Micro-tech Systems', research the percentage of both these metals in solder and the effects that this has on the cooling curve of the mixture. Find out the temperature at which both metals are solid in the mixture and what difference to the shape of the cooling curve would be made if the percentage of lead to tin were changed.

Super cooling

The process of crystallisation for liquid substances begins at the 'melting point'. This process requires a nucleus from which a crystal of the solid form can begin to form.

Generally, this nucleus can be a dust particle, flaw in the molten material or smaller impurity within the liquid phase which then allows 'nucleation' to occur (a vibration in the liquid can also induce nucleation). Once this has started, the full crystallisation and solid phase will develop. In some very advanced industrial processes, nucleation can be started by even random atom or molecule movements.

Super cooling is the process by which a liquid, such as water, can be cooled to temperatures well below its freezing point but can remain in liquid state. In order to super cool a liquid or molten material, you have to remove the impurities which can trigger the nucleation process. This means that attempting to keep pure or distilled water at a liquid state below 0 °C is possible if the conditions are correct.

Intermolecular forces and cooling

Molecular gases, liquids and solids exist as such as a result of the balance between their **intermolecular forces** holding them together and the kinetic energy of their molecules driving them apart. Cooling and heating of liquids and solids can also determine their physical state.

Liquids and solids have the following characteristics.

▶ **Table 2.4:** Characteristics of liquids and solids

Liquids	Solids
Definite volume that is independent of container shape	Do not take the shape of containers well
Less dense than solids	More dense and less compactable than liquids
Intermolecular forces strong to hold atoms together	Intermolecular forces hold molecules together
Attractive forces cannot hold molecules together	Attractive forces provide rigid structure
Diffusion occurs slowly	Diffusion occurs very slowly

Intermolecular forces are the sum of repulsive and attractive forces between molecules. This does not include atomic bonding of the substance, which is termed intramolecular forces.

The strength of the attractive forces and the kinetic energy of the atoms in a substance are the key factors which determine what physical state it will be in. The intermolecular forces influence melting and boiling points, pressures and viscosity.

Link

Go to *Unit 1A: Principles and Applications of Science.*

Greater attractive intermolecular forces means that higher temperatures are needed to reduce the forces of attraction binding the molecules. Changing the state of a solid to a liquid needs a lot of extra heat energy (measured in kJ/mol). Changing the state of a liquid to a gas requires even more heat energy.

PAUSE POINT Produce a generalised diagram using arrows and appropriate labels to identify what happens when a solid becomes a liquid and when a liquid becomes a solid.

Hint Include correct terms and arrow directions, energy changes, molecular forces and structure.

Extend Add a phase change to gaseous state for your diagram.

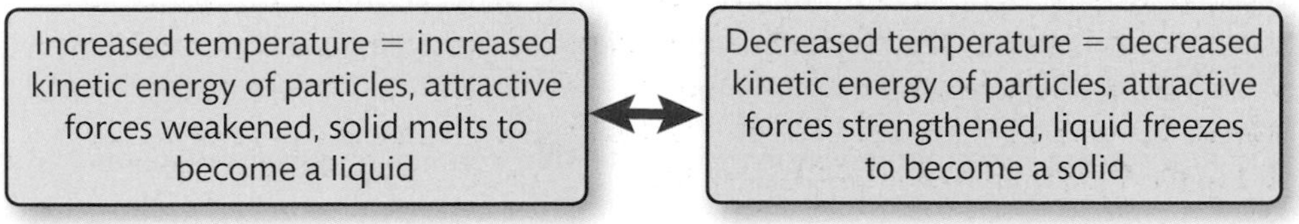

Increased temperature = increased kinetic energy of particles, attractive forces weakened, solid melts to become a liquid

↔

Decreased temperature = decreased kinetic energy of particles, attractive forces strengthened, liquid freezes to become a solid

▶ **Figure 2.15:** The effects of increasing and decreasing temperature

---

Assessment practice 2.2

B.P3 | B.P4 | B.M2 | B.D2

Double glazing works because it traps a thin layer of air between the glass sheets to reduce heat loss.

To test this principle, produce two cooling curves from investigation by using hot water in a) a single glass beaker and b) a glass beaker inside a larger glass beaker. Provide each with a suitable lid.

After producing your cooling curves, analyse the rate of cooling for both investigations to inform your conclusion.

Evaluate the accuracy of your work in relation to the analysis of your cooling curves. Consider your temperature measurements in terms of probable error and the range of values acceptable for the activity. If the value recorded falls outside the acceptable range, outline how you have used this information.

Plan
- What is the task? What am I being asked to do?
- How confident do I feel in my own abilities to complete this task? Are there any areas I think I may struggle with?

Do
- I know what it is I'm doing and what I want to achieve.
- I can identify when I've gone wrong and adjust my thinking/approach to get myself back on course.

Review
- I can explain what the task was and how I approached the task.
- I can explain how I would approach the hard elements differently next time (i.e. what I would do differently).

C Undertake chromatographic techniques to identify components in mixtures

In chemistry, you need to isolate one substance from another to be able to analyse the substance further or to prepare it for use in other ways. **Chromatography** (from the Greek word *khroma*, meaning colour) is a method used to separate mixtures and identify substances. From this, you can find out the purity of a chemical substance. The basic principle is quite simple. It relies on moving a liquid or gas over a stationary paper or powder. This section introduces you to the different techniques in chromatography, the various pieces of laboratory equipment that you will need to use, and the applications of the process related to real-life situations.

Chromatographic techniques

Paper chromatography

In chromatography, substances are separated as they travel in a **mobile phase** which passes over a **stationary phase**. Different substances travel at different speeds, so some move further than others in the time specified.

In paper chromatography, the stationary (non-moving) phase is paper. The mobile phase may be either an aqueous (water-based) liquid or a non-aqueous organic (carbon-based) solvent. An example of an organic solvent is propanone. (Propanone is the main chemical in nail varnish remover.) For each chemical in the sample, the

> **Key terms**
>
> **Chromatography** – a method used to separate chemical mixtures for analysis.
>
> **Mobile phase** – the liquid that transports the substance mixture through the absorbing material which travels along the stationary phase or 'bed' and carries the substance components with it.
>
> **Stationary phase** – the solid material that absorbs the mixture flowing through it.

separation depends on how strongly attracted the chemicals are to the mobile and the stationary phases. This characteristic is unique to each chemical compound and can be used to separate them in a mixture.

It is important to note that the paper itself has a thin 'coating' of water molecules which may have some minor effect on the results. This is partially a result of the manufacturing process and the fact that cellulose fibres, from which the paper is made, absorb moisture from the air.

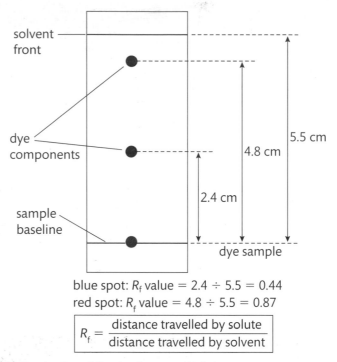

blue spot: R_f value = 2.4 ÷ 5.5 = 0.44
red spot: R_f value = 4.8 ÷ 5.5 = 0.87

$$R_f = \frac{\text{distance travelled by solute}}{\text{distance travelled by solvent}}$$

▶ **Figure 2.16:** Determination of R_f value from a typical chromatogram

Link

Go to *Unit 4: Laboratory Techniques and their Application, Learning aim C* and *Uni 19: Practical Chemical Analysis, Learning aim C*.

Thin-layer chromatography (TLC)

This type of chromatography is used to analyse dyes in fibres, inks and paints. It is also used in the detection of pesticides or insecticides in food products. Thin-layer chromatography is very similar to paper chromatography. Instead of paper, the stationary phase is a thin layer of an unreactive substance (for example, silica or aluminium oxide) supported on a flat, inert surface such as a glass, metal or plastic plate. A small amount of the mixture to be analysed is put near the bottom of the plate in the form of spots.

Each component of the mixture is adsorbed on the solid because of their differences in solubility and in the strength of their **adsorption** to the stationary phase substance. Some components will be carried further up the plate than others. When the solvent has reached the top of the plate, the plate is removed and dried.

For coloured compounds, analysing the results is generally simple. For colourless samples, an ultra-violet (UV) lamp is used because many organic compounds absorb UV light.

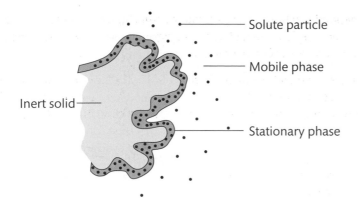

Solute particle

Mobile phase

Inert solid

Stationary phase

▶ **Figure 2.18:** Each solute partitions itself between the stationary phase and the mobile phase

TLC has some advantages over paper chromatography.

▶ The mobile phase moves more quickly through the stationary phase.

▶ The mobile phase moves more evenly through the stationary phase.

▶ There are a range of absorbencies for the stationary phase.

TLC tends to produce more useful **chromatograms** than paper chromatography. TLC chromatograms show greater separation of the components in the mixture than paper chromatograms and are therefore easier to analyse. Of course, how dependable the results are and whether you can repeat those results will be linked to how constant you keep:

▶ the solvent used

▶ the amount of stationary phase (absorbent) used on the plate

▶ the amount of substance spotted onto the plate

▶ the temperature controls.

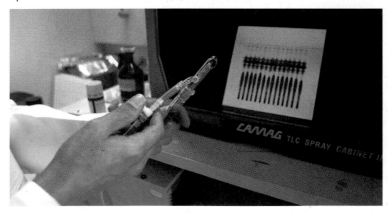

▶ TLC apparatus of the type which is commonly used. You will notice that there are a number of spots indicating multiple chemical substances under test.

Theory into practice

Industrial and research laboratories may have a number of different types of chromatographic devices. The variety is related to their specialist method of separation, although all of them carry out the same basic function.

On entering an establishment as a junior member of the technical staff, a valid and useful exercise would be to list the different chromatographs available in the department and carry out research on their differences and similarities. Identify where each differs in design and materials, and highlight where they are different in 'style' only. You could also provide the uses for which each is best suited.

Key terms

Adsorption – the process by which atoms, molecules or ions from a gas or liquid adhere to a surface. The process is not permanent.

Chromatogram – the resulting paper or plate produced showing the substance separation.

Safety tip

Some solvents can be toxic, flammable and quite costly. The following provides a guide for which solvents to use:

• alcohols (methanol and ethanol)

• acetone

• acetic acid (corrosive and vapours are irritating)

• hexanes (petroleum ethers are flammable)

• diethyl ether (volatile and flammable).

You can use different percentage compositions. An ideal mobile phase is one in which the substances are only partially soluble and have differing solubilities.

Preparing your samples

Many substances requiring sampling in chromatography are not in a suitable form which can be readily used in the process and will need preparation. Samples not prepared effectively can be a significant cause of errors in the sampling process as a result of contamination, for example.

The sample to be analysed must have a high concentration but contained in a small volume to ensure good separation of the solute from the solvent. Many small drops are put into the same spot and the solvent allowed to evaporate to concentrate the sample. There are various methods that you can use to prepare your samples.

Solvent extraction

This process provides a method of separating two compounds from a chemical mixture based on the differences of their solubilities. When petrol is added to water, for example, the two do not mix. They are **immiscible**.

If a chemical compound is already dissolved in the water but dissolves more readily in the petrol, the compound will move to the petrol after the mixture is shaken. When left to stand, the water and petrol will separate with the chemical compound now dissolved in the petrol and *not* the water.

Filtration

This process involves passing a mixture of solids and liquids through filter paper placed in a funnel. The solids are insoluble. The liquid flows through the filter paper. The solid residue is left behind in the filter paper. If the filter paper is fluted (folded), then results are improved. You should rinse any solid left in the filter paper in the liquid again and pass it through the filter paper again to complete the process. Paper filtration is a basic technique using a semi-permeable paper barrier placed perpendicular to the flow of liquid. In scientific laboratories, the filter paper is placed in a filter funnel, Hirsch or Buchner funnel.

Choosing the filter paper that has the correct **porosity** is crucial in the filtering process of the medium and the length of time needed for it to filter. Vacuum filtration is sometimes used to speed up the process. However, since this process is carried out with a reduced pressure, the glassware must be checked thoroughly for signs of cracking and the vacuum apparatus must be secured to prevent movement. There must be a vacuum trap in place and strong rubber tubing must be used. This process is not suitable for low boiling point solvents which can evaporate in the vacuum and the precipitate can block the filter paper pores.

Evaporation

This process involves separating the soluble chemical compound by means of removing excess water from a chemical solution by heating the solution. Removing NaCl and other salts from sea water is an example. The heating process must be carried out carefully and slowly, so that over-evaporation of the solution does not occur. Repeated evaporation cycles can lead to increased concentration of the sample to be separated, which can then be used to determine the identification of the substance with more clarity.

Locating agents

Samples, such as amino acids, are colourless and need some means to make them visible for analysis. Locating agents are chemical substances which are added to the samples by a spray in many cases. They react with the sample and change the sample colour as a result.

Key term

Immiscible – liquids that do not mix together.

Porosity – a measure of the volume of tiny holes (pores, from the Greek 'poros') in a material divided by the total volume of the material.

Ⅱ   PAUSE POINT   It is useful for a science department to have clear and well-presented information about practical investigative techniques. As a means of confirming your knowledge and understanding of chromatography, produce a suitable aid which could be used by all.

       Hint   Perhaps, a large poster, set of informative notes or instruction document – written by learners for learners.

       Extend   Include your own glossary of terms, relevant and detailed diagrams and additional examples of the use of the techniques to those in this student book.

Safety tip

Propanone (acetone) is highly flammable. Treat with caution and keep away from naked flames. The solvent is also an irritant if it gets into your eyes, so you must wear goggles.

Applications of chromatography

Investigation 2.5

Plant pigment extraction

Chlorophyll is responsible for the green colouration of leaves in plants and allows the process of photosynthesis to take place, providing the plants with energy. A leaf contains more than one pigment.

Steps in the investigation	Pay particular attention to...	Think about this...
1. Place a few leaves from the same plant into a mortar.		There is no benefit in adding more than one type of leaf because we need to control the variables in the investigation.
2. Add a small amount of grit or sand (to break the cell walls) and approximately six drops of propanone.	A small amount of grit or sand is needed help tear and expose the inner leaf.	
3. Grind the mixture for a few minutes.		
4. Put a pencil line 3 cm from the bottom on the TLC plate.		
5. Use a micro-capillary tube to put a small spot in the centre of the pencil line.	Practise putting the spots onto paper before doing it 'for real'. This will help you to develop a good technique.	Adding further drops on top of the initial drops will help to concentrate the pigments for better results.
6. Add the solvent to the beaker.	You may need to add a small quantity at first.	
7. Place the TLC plate into the beaker and ensure that the level of solvent is below the spot.	Add more solvent if the initial level is too low.	
8. Allow the solvent to rise until it stops. Mark this point with a pencil line.	You will need to observe the solvent movement over a suitable time period.	Make a small mark to gauge when the solvent has stopped moving up the plate.
9. Calculate the R_f of each pigment.		

Key terms

Polypeptides – a long chain of amino acids (and, therefore peptides) producing proteins of a high molecular weight.

Peptides – a chemical compound made of two or more amino acids.

Amino acid identification

Cell proteins are made of polymers called **polypeptides** which consist of amino acids linked by **peptide** bonds. The exact structure of the polypeptides determines the biological function of the protein. There are 20 amino acids, but the sequence and type of amino acids varies with different polypeptides. All amino acids have a common structure with a central carbon atom (alpha carbon) linked to other molecules, including an R-Group (hydrogen or carbon chain bonded to the central alpha carbon). It is the R-Group that differs between amino acids and allows identification of each.

Investigation 2.6

Finding R_f values for amino acids

Amino acids are the very building blocks of proteins which are large molecules found in all living things. Proteins are needed for growth and repair of body systems, such as the immune system and are essential for metabolic processes. Amino acids can be identified by using chromatography but are colourless compounds. A solution called ninhydrin is added to produce a purple colour.

Steps in the investigation	Pay particular attention to...	Think about this...
1. Put a pencil line across the chromatography paper 3 cm from the bottom.		
2. Place a small pencil dot every 2.5 cm apart for each of the amino acids under test and label them.	Be sure to label the pencil dots before testing.	
3. Pipette 2 ml of an amino acid onto one of the dots. Repeat for the other two, using a different pipette for each.		Using a fresh pipette for each amino acid is needed to avoid contamination of the spots which will result in separations.
4. Repeat this another four times.	Each dot should applied at least four times to develop a suitable concentration of all amino acids.	
5. Allow to dry and roll the paper into a cylinder. Be careful not to contaminate the paper with your fingers.		If you contaminate the paper through poor handling technique, the results will be invalid.
6. Add the solvent to the beaker.	Be careful when adding the solvent to avoid splashing the paper in random areas.	
7. Place the cylinder into the beaker and ensure that the level of solvent is below the line.		
8. Allow the solvent to rise until it stops. Mark this point with a pencil line.		You will need to make clear observations to gauge correctly when the solvent has stopped rising.
9. The technician will now spray ninhydrin onto the paper and heat-develop in an oven.	Ninhydrin is flammable, harmful if swallowed, an irritant on skin and respiratory system, and can cause dizziness if inhaled.	
10. Calculate the R_f of each amino acid.		

Research

- Find out the R_f values of the amino acids: alanine, leucine and aspartic acid. Do they match your values from the investigation?
- What possible reasons can you suggest for any differences between your values and the professional research figures?

Ⅱ   PAUSE POINT   From the investigations you have carried out for plant pigment extraction and amino acid identification, identify and comment on any of the procedures which needed slight changes in order to provide suitable results.

Hint   Look at the preparation of the sample, the effectiveness of the solvent used, the position of the pencil line and the clarity of the separation of colours.

Extend   Suggest amendments or additions to the procedures for future learners and/or laboratory technicians.

Other types of chromatography

Gas-liquid chromatography

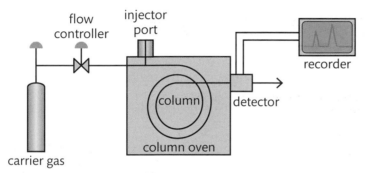

▶ **Figure 2.18:** Gas-liquid chromatography system. Can you identify any differences between this system and TLC?

This is how gas-liquid chromatography works.

- ▶ A liquid sample is injected into the oven.
- ▶ The oven boils the sample to produce a vapour.
- ▶ The vapour is carried by inert gas (such as helium) through a column (steel tube packed with porous rock).
- ▶ Molecules of sample move with the gas (mobile phase) through the system and into contact with a liquid solvent (stationary phase – helium gas) adsorbed onto the solid material.
- ▶ The time taken for the sample to pass through the machine to the detector on the column is the retention time and depends on the solubility of the sample in liquid or gas solvents. There are a number of types of detector in current use. One particular method, Thermal Conductivity Detector, uses the temperature difference of burning between two streams of gas; one stream with no compound and one stream with the compound in it. The temperature difference provides an electrical resistance difference which helps to identify the substance.
- ▶ The temperature of the oven is controlled at stages.
- ▶ The display on the processor shows a series of peaks indicating the retention time and so identifying the sample. The area under the peak is a measure of the amount of the substance present in the sample.

This technique has many useful applications. The process can be used to analyse the concentration of alcohol in the blood of a suspected drink-driver, determine the concentration of animal fats in vegetable oils and also identify the chemical substances in an athlete who has been banned by sports awarding organisations.

Ion-exchange chromatography

This method of chromatography is the preferred process for purification of proteins and other charged molecules including amino acids. The procedure is also used in water analysis.

The method relies on the attraction of oppositely charged ions between the mobile phase and stationary phase. Typically, a low concentration salt mobile phase will interact with the stationary phase ions weakly, thereby **eluting** first. Identification of chemical species in samples shows as sharp spikes on a graph of conductivity/time.

▸ **Cation** exchange – The immobilised or stationary phase in this type of chromatography is charged positively. The ionic interaction between negatively charged ions in the sample and positively charged ions in the stationary phase is strong and the negatively charged ions in the sample bind with the stationary phase. Positively charged ions in the sample will be removed (elute).

▸ **Anion** exchange – The stationary phase in this process is negatively charged. Proteins charged positively will interact with the negatively charged stationary phase molecules and negatively charged molecules will elute from the column.

The pH of the mobile phase is important. If the pH is raised, the number of available hydrogen ions decreases and so the mobile phase becomes less positive.

By lowering the pH of the mobile phase, the proton availability increases. This makes the mobile phase more positively charged.

The advantage of this type of chromatography is its speed at producing results. In environmental analysis, identification of anions, for example, could take less than 20 minutes while other techniques could take more than two days.

Amino acids are sensitive to pH changes. If the pH is adjusted for a given protein, for example, then this may develop a net positive or negative charge allowing it to be eluted in the column, depending on the net charge of the column. So, by adjusting the pH, a variety of proteins can be targeted for separation.

Ion-exchange chromatography (anion exchange)

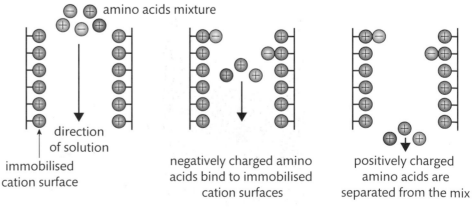

▸ **Figure 2.19:** Negative/positive ion interactions for separating out amino acids. The charge on the solid support will be negative during cation exchange.

Water is sometimes referred to as 'hard' or 'soft'. 'Hard' water contains salts of substances that are less soluble than others, such as calcium and magnesium. These salts dissolved in water produce cations (positively charged ions) of 2+ charge. Sodium and potassium salts dissolved in water also produce cations, but are only of 1+ charge.

As a new member of a water treatment laboratory, find out details of the type of resin used in ion-exchange to 'soften' the water. You will need to look closely at the cations in the water and how ion-exchange removes the hardness salts from the water.

PAUSE POINT   The pH level of the mobile phase in ion-exchange chromatography is important. If the mobile phase were tested as neutral in pH, what effect, if any, would you expect it to have on the resulting chromatographic procedure?

Hint   Consider what raising and lowering the pH does to the ionic interaction between the mobile and stationary phases.

Extend   Suggest an addition to your procedure to ensure that you have a mobile phase at the correct pH which offers the clearest possible results.

Theory and principles behind chromatography

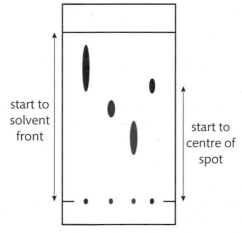

▶ **Figure 2.20:** A TLC chromatogram showing the position of spots and measurement to the centre of each substance evolved

▶ A TLC chromatogram. Compare the traces on this plate with your own investigations. Are they similar?

Polar molecules – molecules without an equal distribution of electrons, causing them to have opposite electrical poles.

Non-polar molecules – molecules with an equal distribution of electrons, resulting in no observable electrical poles.

The way in which the atoms are arranged in the molecules of a substance and by differences in electronegativity between the attoms in a molecule determine whether it is a **polar molecule** or a **non-polar molecule**.

Polar molecules exhibit a positive charge on one end of the molecule and a negative charge on the other. Non-polar molecules do not show this simplicity of opposite charges and so do not have the same characteristic behaviour in terms of their solubility and miscibility.

This characteristic of molecules can be explained if you look at how atoms bond together. Water has the chemical formula H_2O and the hydrogen atoms which are covalently bonded to the outer electron shell of the oxygen atom produce a net positive pole to the negative pole of the electrons in the oxygen. Water is therefore a polar molecule. Ethanol is an example of another polar molecule.

When carbon bonds with two oxygen molecules to form carbon dioxide (CO_2), the carbon atom is essentially positioned between the two oxygen atoms and the electrons are evenly distributed. There are no observable positive and negative poles. The molecule is non-polar. Petroleum is another example of a non-polar substance.

▶ **Figure 2.21:** The polarity of a water molecule

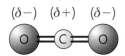

▶ **Figure 2.22:** The carbon atom positioned between the oxygen atoms (a non-polar molecule)

Discussion

Generally, polar molecules mix to form solutions, non-polar molecules mix to form solutions, but polar and non-polar molecules cannot mix to form solutions. Polar molecules are soluble in water. Non-polar molecules are soluble in fats.

Discuss how these facts may explain how you can use chromatography to separate chemical substances, particularly in the mobile phase solvents and the stationary phase molecules. Can you work out which of these molecules is polar and non-polar?

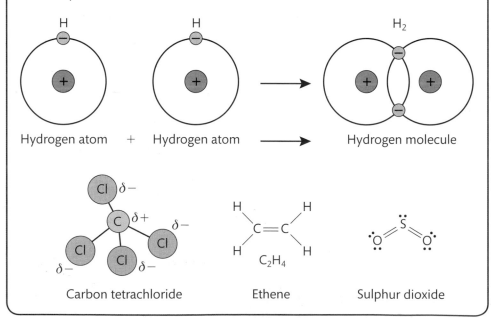

Knowing the polarity of your solvents gives you a clear indication of the speed at which the components in the mixture move up the stationary phase and separate out. More polar solvents allow the components of the mixture to move faster than less polar solvents.

The strength of bonding of the molecules in the solvent to the adsorbent also depends on the interaction with the dipoles, bonding and molecular forces present. With TLC, for example, strongly polar molecules interact well with the polar silicon hydroxide $Si(OH)_4$ adsorbent, while molecules that are less polar move quite quickly through the system.

Molecule size and solubility

The overall size of a molecule plays an important part in its ability to be soluble in water. You have probably already carried out investigations into rates of chemical reactions and the factors involved. One factor is the molecule size. Molecule size has a direct physical influence on the speed at which a substance can dissolve. Smaller molecules dissolve more quickly in a solvent because they are more able to fit into the inter-spacial areas within the larger molecules of the solvent. Temperature is also very important in the solubility of solids. When the temperature increases, solutes in solvents lose their bonding strength as a result of the increase in kinetic energy of the solvent molecules. The solid becomes more soluble.

The alcohols, however, can also be used to explain dissolving in terms of solubility, because they have very similar chemical formulas and are soluble in water. Methanol (CH_3OH) is a small molecule in comparison to other alcohols but has an -OH group attached to the carbon atom. This allows the possibility of a **hydrogen bond** while the O-H bond is also polar and so methanol is totally miscible in water.

The more similar a molecule is to the mobile phase solvent or gas then the more likely it is to transfer into the mobile phase from the stationary and the further along it will travel (e.g. anything with OH is likely to have an affinity to water or an alcohol solvent). Size generally makes molecules slower as they are bigger and less likely to be soluble.

> ### Key term
>
> **Hydrogen bond** – a force of attraction between a very strongly de-shielded hydrogen atom's nucleus and the lone pair of electrons on an electronegative element on another molecule.

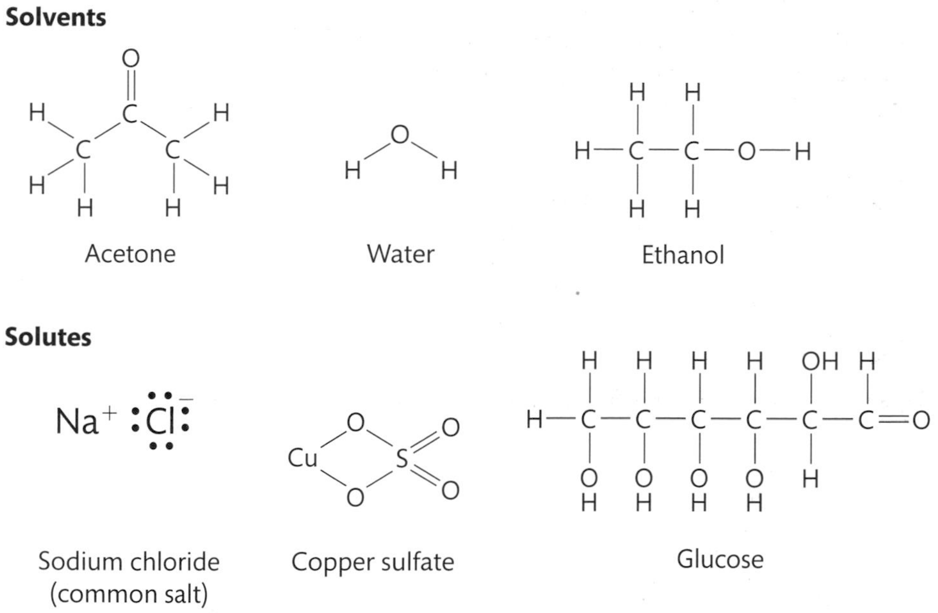

Solvents

Acetone   Water   Ethanol

Solutes

Sodium chloride   Copper sulfate   Glucose
(common salt)

▶ **Figure 2.23:** Which solutes will readily dissolve in which solvent?

Molecule size and mobility

The molecules of substances are different in size and weight. This physical property will allow separation of chemical substances in a process similar to a mechanical sieve. In a system known as 'gel permeation chromatography', the stationary phase is a polymer gel. The gel is porous and smaller molecules can permeate through, being held by the stationary phase. Larger molecules will travel further in the mobile phase.

This system is used in the bio-chemicals industry, for example, to separate small molecules from proteins and enzymes.

Calculating R_f

The **retention factor** (R_f) is defined as the distance travelled by the compound divided by the distance travelled by the solvent.

$$R_f = \frac{\text{distance travelled by solute}}{\text{distance travelled by solvent}}$$

Worked Example

A solvent travels 2.7 cm from the bottom mark to the point at the top where it comes to rest. The compound which separates out of the solvent reaches a point 2.2 cm from the bottom mark. Calculate the R_f of the compound.

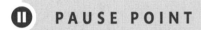

$$R_f = \frac{\text{distance travelled by solute}}{\text{distance travelled by solvent}}$$

$$R_f = \frac{2.2}{2.7} = 0.81$$

If the stationary phase is polar or ionic in character, then generally the more polar a compound, the lower its R_f value, because the substance interacts with the stationary phase. The reverse is also true. The R_f values of two spots on a TLC plate, for example, can provide some evidence as to the identification of a chemical compound.

PAUSE POINT

Using TLC, two compounds (A and B) are tested under identical conditions with a polar absorbent. Compound B has a larger R_f value. What does this tell you about the polarity of compound B?

Hint Decide what polarity will interact with a polar adsorbent.

Extend If both R_f values appeared to be the same, what would you do to check your result? (Consider all factors.)

Case study

Affecting R_f values

Brady works in a busy undergraduate laboratory in the Faculty of Science. His work is based in the Schools of Pharmacy and Chemistry. He is part of a team of technicians responsible for the preparation and clearing of chemicals and materials used in undergraduate teaching sessions and helps in all activities as required.

Brady moves a large number of containers of the faculty's supply of solvents to another store room which is heated with a large radiator, has good shelving and a window with a sill. The window is kept open, even on days when the temperature outside is freezing. One particular solvent – ethyl alcohol – is contained in two identical glass bottles which are clearly and correctly labelled. However, each bottle is placed in a slightly different location in the store room.

Following a practical session involving separation of food dyes using thin-layer chromatography, the tutor recognises that the R_f values obtained were slightly lower for all tests for group A when compared to group B. Each group were given one separate bottle of the solvent.

Check your knowledge

1 What different positions could have been used?

2 What effect does temperature have on the solvent?

3 Which group, A or B, had the bottle of solvent stored close to the radiator?

4 Could either group's set of results be relied upon?

5 What would you suggest to the technician in terms of future storage?

Chromatography – problems in technique

Here are some problems which can arise as a result of poor technique in chromatography.

▶ **Correct spotting of samples** – if your samples are not spotted above the solvent level, they will not travel with the solvent but are more likely to be washed into the solvent before the mobile phase occurs.

▶ **Uneven movement of the solvent in TLC** – using water as the solvent is not advised because of its surface tension which produces a curved solvent front and possible errors in the R_f calculation. Use a large amount of solvent so that the mobile phase can advance to its full limit. Ensure that the plate is positioned flat against the chamber floor and is not tilted. Use a ruler or even spirit level to check this.

▶ **Over-concentration of spot** – if this occurs during your preparation, streaking will happen rather than separations. Make sure that your sample is accurately prepared in accordance with the procedures outlined in this section.

▶ **Excessive spot sizes** – generally, spot sizes must not be larger than 1 mm to 2 mm in diameter. If the spot size is too large, overlapping can occur with spots of similar R_f values. As a result, the differences in R_f cannot be easily calculated.

You also need a constant temperature. Do not move the sample!

Assessment practice 2.3   C.P5 C.P6 C.M3 C.D3

It is often very difficult to keep the necessary factors constant for each chromatography experiment. Carry out paper chromatography of a selection of popular sweets (e.g. Smarties or M&Ms).

You will need to ensure safe working practices and have two experiments running at the same time with the same sweets. This will be used for comparison. Once completed, calculate all R_f values and compare food dyes with each other. Analyse the chromatograms and relate the factors that affect the R_f values to the quality of results obtained. Evaluate your techniques in relation to the results and suggest experimental improvements.

Plan
- What is the task? What am I being asked to do?
- How confident do I feel in my own abilities to complete this task? Are there any areas I think I may struggle with?

Do
- I know what it is I am doing and what I want to achieve.
- I can identify when I have gone wrong and adjust my thinking/approach to get myself back on course.

Review
- I can explain what the task was and how I approached the task.
- I can explain how I would approach the hard elements differently next time (i.e. what I would do differently).

Review personal development of scientific skills for laboratory work

Undertaking employment in all areas of life is not simply a matter of turning up at the correct time, completing the set work to appropriate and acceptable standards and then leaving the place of work at the correct time. All employees are expected to ensure that their practices are **evaluated** in order that possible errors are limited, **skills** are improved and **personal development** and achievement in work can take place.

Key terms

Evaluate – to make a judgement and determine the value, amount, quality or importance of something.

Skills – the abilities required to do something well or expertly.

Personal development – improving yourself through a range of activities.

This foundation ethos is of great importance in industrial settings where scientific principles form the basis of the work completed.

Link

Go to *Unit 4: Laboratory Techniques and their Application, Learning aim A.*

▶ Carrying out a routine investigation. Can you see any positive or negative aspects to the way in which this activity is being performed?

Personal responsibility

Working in the science-related industry is a demanding but very rewarding occupation. It requires a particular set of skills, personal qualities and attitude from potential employees. Look at the following list of abilities which appear in most job advertisements for technical staff.

▶ Follow strict safety procedures and safety checks.

▶ Keep up to date with scientific developments.

▶ Identify ways of saving time and improving reliability in practical activities.

▶ Use computers to perform calculations, graphs, research and communication.

▶ Identify when stocks and supplies need re-sourcing.

▶ Prepare appropriate samples for testing.

▶ Perform essential lab tests to determine, for example, concentration values for standards.

▶ Be able to use all laboratory equipment effectively.

▶ Perform essential maintenance of laboratory equipment and apparatus.

▶ Record results accurately and report to management.

PAUSE POINT   Which of these attributes do you feel you have at the moment?

Hint   Refer back to your scientific investigations and critically assess your performance in at least two of them. Include an example of group work and appraise the performance of another member of your group.

Extend   Provide a document in the form of a formal probationary appraisal of a colleague in no more than one A4 page.

Figure 2.24 shows four personal responsibilities that are the most fundamental attributes required by all people working successfully in the science industry. You should try to develop these qualities throughout your study of this qualification.

▶ **Figure 2.24:** Four personal responsibilities that are fundamental attributes required by all people working successfully in the science industry

Work to appropriate standards and protocols

In order to comply with the fundamental expectations of working in a laboratory or other technical environment, you must ask yourself: How prepared am I? (See Figure 2.25.)

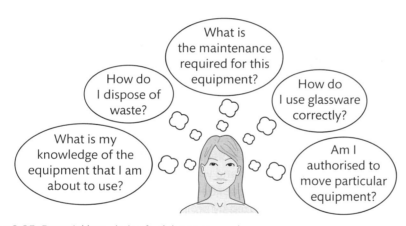

▶ **Figure 2.25:** Essential knowledge for laboratory work

Table 2.5 identifies examples of standard equipment and their uses.

▶ **Table 2.5:** Standard laboratory equipment

Equipment	Use
Top pan balance	Measurement of mass of solids, liquids and other materials
Microscope	To magnify an object or specimen to provide clarity and depth of viewing
pH meter	Measurement of the acidity or alkalinity of a chemical solution
Graduated pipette	Accurately calibrated glassware for measurement of liquids
Burette	Graduated glassware with release tap used in titration standardisation

Standard Operating Procedures (SOPs) are those procedures used to ensure that routine and irregular activities are performed to a set standard which are common in the technical industry. The procedures are repeated and are always carried out in the same way. Examples are: chemical testing, handling of materials waste, operation and calibration of laboratory equipment.

Key term

Standard Operating Procedures (SOPs) – established procedures or methods for the completion of a routine operation.

Getting ready for assessment

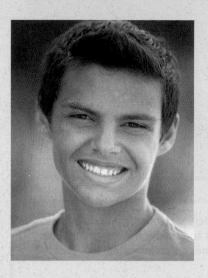

Matthew is working towards his BTEC National Subsidiary Diploma in Applied Science. He was given an assignment at the start of the first term which was based on finding out the concentration of sulfuric acid in a sample. The assignment provided a scenario linked to a wastewater treatment plant which uses hydrochloric and sulfuric acid in the process. Random sampling has found that a lowering of pH of a nearby pond may be a result of sulfuric acid contamination. Matthew has to complete a full report on a 'sample' provided after carrying out suitable tests.

The report must include:

▶ formal sections based on good scientific reporting
▶ safety guidelines
▶ results of a full practical investigation with data and analysis.

Matthew shares his experience below.

How I got started

First I made sure that my laboratory notebook was available and that I had a clear understanding of what I needed to do. I planned the activity, listing all the apparatus and solutions I thought I would need and drew up a table for my expected results. I obtained a $1 \, Mol \, dm^{-3}$ solution of sodium hydroxide as a suitable alkali for the titration of the sample of sulfuric acid and set about checking the sample with a crude pH litmus test.

I performed the titration under strict scientific controls and repeated the activity to make sure of the final result to use in my conclusion and analysis. At all times I made sure that safety aspects were taken into account by wearing goggles and a lab coat. All solutions were labelled using a permanent marker.

How I brought it all together

I used a preliminary test on the solution to guide me with the values expected on the burette and managed to keep to a $25 \, cm^3$ volume of sodium hydroxide in the conical flask. I then:

▶ put all the results into the appropriate table
▶ completed my report to the conclusion stage with clear diagrams and method
▶ evaluated the activity based on the results obtained and after producing a pH curve.

After completing most sections in my report and evaluating the work based on my results and graph, I was able to complete an abstract for the report outlining what the activity had 'told' me about the sample and how my procedures were able to confirm my findings.

What I learned from the experience

I carried out a preliminary test which gave me a good understanding of the expected value of concentration. I was also pleased with the way in which I was able to carry out the procedure, even though I have not completed one for at least two weeks. The activity was made more 'real' because of the realistic setting given to me.

I became a little confused when the number of glass beakers containing acids and alkali waste solutions began to fill the bench top, especially since I was forced to obtain more glassware following a slight breakage. I should have immediately marked them with permanent marker.

Think about it

▶ Have you made a clear note of the agreed submission date of the assignment?

▶ Do you have your previous class notes and laboratory notebook experiments to hand in order to use them for reference with procedures?

▶ Is your final report written in your own words and referenced clearly where you have used quotations or information from a book, journal or website?

Science Investigation Skills 3

Getting to know your unit

Advances in science have produced great benefits for society. These advances depend on research and carrying out scientific investigations. In this unit, you will acquire the skills needed to plan a scientific investigation and record your results appropriately. You will learn how to process and analyse your results, draw scientific conclusions and evaluate your work. Science investigation skills will help you in many scientific or enquiry-based learning courses in higher education, as well as preparing you for employment in a science-related industry.

How you will be assessed

For this assessment, you will be given a method for a practical investigation which you are required to carry out. You will record your results and observations on an observation sheet which will be collected in by your tutor at the end of the practical session. You will not be directly assessed during the practical work, but you will need your results and observations, which will be returned to you for the written test.

The 90-minute written test will be divided into two sections. The main part of this assessment, Section A, will be related to the practical investigation and you will need your observation sheet to complete this section. This section will involve recording, processing, analysing and evaluating both primary and secondary evidence. Section B will involve writing a plan for a scientific investigation. This plan will not be related to the practical investigation carried out for Section A. When you have completed the assessment, your test paper will be sent in to Pearson to be marked by a Pearson examiner.

Throughout this unit you will have plenty of opportunities to practise the skills that you will need to complete the final assessment.

Assessment criteria

This table shows what you must do in order to achieve a **Pass** or **Distinction** grade in this unit.

Pass	Distinction
Demonstrate a sound knowledge and understanding of scientific concepts, procedures, processes and techniques and their application within a practical context. Interpret and analyse your own data and secondary data, leading to reasoned judgements on the qualitative and quantitative data you have collected during your investigation. You will be able to draw links between different scientific concepts, procedures, processes and techniques to make a hypothesis and plan an investigation. You will be able to make evaluative judgements on scientific data, processes and procedures which make reference to scientific reasoning.	Demonstrate a thorough understanding of how scientific concepts, procedures, processes and techniques can be integrated and applied within a practical context. Interpret, analyse and evaluate your own collected data and secondary data, to support judgements and conclusions drawn. You will be able to use and integrate knowledge and understanding of scientific concepts, procedures, processes and techniques to make a hypothesis and plan an investigation which is fully supported by scientific reasoning. You will be able to provide rationalised evaluative judgements on scientific data, processes and procedures which are fully supported by scientific reasoning.

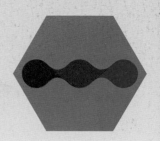

A Planning a scientific investigation

Before undertaking a scientific investigation, it is important to write a detailed plan. In this section, you will learn about what you need to include in your plan and all the factors you must take into account when writing your plan.

Writing a hypothesis for an investigation

When planning a scientific investigation, you need to think about what you are trying to find out from the investigation. You should also think about what type of trend you would expect to see from your results, and make a prediction based on this expected trend. This prediction is your **hypothesis**. In most cases, a hypothesis is an assumption based on your knowledge, understanding of the topic and observations.

In an investigation into which chemical elements are necessary for plant growth, your hypothesis could be: The more nitrogen that is supplied to a plant, the faster it will grow. Your observations will show that other elements are also needed, so your hypothesis could be changed and further investigations carried out.

> **Key term**
>
> **Hypothesis** – a prediction, based on scientific ideas, made as a starting point for further investigation.

Discussion

Suppose you have been asked to plan an investigation to study the effect of temperature on the rate of reaction between magnesium ribbon and hydrochloric acid. What would be your hypothesis for this investigation?

In some cases, you may wish to make a **null hypothesis**. This applies to situations where you do not expect to find a particular trend or pattern in your results. It is often the case that after carrying out an investigation you are able to reject the null hypothesis.

Selection of appropriate equipment, techniques and standard procedures

When planning both qualitative and quantitative scientific investigations, you need to know what equipment to use and how to use it.

> **Key term**
>
> **Null hypothesis** – a prediction which states that there is no relationship between two variables or no difference among groups.

Equipment

When writing a plan for your investigation, you need to be able to choose appropriate equipment to use in your investigation and explain why you have chosen to use this equipment. For example, when doing a quantitive experiment such as an acid-base titration it would not be appropriate to use a measuring cylinder to measure out 25 cm^3 of acid, as the measurement would not be precise enough as in titrations measurements need to be taken to 1 decimal place. For quantitive investigations it would not be necessary to use precise measuring equipment. \

qualitative

Practical techniques

You must also be able to describe any practical techniques that you intend to use in your investigation. For example, when purifying a solid by re-crystallisation, the technique you need to use is shown by the following steps:

Heat the solvent. ▶ Add the minimum amount of hot solvent needed for the solid to dissolve. ▶ Allow the solution to cool. Crystals will appear. ▶ Place in an ice-bath so that more crystals will form. ▶ Using filter paper, separate out the crystals and wash them with a small amount of cold solvent. ▶ Scrape the crystals onto a watch glass. ▶ Leave them in a warm place to dry.

Standard procedures

When planning your investigation, you need to be aware of any standard procedures you need to adhere to.

Standard Operating Procedures (SOP) are in place in many laboratories and can cover many different aspects of the work. These could include the following points.

▶ How tests are carried out.
▶ How chemicals should be handled.
▶ How waste should be disposed of.
▶ How equipment should be used and maintained.

Health and safety issues

When planning an investigation you need to carry out a **risk** assessment. This involves identifying the **hazards** and risks associated with the method you are using for the investigation and then deciding on the best way to minimise the risk.

Example

If you are using a Van de Graaff generator to learn about electrostatics, you would need to consider the following:

▶ hazard – static electricity.
▶ risk – possibility of electric shock.

When a person is being charged by placing their hand on the dome of the Van de Graaff generator, you would need to minimise the risk by making sure they follow the safety tips shown here. To avoid a serious accident, do not allow a person with a heart condition or anyone fitted with a pacemaker or other electronic medical appliance to touch either the Van de Graaff generator or any other person who has been charged by the generator.

PAUSE POINT   In your investigation to study the effect of temperature on the rate of reaction between magnesium ribbon and hydrochloric acid, you need to do a risk assessment before starting your investigation.

Hint   Identify two hazards in this investigation.

Extend   What are the risks associated with these two hazards?

What could you do to minimise these risks?

Variables in the investigation

When planning an investigation, it is important to consider the **variables** that are involved. To make sure the investigation is valid, it is important that you only change one variable and that all other variables are kept constant.

▶ The variable that you are going to change is the *independent* variable.

▶ The variable that may change as a result of changing the independent variable is the *dependent* variable. This is the variable that you will measure.

▶ The variables that you need to keep constant are the *control* variables.

Example

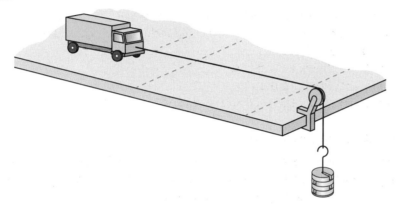

▶ **Figure 3.1:** Investigating how long it takes for a truck to move along a bench

In an investigation to find how long it takes for a truck to move along a bench when different masses are used to accelerate it (see Figure 3.1), the variables are as follows:

▶ independent variable – the mass added

▶ dependent variable – time taken

▶ control variables:
  • the distance travelled by the truck
  • the surface on which the truck travels
  • use the same truck each time.

> **Key term**
>
> **Variables** – factors that can change or be changed in an investigation.

PAUSE POINT   Identify the independent and dependent variables in an investigation to study the effect of temperature on the rate of reaction between magnesium ribbon and hydrochloric acid. In this investigation, strips of magnesium ribbon are placed into excess hydrochloric acid at different temperatures and the time taken for the magnesium ribbon to disappear is recorded.

Hint   Which variables do you need to control in this investigation?

Extend   Why is it important to control all variables apart from the independent and dependent variable?

Method for data collection and analysis

An important part of a scientific investigation is to be able to write a clear, logically ordered method.

This method should include the following.
▶ A list of the apparatus you will use, including a labelled diagram if appropriate.
▶ Step-by-step instructions on how you will perform the investigation.
▶ The number and range of measurements that you will take. For example, in the investigation into the effect of temperature on the rate of reaction between magnesium and hydrochloric acid, a suitable range of temperature would be from 10 °C to 60 °C, as below 10 °C the reaction would be too slow and above 60 °C it would be too fast to time accurately.
▶ The number of repeat readings you will take.

Accuracy and precision

When planning a scientific investigation, you need to understand the importance of obtaining your data to an appropriate degree of **accuracy**, **reliability** and to appropriate levels of **precision**.

If repeat readings are taken and are the same or very similar, they have good precision. It is also likely that the results are reliable and close to the true value and therefore accurate. For example, two titration results that are within 0.1 cm³ of each other are likely to be accurate.

Precision also depends on the apparatus used. For example, a balance reading to 0.001 g is more precise than a balance reading to 0.1 g, because the degree of uncertainty of the measurement is less.

<aside>
Key terms

Accuracy – how close the readings are to the actual values.

Reliability – how trustworthy the data is.

Precision – how close repeat readings are to each other.
</aside>

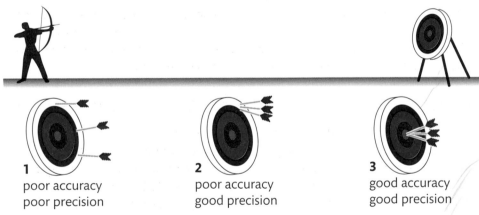

1
poor accuracy
poor precision

2
poor accuracy
good precision

3
good accuracy
good precision

▶ **Figure 3.2:** Archery analogy showing the difference between accuracy and precision

Figure 3.2 uses archery to demonstrate the difference between accuracy and precision. The shooting is more accurate when the arrows are close to the centre of the target. The shooting is more precise when the arrows are close together. Boards 2 and 3 are also reliable.

Variables and data analysis

When planning a scientific investigation, you need to know:
▶ how to control the variables that you need to control, for example, use a water bath to control temperature
▶ how to measure or monitor the dependent variable, for example, use a stopwatch to measure time for a toy car to travel down a ramp
▶ the best way of recording your data
▶ how you are going to analyse the data or information that you have collected.

Assessment practice 3.1

Hydrochloric acid reacts with magnesium to produce hydrogen gas.
The equation for the reaction is:

$$Mg + 2HCl \rightarrow MgCl_2 + H_2$$

You have been provided with different concentrations of hydrochloric acid, and magnesium ribbon.

You are to plan an investigation into how changing the concentration of hydrochloric acid affects the rate of reaction between hydrochloric acid and magnesium.

Your plan should include:

- a hypothesis
- selection and justification of the equipment you are going to use
- hazards and risks associated with the investigation
- independent, dependent and control variables
- a method for data collection to test the hypothesis including:
  - the quantities to be measured
  - the number and range of measurements to be taken
  - how the apparatus may be used.

B   Data collection, processing and analysis and interpretation

In this section you will learn how to collect data and record it appropriately. You will learn the different ways in which you can process your data, including appropriate mathematical techniques and how to plot suitable graphs. You will also learn how to analyse and interpret the data and identify **anomalous results**.

> **Key term**
>
> **Anomalous results** – results that do not appear to fit the trend in the data.

Collection of quantitative and qualitative data

Taking accurate reliable and precise measurements

When you are collecting data, it is important that you take the measurements accurately and to the appropriate level of precision. For example, when reading a volume on a burette or measuring cylinder, it is important to take the reading at eye level and at the bottom of the meniscus every time, to eliminate human error and make sure your measurements are reliable. Figure 3.3 shows an example.

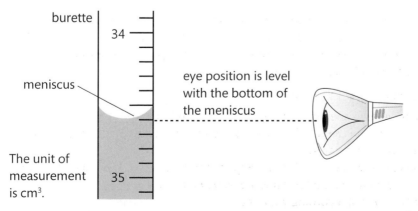

▶ **Figure 3.3:** The reading on this burette is 34.6 cm³

Taking repeat readings and identifying anomalous data

To ensure that the data you are collecting is reliable, you should normally take repeat readings. If the repeat readings are in good agreement with each other, you can usually assume that the data is reliable. Taking repeat readings can also help you to identify any anomalous results. If you think a result is anomalous, you can take a further repeat reading. If you find the result to be anomalous, you can ignore it when you plot a graph of the results or when you calculate a mean.

Recording data in an appropriate table

An important part of data collection is being able to display the data in a clear and logical way. This is normally best done in a table. The table should have correct headings with units where appropriate. For example, if using a balance to measure mass, the table heading should be Mass and the unit g. In the table, **quantitative data** (results that involve numbers) should be given to appropriate levels of precision for the measuring equipment you have used. For example, if a balance reads to one decimal place, then it is good practice to give all balance readings to the same level of precision, so a mass measurement of 24 g should be written as 24.0 g. It would not be appropriate to write 24.00 g as the balance does not read to this level of precision.

Example

Hard water is water that contains either calcium ions or magnesium ions. An investigation was carried out in order to find the height of lather formed when soap solution was shaken with water containing different concentrations of magnesium ions, Mg^{2+}. The results obtained were recorded in a table as shown in Table 3.1.

▶ **Table 3.1:** Concentration of magnesium versus height of lather

Concentration of Mg^{2+} / mol dm^{-3}	Height of lather / mm				
	Run 1	Run 2	Run 3	Run 4	Mean
0.000	84	80	86	80	82.5
0.005	75	74	77	76	75.5
0.010	58	60	78	62	60.0
0.020	15	15	17	17	16.0
0.040	8	8	10	6	8.0
0.060	6	5	5	6	5.5
0.080	4	4	3	3	3.5

Table 3.1 has correct headings with units and shows the data recorded to an appropriate level of precision. Heights were measured with a 15 cm ruler with 1 mm divisions. Repeat readings have been taken and means have been calculated.

The third result for measurements of lather height using a concentration of 0.10 mol dm^{-3} has been ignored when calculating the **mean**. This result has been ignored as it can be considered as an anomalous result because it is much larger than the other 3 and is larger than the mean for the 0.005 mol dm^{-3} concentration.

Making qualitative observations and drawing conclusions

You can also record **qualitative data** in a table. This may make the data easy to analyse and you can draw your conclusions from it.

Example

An investigation was carried out to find the order for reactivity of the halogens by mixing solutions of the halogens with colourless sodium halide solutions. Table 3.2 shows the observations that were made.

▶ **Table 3.2:** Halogens mixed with sodium halide solutions

Solution	Sodium chloride	Sodium bromide	Sodium iodide
Chlorine	X	colourless chlorine solution turns orange	colourless chlorine solution turns brown
Bromine	orange bromine solution stays orange	X	orange bromine solution turns brown
Iodine	brown iodine solution stays brown	brown iodine solution stays brown	X

▶ In which mixtures did a reaction take place?

▶ Why was sodium chloride not mixed with chlorine, sodium bromide not mixed with bromine and sodium iodide not mixed with iodine?

The results show that chlorine is the most reactive halogen as it reacts with both sodium bromide and sodium iodide to produce orange bromine solution and brown iodine solution respectively. Bromine is the next most reactive as it reacts with sodium iodide but not with sodium chloride, and iodine is the least reactive as it does not react with either sodium chloride or sodium iodide.

Processing data

Having collected your data, you need to be able to process it. Depending on the investigation, processing the data can involve statistical analysis, using mathematical relationships, finding percentage errors of measuring equipment and plotting suitable graphs. In this section you will learn about all these processing techniques and have the opportunity to practise using them. Although this can often be done using data analysis functions in Excel, for this unit you need to be able to complete this analysis under exam conditions, using the techniques in this section.

Mean and standard deviation

Mean

In statistical analysis, symbols are used. You need to become familiar with these. These symbols are used when calculating standard deviation.

▶ Σ is the Greek uppercase letter 'sigma' and it means 'sum of'.

▶ $\bar{x}$ is the mean.

▶ n is the number of data values.

▶ x_i represents a particular value.

▶ s represents **standard deviation**.

> **Key term**
>
> **Standard deviation** – a measure of how far data values are from the mean value.

Mean = sum of all results ÷ number of results

Using statistical notation the mean can be expressed as:

$$\bar{x} = \frac{\Sigma x_i}{n}$$

where $\Sigma x_i = x_1 + x_2 + ... + x_n$

Worked Example

Cheerag has built a mini greenhouse for his biology project. He has recorded the temperatures over the last 10 days.

Day	1	2	3	4	5	6	7	8	9	10
Temperature/°C	25	28	25	23	26	29	27	28	26	23

What is the mean temperature recorded in his greenhouse?

$$\bar{x} = \frac{25 + 28 + 25 + 23 + 26 + 29 + 27 + 28 + 26 + 23}{10} = 26\,°C$$

Standard deviation

The standard deviation indicates how closely a set of data values are positioned around the mean. It is calculated using the following equation:

$$s = \sqrt{\frac{\Sigma(x - x)^2}{n - 1}}$$

Use the following steps in your calculation.

▶ Find the mean.

▶ Subtract the mean from each of your data values to get the deviations.

▶ Square each deviation and add them all up.

▶ Divide this figure by one less than your sample number.

▶ The standard deviation (s) is the square root of this value.

Some scientific calculators have a statistical mode which will allow you to calculate standard deviation. This will enable you to do the calculation quickly. Your tutor should be able to show you how to use this function on your calculator.

Worked Example

You are working with a team of biologists who have been investigating a type of herb which could be used to cure a disease. You have just measured the height in centimetres of different specimens of the same herb. The data you have collected are:

10.2, 10.3, 10.4, 10.4, 10.5, 10.6, 10.6, 10.6, 10.8, 10.9.

Calculate the standard deviation.

Step 1: Calculate the mean, $\bar{x}$

$$\bar{x} = \frac{10.2 + 10.3 + 10.4 + 10.4 + 10.5 + 10.6 + 10.6 + 10.6 + 10.8 + 10.9}{10} = 10.53$$

Step 2: Construct a table as shown. Column A shows the data in ascending order, column B is the difference between the data value and the mean, and column C is the square of the value in column B.

A	B	C
Height/cm	$x - \bar{x}$	$(x - \bar{x})^2$
10.2	-0.33	0.1089
10.3	-0.23	0.0529
10.4	-0.13	0.0169
10.4	-0.13	0.0169
10.5	-0.03	0.0009
10.6	0.07	0.0049
10.6	0.07	0.0049
10.6	0.07	0.0049
10.8	0.27	0.0729
10.9	0.37	0.1369

Adding column C gives: $\Sigma (x - \bar{x})^2 = 0.421$

Divide this number by $n - 1$, which, in this example, is 9.

This gives a value of 0.0468.

Now, use this value to calculate the standard deviation by finding its square root.

$$s = \sqrt{\frac{\Sigma(x - x)^2}{n - 1}}$$

$s = \sqrt{0.421 \div 9} = \sqrt{0.0468} = 0.216$

You should give the standard deviation to the same number of decimal places as your data values. So the standard deviation of the height of the herbs is 0.2 cm (to 1 d.p.).

Ⅱ PAUSE POINT

Jasmeen was investigating the resistivity of nichrome wire. She measured the diameter of the wire at different points, using a micrometer. Her measurements in mm are: 0.234, 0.234, 0.235, 0.237, 0.238. Find the mean of these measurements.

Use the mean and the equation for standard deviation to find the standard deviation.

Hint   Remember to give your answer to the same number of decimal places as Jasmeen's measurements.

Extend   Now use a scientific calculator to find the standard deviation. Are your answers the same?

Significant figures (s.f.)

When doing scientific calculations, you need to be aware of the precision to which you should give your answer. Answers to calculations often produce more digits than the accuracy of the original data. The general rule is to give your answer to the same number of significant figures as the original data. Note that you may need to round up or down – this is demonstrated in the following example.

To determine the number of significant figures, count the number of digits from the first one that is not 0. For example, in the pause point on the previous page, Jasmeen's measurements were given to 3 s.f.

Worked Example

Write the following numbers to 2 significant figures:

(a) 6.084

(b) 0.0040254

(c) 24465

Answers

(a) 6.1 If the digit after the 2nd s.f. is 5 or more you need to round the previous digit up, so in this case 0 becomes 1.

(b) 0.0040 Here the first s.f. is the one that is not 0. The 0 after the 4 needs to be there otherwise this answer would only be to 1 s.f.

(c) 24000 Although this may look like 5 s.f. it is still only 2 s.f. as the answer has been rounded down to the nearest 1000.

Key term

Frequency – how often a particular value occurs in a set of values.

Normal distribution

If you plot a graph of **frequency** against standard deviation for a set of data, you usually find that the graph has the shape of a normal distribution as shown in Figure 3.4.

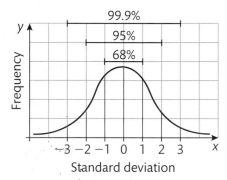

▶ **Figure 3.4:** Graph showing a normal distribution

If you measured the lengths of 100 holly leaves, you would expect to find that 68% would lie within one standard deviation and 95% within two standard deviations from the mean.

Use and interpretation of error bars

When you plot a graph, you should include error bars wherever possible. These error bars represent a measure of uncertainty in the data. This uncertainty could be due to a lack of precision of the instruments used to take the measurements, or a variation in the measurements taken.

Both horizontal and vertical error bars may be plotted, but often only the vertical ones are shown as these represent the uncertainty in the measurements for the dependent variable. For example, if 3 different times are recorded for the time taken for a toy car to run down a ramp as 3.5 s, 3.7 s and 3.8 s, then the mean time of 3.7 s would be plotted on the graph and an error bar would be drawn from 3.5 s to 3.9 s.

Worked Example

Use the data from the investigation to find the height of lather formed when soap solution was shaken with water containing different concentrations of magnesium ions, Mg^{2+}. Plot a line graph of the results showing error bars for the height of lather results.

Concentration of Mg^{2+} / mol dm^{-3}	Height of lather / mm				
	1	2	3	4	Mean
0.000	84	80	86	80	82.5
0.005	75	74	77	76	75.5
0.010	58	60	78	62	60.0
0.020	15	15	17	17	16.0
0.040	8	8	10	6	8.0
0.060	6	5	5	6	5.5
0.080	4	4	3	3	3.5

The graph is plotted with concentration on the x axis and mean height of lather on the y axis. The lowest and highest readings are then plotted for each data set and the vertical error bars are drawn in by joining the upper and lower points for each reading. The anomalous result has been ignored.

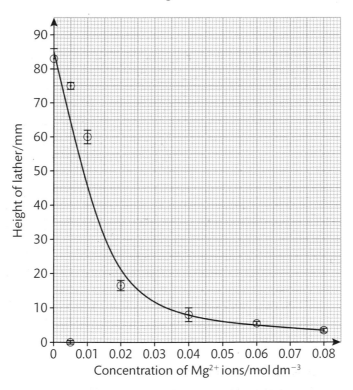

▶ **Figure 3.5** Graph to show height of lather against concentration of Mg^{2+} ions

Using statistical tests

You can use statistical tests to test or support a scientific hypothesis or to see if there is a relationship between two quantities or factors. For example, you may want to compare two sets of experimental data to see whether there is any difference between them.

Key term

Significance level or confidence level (p) – this is used in hypothesis testing. It is a figure used to reject or accept the null hypothesis. Scientists usually use figures ranging from 1% (0.01) to 5% (0.05) significance levels.

The t-test

The t-test is usually used to compare unrelated independent samples of data. The samples are often referred to as 'unmatched pairs'. The data could be from two separate experiments where the two sets of data show a normal-distribution. The means of the two sets are compared and t-test tables are used to determine the significance of the differences in means.

A section of a t-test table is shown here.

Degrees of freedom	20% significance level	5% significance level	1% significance level
1	3.08	12.71	63.66
2	1.89	4.30	9.93
3	1.64	3.18	5.84
4	1.53	2.78	4.60
5	1.48	2.57	4.03
6	1.44	2.45	3.71
7	1.42	2.37	3.50
8	1.40	2.31	3.36
9	1.38	2.26	3.25
10	1.37	2.22	3.17

When comparing two sets of data, the null hypothesis states there is no significant difference between the two sets of data.

Step by step: Carrying out the t-test

9 Steps

The steps for carrying out a t-test for independent samples are shown for the following example.

Five black dog hairs were found on the clothes of a victim at a crime scene. The thickness of these hairs was measured using a micrometer and found to be 46, 57, 54, 51 and 38 μm. A suspect is the owner of a black dog. Five hairs were taken from the suspect's dog and their thicknesses measured and were found to be 31, 35, 50, 35 and 36 μm.

Is it possible that the hairs on the victim were left by the suspect's dog?

1 Calculate the mean for each set of data.

$(46 + 57 + 54 + 51 + 38) \div 5 = 246 \div 5 = 49.2$ and $(31 + 35 + 50 + 35 + 36) \div 5 = 187 \div 5 = 37.4$

▼

2 Calculate the magnitude (size) of the difference between the two means, $\bar{x}_1 . \bar{x}_2$. You only need the value, not its sign (positive or negative). $49.2 - 37.4 = 11.8$

▼

3 Calculate the standard deviation for each set of data, using

$$s = \sqrt{\frac{\Sigma(x - x)^2}{n - 1}}$$

$s_1 = \sqrt{(3.2^2 + 7.8^2 + 4.8^2 + 1.8^2 + 11.2^2) \div 4} = \sqrt{222.8 \div 4} = \sqrt{55.7} = 7.46$

$s_2 = \sqrt{(6.4^2 + 2.4^2 + 12.6^2 + 2.4^2 + 1.4^2) \div 4} = \sqrt{213.2 \div 4} = \sqrt{53.3} = 7.30$

▼

4 Calculate the standard error in the difference between the two samples.

Use the equation: $\sqrt{\dfrac{s_1^2}{n_1} + \dfrac{s_2^2}{n_2}} = \sqrt{(55.7 \div 5) + (53.3 \div 5)} = \sqrt{21.8} = 4.67$ where s_1 and s_2 are the standard deviations for the two sets of data and n_1 and n_2 are the number of measurements in each sample.

5 Calculate the value of t by dividing the difference between the means by the standard error in the difference, which is the answer in step 2 divided by the answer in step 4. So $t = 2.53$.

6 Calculate the number of degrees of freedom = $(n_1 + n_2 - 2)$ where $n_1 + n_2$ is the total number of measurements of the two samples.   Degrees of freedom = $5 + 5 - 2 = 8$

7 Use a t-test table to find the critical value that corresponds to the number of degrees of freedom for the significance level you are working with. This is usually either 1% or 5%. Looking at the t-test table given you can see that for 8 degrees of freedom the critical value at a 5% significance value is 2.31. (You will be able to find a selection of more detailed t-test tables online which you can use.)

8 If the calculated value of t is less than the critical value, there is no significant difference between the two sets of data and the null hypothesis is accepted.

9 If the calculated value of t is equal to or greater than the critical value, the null hypothesis is rejected. This means the two sets of data differ significantly. In this example the calculated value was 2.53 which is greater than 2.31, so the null hypothesis is rejected. There is a significant difference between the two sets of data. This means that you can be 95% confident that the two sets of hairs could not have come from the same dog. You can not be 99% certain based on 1% significance level.

Worked Example

You are an agricultural scientist and you were asked to test two types of fertiliser. You added fertiliser A to eight plots of land and fertiliser B to eight different plots of land. You planted the same number of potato plants in each plot and managed them in the same way.

You recorded the yield of potatoes from each plot as shown.

Plot	Yield of potatoes with fertiliser A /kg	Yield of potatoes with fertiliser B /kg
1	17	18
2	10	9
3	6	8
4	8	11
5	12	14
6	9	10
7	13	15
8	11	17

Considering a 5% significance level, is there a significant difference between the yields of potatoes due to the fertiliser you used?

Step 1: Calculate the two means. $\bar{x}_A$ = 10.75 and $\bar{x}_B$ = 12.75 (you can use the formula for mean to check these values).

Step 2: Find the difference between the means. 12.75 – 10.75 = 2.00. Why is this value given to 2 decimal places?

Step 3: Find the standard deviations for the two sets of data. s_A = 3.37 and s_B = 3.77 (you can use the formula for standard deviation to check these values).

Step 4: Calculate the standard error in the difference using the equation: $\sqrt{\dfrac{s_1^2}{n_1} + \dfrac{s_2^2}{n_2}}$ $\sqrt{(3.37^2/8 + 3.77^2/8)}$ = 1.79

Step 5: Calculate the value of t: t = difference between means standard error in difference = 2.00/1.79 = 1.12

Step 6: Calculate the degrees of freedom: $(n_1 + n_2 - 2)$ = 8 + 8 – 2 = 14

Step 7: Use a t-test table to determine the critical value at a significance level of 5%. For 14 degrees of freedom at a significance level of 5%, the critical value is 2.15.

Step 8: Because t = 1.12 it is less than the critical value of 2.15 so there is no significant difference between the yields of potatoes.

You can therefore be 95% confident that there is no significant difference between the effects of the two fertilisers.

In this case, the null hypothesis is accepted.

Note: You can make the test more reliable by increasing the number of data points. Only eight were used in this investigation.

Example

Two sets of ten water fleas, A and B, were placed in cool river water. The water in which set A were placed contained 0.01% caffeine solution. The heart rates per minute of the two sets were measured by microscopic analysis.

Table 3.3 shows the results.

▶ **Table 3.3:** Heart rates of set A and set B

Heart rate per minute of set A	Heart rate per minute of set B
113	68
111	56
136	62
121	78
108	82
109	64
117	66
122	78
132	77
116	81

For 18 degrees of freedom, (10 + 10 – 2), t-test tables show that at a 1% significance level the critical value is 2.88.

Ⅱ PAUSE POINT   Study the data on the previous page. Considering a 1% significance level, is there a significant difference between the heart rates of the two sets of water fleas?

Hint   Find the means and the standard deviations for the two sets of water fleas. Find the difference between the two means.

Using steps 4 and 5 in the worked example, find the value of t.

Is this value smaller than or larger than the critical value?

Extend   Does adding caffeine to river water make a significant difference to the heart rates of the water fleas? Is the null hypothesis accepted or rejected in this case?

The chi-squared test

The chi-squared test (χ^2 test) is another statistical test which is used to compare the frequencies of individuals in particular categories. The chi-squared test is used to see how the observed frequency compares with the expected frequency.

▶ If there is a significant difference, you can reject the null hypothesis.

▶ If there is no significant difference, you can accept the null hypothesis.

The steps in a chi-squared test are shown for the following example.

It is expected that 5% of the population will be colour blind and the null hypothesis states that there is no difference between the occurrence of colour blindness in males and females. One thousand people were tested for colour blindness: 500 males and 500 females. It was found that 37 of the males were colour blind and 13 of the females.

Step by step: Carrying out the chi-squared test   6 Steps

1 Record your observed results (O) and expected results (E) in a table.

▼

2 Expected number E is 5% of 500 which is 25, for both males and females.

	Number of colour-blind males	Number of colour-blind females
Observed	37	13
Expected	25	25
$(O - E)^2 \div E$	$(37 - 25)^2 \div 25 = 5.76$	$(13 - 25)^2 \div 25 = 5.76$

▼

3 For each pair of values, calculate $(O - E)$. Square this value to find $(O - E)^2$ and divide this by E. Add these values to your table as shown above.

▼

4 Calculate χ^2 by adding all the values of $(O - E)^2 \div E$ together:

$$\chi^2 = \frac{\Sigma(O - E)^2}{E} = 5.76 + 5.76 = 11.52$$

▼

5 Calculate the degrees of freedom (n) in the data. For example, if there are four columns of data, n = 4 − 1 = 3. In the above example there are only 2 columns of data so n = 2 − 1 = 1

▼

6 Use a χ^2 table to find the critical value for the degrees of freedom at a particular confidence level (p). (If you do a Google search you will be able to find a selection of χ^2 tables which you can use.) For the above example the critical value at a confidence level of 1% is 6.64.

If the calculated value of χ^2 is greater than the critical value, then the observed data differ significantly from the expected data and you can reject the null hypothesis.

If the calculated value of χ^2 is less than the critical value, then there is no significant difference between the observed and expected data and you have to accept the null hypothesis. In this example the critical value of 6.64 is much lower than our χ^2 value of 11.52 so we can say with 99% confidence that there is a significant difference between the occurrence of colour blindness in males and females and the null hypothesis can be rejected.

Worked Example

A genetic model suggests that if a red tropical flower self-pollinates the expected outcome of red, pink and yellow flowers is in the ratio 1:3:2.

Mr Gardener is a botanist. He grew 300 plants from the self-pollinated seeds of the tropical plant. Of the flowers he grew:

- 45 were red
- 160 were pink
- 95 were yellow.

Use the chi-squared test to decide whether these results are consistent with those suggested by the genetic model.

Step 1: Calculate the expected result (E) for each flower colour.
The number of flowers in the sample is 300. As the expected ratio is 1:3:2, the expected results would be:
$1 \times 300 \div 6 = 50$ red flowers, $3 \times 300 \div 6 = 150$ pink flowers and $2 \times 300 \div 6 = 100$ yellow flowers.

Step 2: Calculate $(O - E)^2 \div E$ for each flower colour and put all these results in a table as shown:

Number of each flower colour	Red	Pink	Yellow
Observed	45	160	95
Expected	50	150	100
$(O - E)^2 \div E$	$(45 - 50)^2 \div 50 = 0.50$	$(160 - 150)^2 \div 150 = 0.67$	$(95 - 100)^2 \div 100 = 0.25$

Step 3: Add these values to get χ^2:
$\chi^2 = 0.50 + 0.67 + 0.25 = 1.42$

Step 4: Calculate the number of degrees of freedom, n. There are three columns of data, so:
$n = 3 - 1 = 2$ degrees of freedom.

Step 5: Use a χ^2 table to find the critical value. From the χ^2 table, for $n = 2$ at a confidence level (p) of 0.05 (5%) the critical value is 5.99.

As $1.42 < 5.99$, you can accept the null hypothesis. This means that there is no significant difference between the observed and expected data, so Mr Gardener can accept the genetic model. As the critical value is more than three times larger than the value of χ^2, he can be confident that the genetic model is correct.

Example

A section of a river was trawled for four types of freshwater fish. It was expected that there would be equal numbers of each type of fish collected.

A total of 40 fish were collected, of which:

▶ 15 were roach

▶ 15 were perch

▶ 6 were pike

▶ 4 were bream.

The null hypothesis states that there is no significant difference between the observed frequency and expected frequency of fish.

The critical value for 3 degrees of freedom at a confidence level of p = 0.05 (5%) is 7.82.

You can set up a table for a χ^2 test as shown in Table 3.4.

▶ **Table 3.4:** Observed and expected frequencies of fish

Number of fish	Roach	Perch	Pike	Bream
Observed				
Expected				
$(O - E)^2 \div E$				

Ⅱ **PAUSE POINT**   Study the data above. Use a chi-squared test to decide whether the null hypothesis should be accepted or rejected.

Hint   Copy out the table above, fill in the observed and expected numbers of fish and use these values to find $(O - E)^2 \div E$ for each type of fish. Find a value for χ^2. Is this value smaller or larger than the critical value for a confidence level of $p = 0.05$? Is the null hypothesis accepted or rejected in this case?

Extend   Using your results from the chi-squared test, if you trawled another section of the river, explain why you would not expect to find equal numbers of the four types of fish.

Correlation analysis

When carrying out a scientific investigation, you will often look to see how changing one variable (the independent variable) affects another variable (the dependent variable). Correlation analysis helps you to look for relationships that may exist between the two variables.

Types of correlation

There are various types of correlation.

▶ **No correlation:** When there is no clear pattern between the data (see Figure 3.6 (a)).

▶ **Positive correlation:** When x increases, y increases, so the **line of best fit** has a positive slope (see Figure 3.6 (b) and (d)).

▶ **Negative correlation:** When x increases, y decreases, so the line of best fit has a negative slope (see Figure 3.6 (c)).

▶ **Strong correlation:** When most of the data points are close to the line of best fit (see Figure 3.6 (d)).

▶ **Weak correlation:** When the data points are more widely scattered around the line of best fit (see Figure 3.6 (b) and (c)).

Key term

Line of best fit – a straight line or smooth curve drawn to pass through as many data points as possible.

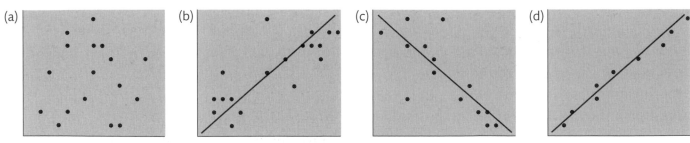

▶ **Figure 3.6:** Graphs showing different types of correlation: (a) no correlation, (b) weak positive correlation, (c) weak negative correlation, (d) strong positive correlation

Note: When you plot a graph, the independent variable is always on the x-axis and the dependent variable on the y-axis.

Formulae

In this unit, you will need to use some common formulae in order to carry out calculations. You will also need to know how to rearrange formulae and give your answers in the correct **SI units**.

Example

There are two formulae that you can use to find power, as Table 3.5 shows.

▶ Equation 1 refers to electrical power.

▶ Equation 2 refers to mechanical power.

In both equations, the unit of power is the watt.

▶ **Table 3.5:** Equations 1 and 2

Equation 1	Equation 2
Power = Voltage × Current (Power = VI)	Power = Work done ÷ Time
Power is measured in watts (W)	Power is measured in watts (W)
Voltage (potential difference) is measured in volts (V)	Work done = energy supplied or transformed and is measured in Joules (J)
Current is measured in amps (A)	Time is measured in seconds (s)

Worked Example

A toaster has a power rating of 1200 W. The voltage supplied by the mains is 240 V. What is the current flowing through the wires to the toaster?

Answer

The formula you need to use is equation 1:

Power = Voltage × Current

Because you want to find the current, you need to rearrange the formula:

Current = Power ÷ Voltage

Current = 1200 W ÷ 240 V = 5 A

Conversion of units

Sometimes you need to convert units so that they are in the correct form to use in the given formula.

In the previous example, if the power rating had been given as 1.2 kilowatts (kW), you would need to convert this value to watts before doing the calculation. The prefix 'kilo' refers to 1000, so 1.2 kW = 1200 W.

Table 3.6 shows some of the prefixes you may come across.

▶ **Table 3.6:** Prefixes and how to convert them

Prefix	Symbol	Factor	Example
Mega	M	1 million (1 000 000)	5 MW = 5 000 000 W
Kilo	k	1 thousand (1000)	6 kV = 6000 V
Centi	c	1 hundredth (0.01)	15 cm = 0.15 m
Milli	m	1 thousandth (0.001)	8 mg = 0.008 g
Micro	μ	1 millionth (0.000 001)	2 μA = 0.000 002 A

Use of standard form

When you are working with either very large or very small numbers, it is easier to use **standard form** than to write the numbers out in full.

Examples

The distance between the Sun and the Earth is 149 million kilometres. When this is written out using the metre as the standard SI unit, it is 149 000 000 000 m. To convert this number to standard form, imagine there is a decimal point after the final 0. Now move this point to the left and count how many times you need to move it until you get to a value of 1.49. In this example, you would have to move it 11 times, so in standard form you would write the answer as 1.49×10^{11} m (where 10^{11} means $10 \times 10 \times 10 \times 10 \times 10 \times 10 \times 10 \times 10 \times 10 \times 10 \times 10$).

The influenza virus is very small (see Figure 3.7 for comparison with the distance between the Sun and the Earth), with a diameter of approximately 0.000 000 08 m. To convert this number to standard form, move the decimal point to the right and count how many times you need to move it to get a value of 8.0. In this example you would have to move it 8 times, so in standard form you would write the answer as 8.0×10^{-8} m (where 10^{-8} means $1/10^8$, in other words 1 divided by $10 \times 10 \times 10 \times 10 \times 10 \times 10 \times 10 \times 10$).

Key term

Standard form – a way of writing down small and large numbers easily using powers of ten.

(a) 8.0×10^{-8} m

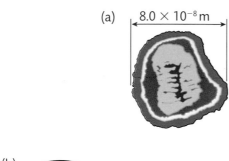

(b)

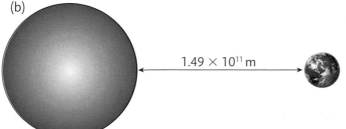

1.49×10^{11} m

▶ **Figure 3.7:** (a) Influenza virus (b) The distance between the Sun and the Earth

You can use standard form on a scientific calculator. If you do not know how to do this, your tutor will be able to show you how.

The formula which connects voltage, current and resistance is:
Voltage = Current × Resistance ($V = IR$). A standby light on a laptop computer is powered by a 15-volt supply and has a resistance of 1000 ohms. Find the current in amps which flows through the light.

Hint   You will need to rearrange the equation.

Extend   Convert your answer in amps into standard form. What is the current in milliamps (mA)?

Percentage error of measuring equipment

Different types of measuring equipment can measure to different degrees of precision.

For example, a 30 cm ruler will normally have a scale showing divisions of 1 mm (0.1 cm), so you could use it to measure to the nearest ± 0.05 cm. This is the maximum absolute error or uncertainty of the ruler.

Similarly, a balance reading to 0.01 g could measure the mass of a sample to the nearest ± 0.005 g. For example, if a strip of magnesium ribbon is found on this balance to have a mass of 0.15 g, then the actual mass could be anywhere between 0.145 g and 0.155 g.

When carrying out scientific investigations, percentage error is more important than absolute error or uncertainty of the measuring equipment. Percentage error depends on the magnitude of the readings taken as well as the precision of the measuring equipment. For the example above the percentage error of the balance would be found by multiplying the uncertainty of the balance by 100 and dividing by the balance reading.

So, percentage error of the balance = (± 0.005 × 100) ÷ 0.15 = ± 3.3%. When you use several different types of measuring equipment in an investigation, it is useful to know which one is likely to result in the greatest percentage error. You can consider using a more precise instrument to take the measurement – for example, using a burette in place of a measuring cylinder to obtain a more precise volume measurement.

Worked Example

Lucy is investigating the reaction between sodium hydroxide and hydrochloric acid. She measures out 25 cm^3 of sodium hydroxide solution of concentration 2.0 mol dm^{-3} into a polystyrene cup, using a measuring cylinder. She records the initial temperature of the solution. She then adds 25 cm^3 of hydrochloric acid of concentration 1.0 mol dm^{-3} and records the highest temperature reached. She repeats the experiment using the same concentration and volume of sodium hydroxide solution with 25 cm^3 of hydrochloric acid of concentration 2.0 mol dm^{-3}.

Here are her results.

Concentration of hydrochloric acid / mol dm^{-3}	Initial temperature of sodium hydroxide solution / °C	Final temperature of mixture / °C	Temperature rise / °C
1.0	22	29	7
2.0	22	35	13

The measuring cylinder measures to the nearest 1 cm^3 and the thermometer to the nearest 1 °C.

What are the maximum percentage errors of the measuring equipment?

How could these errors be reduced?

Answer

Percentage error of the measuring cylinder.

The volume measured is 25 cm^3. To find the percentage error you need to multiply uncertainty of the measuring cylinder, which here is ± 0.5 cm^3 by 100, and divide by the volume measured.

% error = (± 0.5 × 100) ÷ 25 = 2%

Percentage error of the thermometer.

The thermometer reads to the nearest 1 °C so the uncertainty of the thermometer is ± 0.5 °C. When a temperature rise is measured two separate temperature readings are taken and so both readings could be out by ± 0.5 °C giving a maximum absolute error of ± 1 °C. To find the maximum possible percentage error, you need to use the smallest temperature rise, which in this example is 7 °C.

% error = (± 1 °C × 100) ÷ 7 = 14%

In order to reduce the possible percentage error in this investigation, Lucy should use a more precise thermometer, for example, one which reads to the nearest 0.2 °C, which would give a percentage error of ± 2.8%.

There would not be much point in using a burette to measure the volume, as even with a more precise thermometer, the percentage error of the measuring cylinder is still less than that of the thermometer.

Timing equipment

Sometimes where timing is involved in an experiment, human reaction time will be a more likely cause of error than the timing instrument.

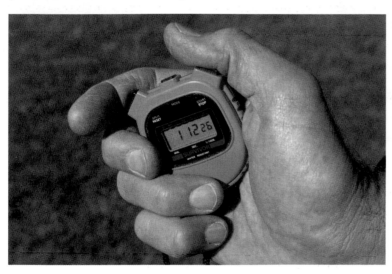

▶ The stopwatch measures to the nearest 0.01 of a second, giving an uncertainty of ±0.005 seconds. Human reaction time is about 0.2 seconds, resulting in a much larger percentage error.

> **Research**
>
> Use the internet to find out ways of measuring your reaction time. A useful web site is **http://www.mathsisfun.com/games/reaction-time.html**

Example

You were asked to find the density of a block of aluminium. You measured the length, height and width of the block using a 30 cm ruler which measured to the nearest mm (0.1 cm). You then found the mass of the block using a balance reading to 0.1 g. You found that the block measured 5.0 by 5.0 by 5.0 cm, giving a volume of 125 cm³, and had a mass of 336.2 g.

PAUSE POINT

Using the data above find the percentage errors of the ruler and the balance.
Use the formula: Density = Mass ÷ Volume to find the density of aluminium in $g\,cm^{-3}$ and then convert it to $kg\,m^{-3}$.
Give your answers to 2 significant figures.

Hint

You used the ruler three times to take the measurements. What do you think is the percentage error of the volume?
Look up the density of aluminium in a data book or on the Internet. How close is your value to the data book value?

Extend

Explain why there would be no point in using a balance reading to 0.01 g.

Different ways of displaying data

In this section you will learn how to choose the correct way to display the data you have collected.

Results tables

Both quantitative and qualitative data are best displayed in a results table, with correct headings and units for quantitative data. With quantitative data, you can display the results in an appropriate chart or plot them on a suitable graph. All charts and graphs should have a heading with the title of the investigation and the variables plotted. For example, if you were investigating the relationship between voltage and current for a filament bulb your heading would be: 'Graph showing the relationship between voltage and current for a filament bulb.'

Frequency tables

A frequency table or tally chart is one way of organising data so that it is easier to analyse.

Example

Twenty agar plates were left exposed for 24 hours under identical conditions. The numbers of colonies of bacterial growth on the 20 plates were as follows:

1, 2, 4, 3, 6, 7, 6, 8, 3, 9, 6, 7, 7, 6, 5, 4, 5, 6, 5, 8.

You can construct a tally chart to show these results more clearly (Table 3.7).

▶ **Table 3.7:** Tally chart of colony counts on agar plates

Plate number	Tally of colony count	Frequency
1	I	1
2	I	1
3	II	2
4	II	2
5	III	3
6	JHT	5
7	III	3
8	II	2
9	I	1
		20

Key term

Mode – the data value that occurs most often.

The colony count which occurs most often in this table is 6. This is the **mode**.

▶ The mean number of bacterial colonies is found by dividing the total number of bacterial colonies by the total number of plates.

- This can be worked out from the numbers in the tally chart.
- Mean = sum of number of colonies (tally) ÷ number of plates
  = 20 ÷ 9 = 2.2

Types of variable

In order to decide what type of chart or graph to plot, you need to know the type of variable involved. Variables can be discrete, continuous or categoric, as shown in Table 3.8.

▶ **Table 3.8:** Types of variable

Type of variable	Description	Examples	Type of graph or chart
Categoric	Data with specific labels	• Percentages of gases in the air • Melting points of the alkali metals	• Pie chart or bar chart • Bar chart
Discrete	Whole number data	• Number of prickles on a holly leaf	• Bar chart
Continuous	Data that can be any number	• Heights of learners in a class • Change in temperature of a reaction mixture over time	• Histogram • Line graph

Pie charts

Pie charts are a visual way to display data. In a pie chart, the data are plotted in a circle with the proportions of each item being displayed as a segment of the circle. Each segment has a specific angle, where the total of all angles adds up to 360°.

Example

In a group of 100 British blood donors, blood groups were found to be as shown in Table 3.9. Figure 3.8 uses a pie chart to show the proportion of blood donors from each blood group.

▶ **Table 3.9:** Blood groups of donors

Blood group	Number of blood donors	Angle (out of 360°)
A	40	144°
B	10	36°
O	46	166°
AB	4	14°

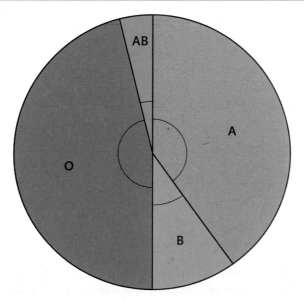

▶ **Figure 3.8:** Pie chart showing the proportion of blood donors from each of the four blood groups

Bar charts

You can use bar charts to display either discrete or categoric data.

Example

The bar chart in Figure 3.9 shows the densities of different metals. The bars should be drawn separately and should not be touching. The x axis is labelled with the categoric variable, which in this case is the symbol of the metal. The y axis is labelled with the quantity you are measuring, which in this case is the density.

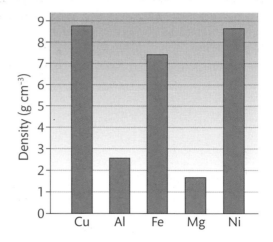

▶ **Figure 3.9:** Bar chart showing the densities of five different metals

Histograms

Histograms are plotted for continuous variables when a large amount of data is being considered. Before plotting your histogram, you would normally put your data into a frequency table. This may involve organising your data into class intervals. For example, if you were measuring the heights of a group of learners, one class could be those with a height between 150 cm and 159 cm and another class those with a height between 160 cm and 169 cm.

A histogram is similar to a bar chart, but because the variable being measured is continuous, the bars should be touching. In most cases you will find that the histogram follows the shape of a normal distribution curve (see Figure 3.10).

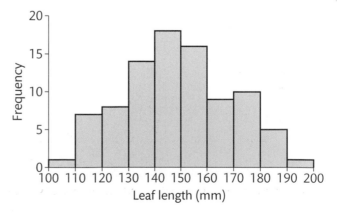

▶ **Figure 3.10:** Histogram showing leaf length in sweet chestnut

The x axis shows the leaf length in mm in class intervals, the first one going from 100 mm to 109 mm and the next from 110 mm to 119 mm, etc. The y axis shows the frequency, that is, the numbers of leaves in each class interval.

- What is the mode for leaf length in this sample of sweet chestnuts?
- How many leaves were in this class interval?
- How many leaves in total were measured for this investigation?

Line graphs

Line graphs are used when plotting continuous variables. They can be straight lines or smooth curves. Axes should be labelled with the appropriate variables and units. The scale should be uniform and chosen so that as much of the graph paper as possible is used. You should also give your graph a heading. Figure 3.11 shows a line graph.

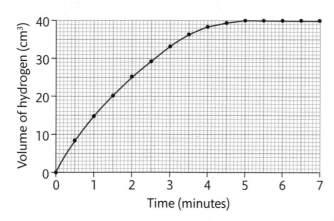

▶ **Figure 3.11:** Line graph showing the relationship between volume of hydrogen gas collected in cm^3 and time in seconds, when magnesium is reacted with hydrochloric acid

Lines of best fit and anomalous points

Experimental data often does not exactly fit the line or curve, so you need to draw a line of best fit.

The line of best fit should be drawn so that there are equal numbers of points either side of the line. Any point that is clearly a long way off the line is likely to be an **anomaly** (see Figure 3.12 for an example of this) and should not be taken into account when drawing the line of best fit.

Key term

Anomaly – a data point that does not fit the overall trend in the data.

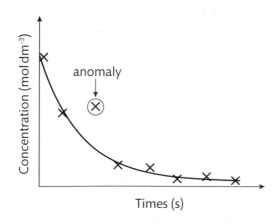

▶ **Figure 3.12:** This line graph shows a line of best fit and has an anomalous point

Example

Table 3.10 shows the time taken for the reaction to complete when 2 cm strips of magnesium ribbon are added to a particular concentration of hydrochloric acid heated to different temperatures.

▶ **Table 3.10:** Time taken for reaction

Temperature / °C	Time / s
12	100
21	52
30	27
44	14
50	9

Ⅱ PAUSE POINT Using Table 3.9, draw a bar chart to display the data showing the number of blood donors with each type of blood group.

Hint Make sure you add the appropriate labels.

Extend Draw a line graph using the data in Table 3.10.

Checklist

Look at the line graph you just plotted.

▶ Did you choose suitable uniform scales which cover more than half the graph paper?

▶ Did you plot all points correctly and draw a smooth curve (line of best fit) through the points?

▶ Did you label both axes, with time on the *y* axis and temperature on the *x* axis, and did you include units?

▶ Did you write a suitable heading for your graph?

Assessment practice 3.2

Nick and Manisha were investigating how changing the height of an 80 cm ramp affected the time it took for a toy car to reach the bottom of the ramp.

- They set the top of the ramp at a height of 3 cm and placed a toy car at the top of the ramp and then timed how long it took for the car to reach the bottom of the ramp.
- They repeated this two more times.
- They then repeated the whole experiment for five more different ramp heights.
- Their results are shown in the following table.

Ramp height / cm	Time / s			Mean time / s
3.0	2.46	2.53	2.43	
4.0	2.37	2.43	2.24	
5.0	1.73	1.96	1.93	
6.0	1.46	1.64	1.71	
7.0	1.41	1.58	1.32	
8.0	1.34	1.28	1.28	

1 Copy out the table and calculate the six mean times.

2 Plot a graph of mean time against ramp height.

3 Add vertical error bars to your graph.

4 The ruler you used measured to the nearest 0.1 cm and the timer to the nearest 0.01 s. Calculate the percentage errors on the measuring equipment.

5 The percentage error of the timing equipment is very small compared to that of the ruler. Explain why this is not really a true measure of the timing error.

Drawing conclusions and evaluation

Once you have planned your investigation, collected and processed your data, you now have to interpret, analyse and evaluate your data. In this section you will learn how to identify trends and patterns in your data. This will allow you to draw relevant and valid conclusions. You will learn how to evaluate your investigation and to identify strengths and weaknesses in the method. You should then be able to suggest how your investigation could be improved.

Interpretation and analysis of data

Having processed your data using suitable tables, charts or graphs, you should be able to identify trends and patterns in your data. For quantitative data, you need to ask this question.

▶ Is there a relationship between the variables?

▶ If there is a relationship between the variables, you need to ask these questions.
  - Is there a positive or negative correlation?
  - Is it a weak or strong correlation?
  - Are the variables directly proportional? (If the graph plotted for the two variables is a straight line passing through the origin, then the variables are in direct proportion.)
  - Are the variables inversely proportional? This is the case if one variable doubles, the other halves and the graph plotted for the two variables is a curve sloping downwards.

For both quantitative and qualitative data, you also need to ask yourself these questions.

▶ Do your results support your hypothesis or does the hypothesis need to be amended?

▶ Does secondary data collected support or contradict your primary data?

Having interpreted your data, you should now be able to amend your hypothesis if necessary and draw a relevant and valid conclusion. For example, when doing an investigation which requires using a statistical test, you should be able to use tables of critical values at a 5% significance level and draw a conclusion as to whether the null hypothesis should be accepted or rejected.

Evaluation

In your evaluation, you need to be able to:

▶ explain the reasons for any anomalous data

▶ suggest improvements to your investigation which would make the data more reliable and help to eliminate anomalies.

You should be able to discuss any qualitative or quantitative sources of error. Determining the percentage errors of measuring equipment used should help you to decide whether or not more precise measuring equipment is needed.

Investigating temperature changes

Suzie is investigating the temperature changes when pieces of magnesium ribbon are added to different concentrations of hydrochloric acid in a 100 cm³ beaker. She found that the thermometer she used had a much larger percentage error than the balance used to find the mass of magnesium and the measuring cylinder used to measure out the hydrochloric acid.

In her evaluation, she suggested that it would be better to use a temperature probe and data logger which measures temperatures to the nearest 0.1 °C in order to improve the reliability of her results.

Other improvements she suggested included using a polystyrene cup instead of a beaker to reduce heat loss,

and using a wider range of concentrations of acid in order to see a clearer pattern of results.

Check your knowledge

1   Why would there be little point in using a more accurate balance or a burette instead of a measuring cylinder?

2   One problem with the method was that the piece of magnesium ribbon kept floating to the top of the acid. How could Suzie stop this from happening?

3   Can you think of any other changes Suzie could make to improve this investigation?

Reliability of data

In your evaluation, you should give evidence about the reliability of the data collected during the investigation. You need to ask yourself the following questions.

▶   Is there an easily identifiable pattern in the data? For instance, are all the points on a graph close to or on the line of best fit?

▶   Was the method repeatable? Did you obtain similar results every time you repeated the experiment under the same conditions?

▶   Were other people able to repeat the experiment and obtain similar results?

▶   Did the secondary data you collected support your primary data?

If your answer to all these questions is yes, you can assume that your data is reliable.

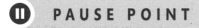 **PAUSE POINT**   In content areas A, B and C you have covered the skills that you may need to use when carrying out a science investigation. Without looking back through sections A, B and C, make a list of the different stages involved in carrying out a science investigation.

Hint   You could use flow diagrams to help you put the different stages in a logical order.

Extend   Why would it not be appropriate to include all the different mathematical techniques described in content area B in your list?

Assessment practice 3.3

Use the information and your answers to the assessment practice activity at the end of section B to answer the following questions.

1   Write a conclusion to explain how the height of the ramp affects the time taken for the toy car to reach the bottom of the ramp.

2   Give reasons why there are quite large variations in the times when repeat results are taken.

3   Suggest improvements you could make to the method to make the results more reliable.

4   Explain how you could extend this investigation to provide further evidence to support your conclusion.

Enzymes in action

In this section, you will learn about the structure of enzymes. You will learn how enzymes work as biological catalysts in chemical reactions. You will also learn about the factors that affect enzyme activity, and how you can plan and carry out scientific investigations to study these factors.

Protein structure

Enzymes are protein molecules. In order to understand how enzymes work, you need to know something about the structure of proteins.

Proteins are made up of amino acid residues joined together by **peptide links**. There are many amino acids residues in one protein molecule. Figure 3.13 shows the structure of an amino acid.

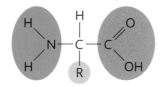

amino group   carboxyl group

▶ **Figure 3.13:** Structure of an amino acid

All **alpha amino acids** have the following four groups which are attached to the central carbon atom:
▶ a hydrogen atom
▶ an amino group
▶ a carboxyl group
▶ a variable R group.

There are 20 different amino acids which make up proteins. Each of these amino acids has a different R group. The R group could be H, CH_3 or some other group containing oxygen, nitrogen or sulfur, for example, COOH. All the proteins in the human body are made up of these 20 amino acids.

When two amino acids are joined together by a peptide link, a water molecule is removed and a dipeptide is formed (see Figure 3.14).

Link

You will learn about protein structure in more detail if you study *Unit 10: Biological Molecules and Metabolic Pathways*.

condensation reaction

peptide bond (covalent) formed

water released

amino acid   amino acid

dipeptide molecule   water

▶ **Figure 3.14:** A dipeptide showing a peptide link

Reflect

If you have access to molecular model kits in your school or college you could build models of two amino acids and then join them together to form a dipeptide.

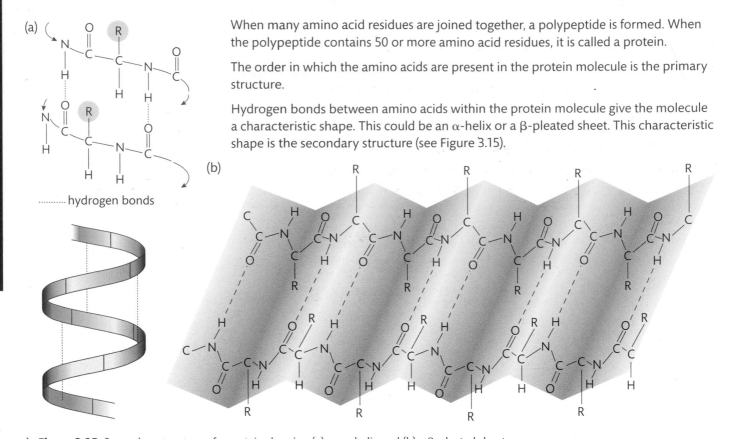

(a)

(b)

········ hydrogen bonds

When many amino acid residues are joined together, a polypeptide is formed. When the polypeptide contains 50 or more amino acid residues, it is called a protein.

The order in which the amino acids are present in the protein molecule is the primary structure.

Hydrogen bonds between amino acids within the protein molecule give the molecule a characteristic shape. This could be an α-helix or a β-pleated sheet. This characteristic shape is the secondary structure (see Figure 3.15).

▶ **Figure 3.15:** Secondary structure of a protein showing (a) an α-helix and (b) a β-pleated sheet

The way in which the secondary structures fold themselves into a three-dimensional shape is the tertiary structure (see Figure 3.16).

This tertiary structure is important in determining how enzymes work.

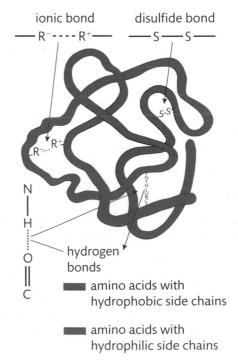

▶ **Figure 3.16:** Tertiary structure of a globular protein

Research

Try to find out some more information about primary, secondary and tertiary structures of proteins, and the importance of hydrogen bonding and disulfide bridges in maintaining the 3D structure of protein molecules.

Active sites and denaturation

An enzyme is a protein molecule with an **active site**.

If the active site of an enzyme is altered in any way, it will not bind with the **substrate** and so will not be able to function. The enzyme has been **denatured**.

> **Key terms**
>
> **Active site** – the area of an enzyme that the substrate binds on to.
>
> **Substrate** – the molecule that is affected by the action of an enzyme.
>
> **Denature** – a change in the tertiary structure of a protein molecule.

Enzymes as biological catalysts in chemical reactions

Enzymes are biological **catalysts**. They speed up reactions in the human body. There are thousands of different enzymes in human cells, each controlling a different chemical reaction.

> **Key term**
>
> **Catalyst** – a substance that speeds up a chemical reaction but remains unchanged at the end of the reaction.

Collision theory

In order to understand how chemical reactions work, and what affects the rate of these reactions, you need to know something about collision theory.

For chemical reactions to occur, the reactants must collide with energy greater than or equal to the **activation energy** and the correct orientation (collision geometry). In practice, only a small minority of collisions that take place lead to a chemical reaction.

> **Key term**
>
> **Activation energy** – the minimum energy required for collisions to break the bonds in the reactants and lead to a reaction.

There are four ways of increasing the rate of a chemical reaction:

1 increasing the concentration of reactants
2 increasing the surface area of a solid reactant
3 increasing the temperature
4 adding a catalyst.

Increasing concentration and surface area will mean that more frequent collisions will occur between reacting particles, leading to a faster reaction.

Increasing the temperature will lead to more frequent and more successful collisions, as the particles will have more energy and will move faster. There will therefore be more collisions with energy equal to or greater than the activation energy.

Adding a catalyst lowers the activation energy for a reaction and so more of the collisions will have enough energy and so again there will be more successful collisions (see Figure 3.17).

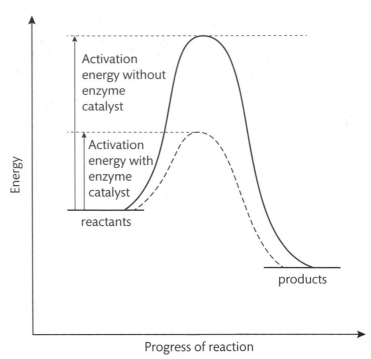

▶ **Figure 3.17:** Energy profile diagram showing how a biological catalyst (enzyme) lowers the activation energy of a chemical reaction

Ⅱ **PAUSE POINT**   Use the ideas of collision theory to answer the following questions. Close the book and list the four ways of increasing the rate of a chemical reaction. Explain how each of the ways in your list increases the rate of reaction.

(Hint)   You should refer to reacting particles and collisions in your answers.

(Extend)   Sometimes reactions need to be slowed down and negative catalysts or inhibitors are used to do this. Draw an energy profile diagram to show what a negative catalyst would do to the activation energy of a reaction.

Formation of enzyme-substrate complexes

The structure of an enzyme is such that it will only catalyse one particular chemical reaction with one particular substrate. This property of an enzyme is known as its specificity. For example, the enzyme peptidase, which increases the rate of protein breakdown in food, will not catalyse the breakdown of starch. Starch breakdown is catalysed by a different enzyme called amylase, which is present in saliva. Try chewing a piece of bread for a long time. You will find that it will begin to taste sweet as the starch in the bread is being converted into sugar by the amylase in your saliva.

The way enzymes work is that the active site on the enzyme binds with the substrate to form an enzyme-substrate complex (see Figure 3.18). The substrate has a specific shape which fits exactly into the active site on the enzyme molecule, in the same way that a key will only fit into one particular lock. This is why one enzyme will only react with a specific substrate.

A chemical reaction then takes place in the enzyme-substrate complex, and the substrate is converted into a product. The product leaves the active site of the enzyme, leaving the enzyme intact to bind to another substrate molecule.

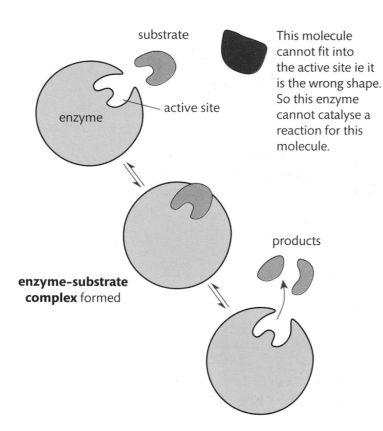

This molecule cannot fit into the active site ie it is the wrong shape. So this enzyme cannot catalyse a reaction for this molecule.

Key term

Enzyme-substrate complex – a transition state where the enzyme and substrate are joined together, before the enzyme converts the substrate into a new producer or products.

▶ **Figure 3.18:** The formation of an enzyme-substrate complex (lock and key mechanism)

Changing substrate concentration will change the rate at which substrate molecules will join the active site.

Factors that can affect enzyme activity

The factors that affect enzyme activity are substrate and enzyme concentration, temperature and pH. In this section, you will look at each of these factors separately.

Substrate and enzyme concentration

Increasing either the substrate or enzyme concentration will mean that there will be more particles in a given volume of solution. The particles will therefore be closer together and so will collide more often. This means more substrate molecules will bind with the enzyme molecules in a given time. This will increase the rate of reaction.

If the substrate concentration is much greater than the enzyme concentration, then there will not be enough enzyme molecules for all the substrate to bind with. The enzyme concentration then becomes a limiting factor, and the rate of reaction is no longer dependant on the substrate concentration at this point.

You will have the opportunity to plan and carry out an investigation into the effect of changing substrate concentration on the rate of an enzyme-catalysed reaction.

Temperature

In normal chemical reactions, as the temperature increases, the rate of reaction increases. However, this is not the case for enzyme-catalysed reactions. This is because each enzyme has a temperature at which it works best (see Figure 3.19). This is the optimum temperature.

For enzymes in the human body, this temperature is about 40 °C, which is just above normal body temperature. As the temperature is increased up to this optimum value, the rate of the enzyme-catalysed reaction will increase, as more molecules will collide with energy greater than or equal to the activation energy.

Above the optimum temperature, the reaction rate starts to decrease. This is because the heat energy starts to break the hydrogen bonds which form the secondary and tertiary structure of the enzyme, so the enzyme loses its shape and the substrate can no longer fit into the active site. At very high temperatures, this change is permanent and the enzyme is denatured. When this happens, the enzyme can no longer catalyse the reaction.

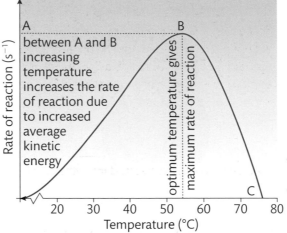

▶ **Figure 3.19:** The effect of temperature on the rate of an enzyme-catalysed reaction

pH

Enzymes have an optimum pH at which they work fastest. In humans, this is usually around pH 7 or 8, which is the pH of your body cells. Some enzymes can work at more extreme pH values. For example, protease enzymes in the stomach need to work in acidic conditions, so they have an optimum pH of 1. For most enzyme-catalysed reactions, if the pH is either very low (strongly acidic) or very high (strongly alkaline), the enzyme can become denatured. This means that:

▶ the enzyme will permanently lose its shape

▶ the substrate will not be able to fit into the active site

▶ the enzyme can no longer catalyse the reaction.

The importance of measuring initial reaction rates

When investigating reaction rates, it is important to consider the initial rate of a reaction. This is because as a reaction proceeds, reactants are being used up and products are being formed.

In an enzyme-catalysed reaction, the concentration of the enzyme remains the same but, as the reaction proceeds, the substrate is becoming less concentrated, which will cause the reaction to slow down. For example, if you are investigating the effect of temperature on the rate of an enzyme-catalysed reaction you should find the rate of reaction for each temperature soon after the start of reaction, before the substrate concentration has decreased significantly.

You will have the opportunity to plan and carry out investigations into the effects of temperature and pH on enzyme-catalysed reactions.

⏸ PAUSE POINT   Close the book and list the factors which affect enzyme activity. For each of these factors, explain how they affect enzyme activity.

Hint   You should refer to active sites, and lock and key theory in your answers.

Extend   Draw a sketch graph, similar to Figure 3.19 to show the effect of pH on an enzyme-catalysed reaction for a normal body cell. How would the graph be different for protease enzymes?

Fermentation

Fermentation is an important application of an enzyme-catalysed reaction. Without fermentation, bread would not rise and there would be no alcoholic drinks. Fermentation can also be used to produce ethanol for use as a fuel. In Brazil, where they grow a lot of sugar cane, fermentation is used to produce ethanol for use as a fuel in cars.

Key term

Fermentation – the process by which glucose is converted into ethanol and carbon dioxide in the presence of yeast.

Fermentation takes place when a sugar solution in the presence of yeast is left in **anaerobic** conditions at an optimum temperature of around 35 °C for several days. Yeast is a micro-organism which contains the enzyme zymase. The zymase in yeast converts the sugar into ethanol and carbon dioxide gas.

glucose → ethanol + carbon dioxide

$$C_6H_{12}O_6 \rightarrow 2C_2H_5OH + 2CO_2$$

Key term

Anaerobic – without the presence of oxygen.

In the production of alcoholic drinks, different foodstuffs can be fermented to give characteristic flavours to the drinks. For example, barley can be used to brew beer, and grapes are fermented to produce wine. When making spirits, the alcohol needs to be distilled off after fermentation to make it more concentrated.

In baking bread, the starch in the flour is broken down into glucose. This is then fermented by the zymase enzyme in yeast to produce the carbon dioxide gas which makes the bread rise. The ethanol produced is evaporated off when the bread is baked, so eating too much bread will not make you drunk!

Investigations for the enzymes in action topic

Before you plan your first investigation, it may be a good idea to look at the following example to give you an idea of what you need to include in your plan.

The effect of temperature on the activity of the enzyme lipase

Lipase is an enzyme that catalyses the breakdown of fat in milk to produce fatty acids. A mixture of milk, lipase and sodium carbonate solution is alkaline with a pH of about 10. After lipase catalyses the reaction, the pH drops as acids have been formed. Phenolphthalein is an indicator which is pink in alkaline solution but goes colourless when the pH drops below 8.2. By adding phenolphthalein to the reaction mixture, you can see how long it takes for the pH to drop below 8.2 at different temperatures.

Hypothesis

The higher the temperature, the faster the reaction until the optimum temperature is reached when the rate will decrease.

Hazard

Phenolphthalein solution is an irritant and is highly flammable.

Risk

The solution may cause irritation to eyes. There is a possibility of fire.

Safety tips

- Wear eye protection.
- Keep phenolphthalein solution away from naked flames.

Variables

▸ Independent variable – temperature
▸ Dependent variable – time taken for enough acid to form, so that pH drops below 8.2
▸ Control variables:
  ▸ Volume of milk
  ▸ Volume and concentration of sodium carbonate solution
  ▸ Volume and concentration of lipase solution
  ▸ Rate of stirring

Equipment

▸ Water baths at different temperatures
▸ 2 cm^3 syringe
▸ 10 cm^3 measuring cylinder
▸ 10 test tubes
▸ Thermometer
▸ Stirring rod
▸ Stop clock

Solutions

▸ Milk
▸ Sodium carbonate
▸ Lipase
▸ Phenolphthalein

Steps in the investigation	Pay particular attention to...	Think about this...
1. Set up five water baths at temperatures of 20, 30, 40, 50 and 60 °C.	The temperatures do not have to be exact but they should be at roughly equally spaced intervals.	More meaningful graphs can be plotted when a wide range of results are obtained.
2. Place a test tube of lipase solution in each water bath.	Make sure the whole of the solution is immersed in the water.	
3. In a separate test tube, add four drops of phenolphthalein indicator.	Only very small amounts of indicators are used in science experiments.	
4. Use a 10 cm³ measuring cylinder to measure out 5 cm³ of milk and add it to the test tube with the indicator.	A 10 cm³ measuring cylinder is used as only a small quantity of milk is being measured.	Using a larger measuring cylinder would give a greater uncertainty of the volume measurement.
5. Use a 10 cm³ measuring cylinder to measure out 8 cm³ of sodium carbonate solution and add it to the test tube. The solution will go pink because it is alkaline.	Use a different measuring cylinder for each different solution.	You do not want the different solutions to contaminate each other.
6. Put a thermometer in the test tube and place it in the water bath at 20 °C.	Make sure the reaction mixture reaches the same temperature as the water bath.	The independent variable in this investigation is the temperature.
7. When the temperature reaches 20 °C remove the thermometer from the test tube and use a 2 cm³ syringe to measure out 1 cm³ of lipase solution from the 20 °C water bath.	A syringe will give a more precise measurement than a measuring cylinder for such a small quantity of solution.	
8. Add the lipase solution to the test tube and start the stop clock.	Make sure you add the solution and start the clock at the same time.	
9. Stir the contents with a glass rod until the solution turns colourless.	Ensure that you stir thoroughly to mix the solutions.	If your rate of stirring is not the same each time you do the experiment, this will affect the accuracy of your results.
10. Stop the clock and record the time.		Time for the solution to turn colourless is the dependent variable in the investigation.
11. Repeat steps 3 to 10 for each temperature.	Make sure the solutions reach the temperature of the water bath each time.	
12. Repeat the whole experiment three more times.		Scientific investigations produce more accurate results if they are carried out a number of times.

Recording and processing your data

Once you have collected your data you should record it in an appropriate table, take averages and plot a line graph of your results.

You should be able to write a conclusion and comment on the validity of your hypothesis.

Investigations for you to plan and carry out

1 The effects of temperature and pH on the protein in egg albumen. Egg albumen is denatured by strong acids and alkalis and high temperatures, causing it to solidify and become opaque. (Denaturation has been explained in the theory section.)
2 The effect of temperature on the action of protease on milk.
3 The effect of substrate concentration on the enzyme-catalysed reaction of catalase on hydrogen peroxide solution. Hydrogen peroxide solution decomposes to give oxygen and water in the presence of a suitable catalyst.

   hydrogen peroxide $\rightarrow$ oxygen + water

   $2H_2O_2 \rightarrow O_2 + 2H_2O$

Catalase is a suitable catalyst for this reaction and can be found in several foodstuffs such as liver, potatoes and celery. The rate of the reaction can be increased when there are more hydrogen peroxide substrate molecules to bind with the active site of the catalase enzyme molecules.

As oxygen gas is produced in this reaction, you can time how long it takes to produce a fixed volume of oxygen gas for each concentration of hydrogen peroxide solution.

▶ The action of the enzyme catalase in raw liver on hydrogen peroxide solution

Assessment practice 3.4

Casein is a protein in milk which causes the milk to be white and opaque. Protease breaks down the casein in milk into amino acids. This causes the milk to become clear. The reaction can be followed by looking at a cross through a beaker containing the milk and protease. As the milk starts to clear there will be a point where the cross starts to become visible. You can time how long it takes for the cross to become visible for different reaction temperatures.

You are to plan an investigation as to how temperature affects the action of the enzyme protease on milk.

Your plan should include:
- a hypothesis
- selection and justification of the equipment you are going to use
- hazards and risks associated with the investigation
- independent, dependent and control variables
- a method for data collection to test the hypothesis including:
  - the quantities to be measured
  - the number and range of measurements to be taken
  - how the apparatus may be used.

E Diffusion of molecules

In this section, you will learn about the factors that affect the rate of **diffusion** of molecules. You will also learn about **kinetic theory**. Understanding the concepts covered in this topic will enable you to plan and carry out investigations to study some of the factors that affect the rate of diffusion.

Factors affecting the rate of diffusion

If you spray air freshener in a corner of a room, why can you soon smell it in other parts of the room?

This is due to the diffusion of the molecules in the air freshener. They have moved from an area where there is a high concentration of air freshener molecules to areas of the room where there are little or no air freshener molecules. Eventually the air freshener molecules should mix completely with the air molecules in the room and be spread out evenly throughout the room (see Figure 3.20).

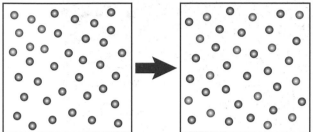

▶ **Figure 3.20:** How air freshener molecules would spread out by diffusion to mix completely with the gas molecules in the air in the room due to the random movement of particles

Several factors affect the rate of diffusion. In this section, you will look at each one in turn.

Concentration gradient

When molecules diffuse, they are moving along a **concentration gradient**.

The greater the concentration gradient (that is, the bigger the difference in concentration between where the molecules are and where they are moving to), the faster the rate of diffusion.

Shape and size of molecules

Smaller molecules will diffuse quicker than larger molecules, and molecules with a more streamlined shape (long thin molecules) will diffuse quicker than less streamlined molecules (fat bulky molecules) of a similar molecular mass.

Temperature

The higher the temperature, the more energy the molecules will have, and the faster they will move, therefore increasing the rate of diffusion.

Distance

The further the molecules have to travel, the longer it will take, so the rate of diffusion is quicker over short distances than over long distances.

Surface area

When diffusion takes place through a **semi-permeable membrane**, such as a cell membrane, the greater the surface area of the membrane, the faster the rate of diffusion of molecules through the membrane. This is important for gas exchange in body cells, as oxygen needs to enter the cells for respiration to take place and carbon dioxide needs to be removed from the cells. If the cell membranes have a larger surface area, this process can take place more quickly and efficiently.

Key term

Concentration gradient – the change in concentration from an area of high concentration of molecules to an area of low concentration.

Key term

Semi-permeable membrane – a membrane that will allow small molecules such as water, carbon dioxide and oxygen to pass through it, but will not allow large molecules to pass through it.

PAUSE POINT

Close your book and list all the factors that can affect the rate of diffusion. Explain how each of these factors affect the rate of diffusion.

Hint You could include particle diagrams in your explanations.

Extend How many everyday examples of diffusion can you think of?

Diffusion demonstrations

The following experiments can be used to demonstrate the diffusion of gas molecules. Due to the hazardous nature of the substances involved, you will not be able to do these experiments yourself.

The diffusion of bromine in air

A gas jar of air is placed above a gas jar of bromine. The bromine starts to diffuse into the gas jar of air even though it is denser than air. After some time, the bromine will be spread evenly throughout the two gas jars (see Figure 3.21).

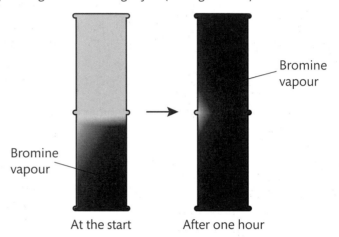

▶ **Figure 3.21:** The diffusion of bromine in air (a) at the start (b) after one hour

Explanation

Both bromine molecules and air molecules are constantly moving and colliding with each other and the walls of the gas jars. This will result in some bromine molecules moving into the upper gas jar and some air molecules moving into the lower gas jar. This process will continue until there is a uniform mixture of bromine and air in both gas jars. The bromine has moved along a concentration gradient from an area of high concentration to an area of low concentration.

The diffusion of hydrogen chloride and ammonia gases

Two pieces of cotton wool, one soaked in concentrated hydrochloric acid, and one soaked in concentrated ammonia solution, are placed at the ends of a long tube, clamped so that it is horizontal.

After several minutes a white ring of ammonium chloride starts to form nearer the end of the tube which contained the cotton wool soaked in hydrochloric acid (see Figure 3.22).

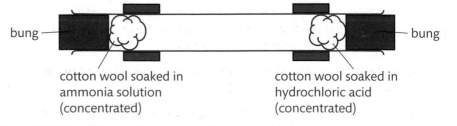

▶ **Figure 3.22:** The diffusion of ammonia and hydrogen chloride gas

Explanation

The concentrated hydrochloric acid gives off hydrogen chloride gas and the concentrated ammonia gives off ammonia gas. Both diffuse along the tube and, when they meet, they react to form ammonium chloride which is a white solid. Because ammonia has smaller molecules and is less dense than hydrogen chloride, it diffuses quicker, so the ammonium chloride is formed closer to the hydrochloric acid end.

ammonia + hydrogen chloride $\longrightarrow$ ammonium chloride

$NH_3(g) + HCl(g) \longrightarrow NH_4Cl(s)$

Case study

Trainee science technician

Richard is a trainee science technician.

He has been asked to set up the demonstrations for the two diffusion experiments shown in Figures 3.21 and 3.22.

You are the senior laboratory technician in the college and it is your job to teach Richard how to set up the two experiments taking all the necessary safety precautions.

The first thing you need to do is consult the hazard cards for the three chemicals involved and write a risk assessment for the three chemicals so that you can teach Richard how to handle them safely.

You then need to show him how to set up the apparatus in the laboratory ready for the demonstration.

Use the following steps to help you to show Richard what he needs to do.

Check your knowledge

1 Look at the hazard cards for bromine, concentrated hydrochloric acid and ammonia, and note down the hazards and the risks. If you do not have these cards in your school or college, you should be able to find them on the CLEAPSS (Consortium of Local Education Authorities for the Provision of Science Services) website.

2 Use this information to write a risk assessment for each of the practical demonstrations.

3 Write a list of the precautions that Richard must take when using the three chemicals.

4 Write out step-by-step instructions as to how Richard is to set up the two demonstrations.

Arrangement and movement of molecules

Chance discoveries have played an important part in the development of scientific ideas. This happened with Brownian motion. In 1827, a botanist called Robert Brown was observing pollen grains in water under a microscope. He noticed that the pollen grains moved around jerkily in a random fashion.

In 1905, Albert Einstein came up with the theory that the pollen grains were moving because they were constantly being bombarded by the much smaller water molecules. This was the first evidence to show that molecules in liquids and gases are constantly moving in a random fashion.

You can observe Brownian motion in the lab by looking at illuminated smoke particles under a microscope (see Figure 3.23).

The smoke particles appear as small specks of light which are moving around randomly, as Figure 3.24 shows.

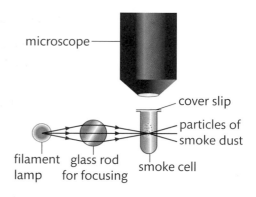

▶ **Figure 3.23:** Observing Brownian motion

185

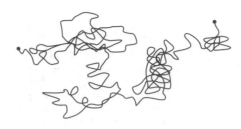

▶ **Figure 3.24:** The random zigzag motion of a smoke particle

The smoke particles move in this way because they are being constantly bombarded by the much smaller molecules in the air, which are in rapid constant random motion.

Observing Brownian motion led to the development of the **kinetic model of matter**.

The arrangement of particles in solids, liquids and gases

The arrangement of particles in solids, liquids and gases is as follows.

▶ Particles in solids are touching each other and in fixed positions (see Figure 3.25).

▶ Particles in liquids are close together, rolling over each other and arranged randomly.

▶ Particles in gases are far apart and arranged randomly.

Key term

Kinetic model of matter
– all matter is made up of very small particles (atoms, molecules or ions) which are in constant motion.

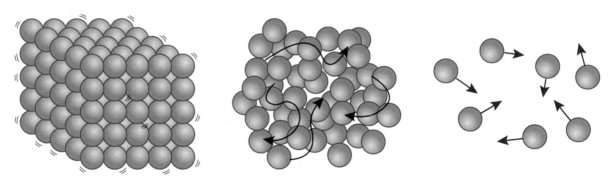

▶ **Figure 3.25:** The arrangement of particles in a solid, a liquid and a gas

Kinetic theory and the random movement of molecules

Kinetic theory is concerned with the motion of particles in solids, liquids and gases (see Figure 3.26).

The particles in solids, liquids and gases at any temperature above **absolute zero** will all exhibit some form of movement. At absolute zero, all movement of particles stops.

Key term

Absolute zero – the lowest possible temperature, which is 0 K on the Kelvin temperature scale and -273 °C on the Celsius temperature scale.

▶ In a solid, the particles can vibrate in all directions but cannot move out of their fixed positions.

▶ In a liquid, the particles move randomly and can slide past each other, but they do not move far.

▶ In a gas, the particles move around quickly and can travel large distances in all directions. The molecules in a gas are constantly hitting each other and the walls of their container, causing them to change direction.

In all states of matter, if the temperature is increased, the particles will gain more kinetic energy and vibrate or move more quickly.

▶ **Figure 3.26:** The random motion of particles in (a) a solid, (b) a liquid and (c) a gas

Diffusion in gases is much faster than in liquids, because molecules have much more kinetic energy and move a lot faster.

PAUSE POINT   Close the book and describe the arrangement of particles in solids, liquids and gases.

Hint   You may use diagrams to help with your descriptions.

Extend   Use your knowledge of kinetic theory to describe the movement of particles in solids, liquids and gases. Why do molecules in liquids diffuse more slowly than molecules in gases?

Dynamic equilibrium

Look back to the diffusion of bromine demonstration. At first, the lower gas jar contains bromine vapour and the upper gas jar only contains air molecules. When the two jars are together:

▶ the bromine molecules start to move into the upper gas jar

▶ some air molecules will move down into the lower gas jar.

Also, when there is bromine in the upper jar, some bromine molecules will move back down into the lower jar. However, as the concentration of bromine is much greater in the lower jar, many more bromine molecules will move up to an area of lower concentration than will move down.

This process will continue until there are equal numbers of bromine molecules in both jars. Now there is no longer a concentration gradient, and so for every bromine molecule that moves up one will move down. At this stage, a **dynamic equilibrium** has been reached. At dynamic equilibrium, bromine molecules move between the two gas jars at the same rate, so there is no net change of concentration of bromine in the two jars.

Key term

Dynamic equilibrium – when two processes take place at the same rate so there is no further change in concentration of the substances involved.

Before dynamic equilibrium has been reached

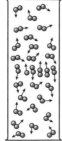

After dynamic equilibrium has been reached

Key:
🔗 a moving bromine molecule

▶ **Figure 3.27:** Particle diagrams showing the movement of bromine molecules (a) before dynamic equilibrium has been reached and (b) after dynamic equilibrium has been reached

Investigations for the diffusion of molecules topic

The effect of temperature on the diffusion of coloured ice cubes in water. This is an investigation which you should be able to write a hypothesis for and then plan, carry out, process, analyse and evaluate your method and results.

Investigation 3.2

The effect of concentration on the diffusion of food colouring through agar

Equipment list

▶ 6 Petri dishes, with lids, containing solidified agar jelly, at least 5 mm deep
▶ 6 concentrations of red or blue food colouring between 20% and 70%
▶ 5 mm cork borer
▶ Cocktail stick
▶ 1 cm^3 graduated syringe
▶ 30 cm ruler with millimetre divisions
▶ Marker pen

Step-by-step method

Steps in the investigation	Pay particular attention to...	Think about this...
1. Use a 5 mm cork borer to cut out three evenly spaced discs in the agar jelly.	It is important that the wells cut in the agar are the same size.	The wells need to be evenly spaced so there is enough room for the food dye to diffuse through the agar.
2. Remove the discs with a cocktail stick to leave wells in the jelly and discard the discs.	The cocktail stick enables the disc to be removed easily.	
3. Use the 1 cm^3 syringe to add 0.1 cm^3 of a 20% concentration of food colouring solution to each of the wells.	Only a small amount of food colouring is needed and a 1 cm^3 syringe is used to obtain an accurate measurement.	Having three wells in the dish will give three sets of data for each concentration. Scientific investigations produce more accurate results if repeat measurements are taken.
4. Place a lid on the Petri dish and label it with the correct concentration using the marker pen.		Covering the Petri dish with a lid will stop the food colouring solution being lost by evaporation.
5. Repeat steps 1 to 4 for the other five Petri dishes with five different concentrations of food colouring.	Choose evenly spaced concentrations, such as 30%, 40%, 50%, 60% and 70%.	More meaningful graphs can be plotted when a wide range of results are obtained.
6. Leave the Petri dishes undisturbed at room temperature for 24 hours.	This should be long enough for the food dye to spread out enough for accurate distance measurements to be taken.	
7. Measure the distances the food colouring has diffused for the three wells on each Petri dish.	Use a transparent 15 cm or 30 cm ruler to measure the diameters of the food colouring circles.	
8. Calculate the mean distance for each concentration and plot a graph of distance travelled by the food colouring against concentration.	You could include error bars on your graph.	

Assessment practice 3.5

Use the information on the diffusion of food colouring through agar investigation to help you answer the following questions.

1 What would be your hypothesis for this investigation?

2 What are the independent and dependent variables in this investigation?

3 What variables would you need to control in this investigation?

4 When you have completed this investigation, calculate the percentage errors of the measuring equipment used.

5 Which piece of equipment is most likely to affect the reliability of your results?

6 Was your hypothesis correct?

7 How could you improve this method to obtain more accurate and reliable results?

8 How could you extend this investigation to obtain further evidence to support your hypothesis?

F Plants and their environment

In this section you will learn about the factors that affect the growth and distribution of plants. You will also learn about the different environmental sampling techniques, so that you will be able to plan and carry out investigations using these techniques.

Factors that can affect plant growth and/or distribution

There are many factors that can affect plant growth and distribution. If any single environmental factor is less than ideal, it will become a **limiting factor** in determining how well the plants will grow. For example, there may be enough carbon dioxide and water available for photosynthesis but if light levels are limited, light intensity becomes a limiting factor and the rate of photosynthesis is reduced.

> **Key term**
>
> **Limiting factor** – a factor that limits the rate of a reaction.

Human effects

There are many ways in which human activities can affect plant growth and distribution. These include the following.

▸ **Trampling:** In a particular area of land trampling by humans or cattle can lead to an uneven distribution of plants.

▸ **Habitat destruction:** This could include deforestation or clearing of land for development or agriculture.

▸ It is sad when areas of natural beauty such as this are destroyed, in order to provide land for building on

- **Pollution:** Acid rain can cause soils to become too acidic, killing some plants and inhibiting the growth of others.
- **Use of chemicals:** Chemicals used include pesticides, fungicides, herbicides, fertilisers and liming. These all affect the growth and distribution of plants, and are likely to reduce **biodiversity** in a natural habitat.
- **Over-harvesting:** This depletes the soil of nutrients, making it difficult for plants to grow.
- **Monocultures:** This involves growing only one type of plant in a particular field, which keeps depleting the soil of the same minerals, which are essential for healthy plant growth.

> **Key terms**
>
> **Habitat** – a place with suitable conditions for a variety of different plants and animals to live in. There are many different types of habitat, e.g. woodland, tropical rainforest, freshwater ponds.
>
> **Biodiversity** – the variety of life in a particular habitat. It includes all the plants, animals and microorganisms that live there.

▶ Monoculture of oil palm plantation showing no biodiversity.

Soil pH and aeration

Plants are sensitive to changes in pH. Most plants grow best in neutral or slightly acidic soil. The pH range of most soils is between 4.5 and 7.5, although there are a few plants that grow better in more extreme conditions, in soils as acidic as pH 3 or as alkaline as pH 9.

▶ Sundews grow in peat bogs where the soil is damp and acidic

Aeration of the soil is also important for the healthy growth of plants. This is because oxygen is needed:

▶ for plants to respire

▶ for microorganisms to respire, as these are needed to decompose organic matter and for nitrification of the soil

▶ to help plants to absorb water and nutrients

▶ to help prevent toxins forming in the soil

▶ to help prevent plants from contracting diseases.

Light intensity

Sunlight is necessary for plants to be able to produce plant food by **photosynthesis**. Plants grow more quickly in summer when there is more sunlight. Usually plants grow better in unshaded areas than in shaded areas.

Key term

Photosynthesis – the process by which plants make food, using carbon dioxide, water and the energy from sunlight.

$$\text{carbon dioxide} \quad + \quad \text{water} \quad \rightarrow \quad \text{glucose} \quad + \quad \text{oxygen}$$
$$6CO_2 \quad + \quad 6H_2O \quad \rightarrow \quad C_6H_{12}O_6 \quad + \quad 6O_2$$

Temperature

Some plants, such as broccoli and spinach, grow well in cooler climates, and others, such as oranges and bananas, grow better in warmer climates.

However, extremes of temperature lead to lack of plant growth.

High temperatures can cause **respiration** to take place more quickly than photosynthesis, causing the products of photosynthesis to be used for respiration. If this happens, the plants cannot grow.

$$\text{glucose} + \text{oxygen} \rightarrow \text{carbon dioxide} + \text{water}$$
$$C_6H_{12}O_6 + 6O_2 \quad \rightarrow \quad 6CO_2 \quad + 6H_2O$$

Low temperatures result in poor growth as photosynthesis is too slow at low temperatures. When temperatures fall below the freezing point of water, 0 °C, plant cells and tissues can be destroyed. This can kill many plants.

Key term

Respiration – the process by which glucose in living cells is converted into carbon dioxide and water, releasing energy.

Presence of water

Water is important to all living organisms on the planet. Plants cannot survive without it.

▶ Moisture in the air is absorbed by the leaves of a plant for photosynthesis.

▶ Water in the soil is taken up by the roots. This collects in the leaves for photosynthesis.

▶ Water is also important as it dissolves the minerals in the soil which can be taken up by the roots to all parts of the plant.

▶ If soils become water-logged, this water can have an adverse effect on some plants because it causes the roots to rot, killing the plant.

▶ Plants also lose water from their leaves by **transpiration**, so this water needs to be replaced.

▶ Different plants need varied amounts of water, e.g. in desert areas where there is very little rainfall, plants are adapted to store as much water as possible.

Key term

Transpiration – evaporation of water from the surface of the leaves of plants.

▶ The leaves of this cactus have been reduced to spines to give them a very small surface area to reduce water loss. The fat green body of the cactus is a stem, which is full of water-storing tissue.

Mineral ions

There are several mineral ions which are essential for the healthy growth of plants. These are present in the soil and are taken up by the roots and distributed throughout the plant.

Table 3.11 shows the mineral ions needed for healthy plant growth and the consequences of a lack of these mineral ions.

▶ **Table 3.11:** Mineral ions and the effects of their deficiency

Key term

Chlorophyll – the green pigment found in the leaves of plants, which is needed for photosynthesis.

Mineral ion	Effect of deficiency of this ion
Calcium, Ca^{2+}	Tissues become soft and the plant is likely to wilt.
Magnesium, Mg^{2+}	Magnesium is an essential part of the **chlorophyll** molecule. Without chlorophyll, the plant is unable to photosynthesise so cannot grow.
Iron, Fe^{3+}	Leaves become bleached, leading to deficiency in chlorophyll and reduced photosynthesis.
Potassium, K^+	Potassium is essential for the formation of healthy flowers and fruit. Leaves lose colour at the tips and may curl and crinkle.
Nitrate, NO_3^-	Nitrates are needed for healthy growth. Plants become short and spindly and deficient in chlorophyll. They may wilt and die.
Phosphate, PO_4^{3-}	Plants grow more slowly, leading to dwarfed or stunted plants.
Sulfate, SO_4^{2-}	Veins in the leaves take on a reddish colour, leading to a deficiency in chlorophyll. Leaves may also become twisted and brittle.

❚❚ PAUSE POINT

Hiromi is an ecologist who works at an experimental farm and is studying the growth and distribution of plants in a meadow. Make a list of all the factors that could affect the growth and distribution of these plants. Which of these factors do you think Hiromi would be able to control?

Hint

Think about what substances and equipment farmers might have to control some of these factors.

Extend

Assuming Hiromi has access to all the necessary farming machinery, explain how she could control these factors.

Sampling techniques

Sampling techniques are used to study the distribution of plants in a particular habitat. Obviously it is impossible to count all the plants in the habitat, so instead you select a small portion of the habitat and study it carefully.

The importance of random sampling

If you wanted to study the distribution of the plants in a meadow, you might be tempted to sample the areas where you could see a large number of different plants, but this would not give you a correct picture of the overall distribution. To avoid this, you must use random sampling. There are several ways of making sure your sample is random. You can:

▶ take samples at regular distances across the habitat

▶ use a computer to generate random numbers to plot co-ordinates in the habitat at which to take samples

▶ select co-ordinates on a map and use a GPS system to find the exact position in the habitat at which to take samples.

Selecting the appropriate sampling technique

The sampling technique you choose to use depends on the type of habitat you are investigating. The number of samples you take depends on the size of the habitat. The different ways of sampling are explained below.

Line transects

Line transects are used for large habitats or at the edge of footpaths. A long tape measure is stretched across the habitat and the plants touching the tape measure are recorded at regular intervals.

Quadrats

Quadrats are used for smaller habitats. A quadrat is a square frame that is placed at the randomly selected areas of the habitat. It is usually either a 1 m by 1 m square or a 50 cm square. There are two types of quadrats: open and gridded. A gridded quadrat is divided into a number of smaller squares, usually 100 for a 1 m^2 quadrat or 25 for a 50 cm^2 quadrat.

▶ Using a line transect

▶ Using a gridded quadrat

Having positioned the quadrat correctly, you will need to count the numbers of each type of plant within it (see Figure 3.28).

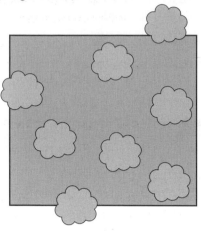

Everything in or touching the quadrat = 8 plants

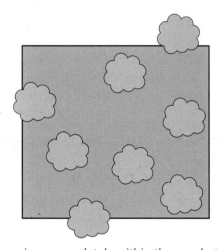

Only organisms completely within the quadrat = 5 plants

▶ **Figure 3.28:** When counting the numbers of a particular plant in a quadrat, you can choose to count (a) every plant inside or touching the quadrat (= 8) or (b) only plants completely within the quadrat (= 5)

Point frames

These are frames with a number of long needles. The frame is lowered into an open quadrat and any plants touching the needles are recorded. If you have a frame with 10 needles, you would move it 10 times in each quadrat to give 100 readings. Each plant recorded as touching the needle will have 1% cover, so you can estimate the percentage cover of each plant from the results.

Sampling sizes

The size of the sample you collect depends on the size of the habitat you are studying. The larger the habitat, the more areas you will need to sample in order to obtain enough results to draw valid conclusions.

In theory, the more areas you can sample the better, but in practice, time is often restricted and you will need to compromise on how many samples you take. If you are working in groups, sometimes pooling your results for analysis is a good way of obtaining more samples.

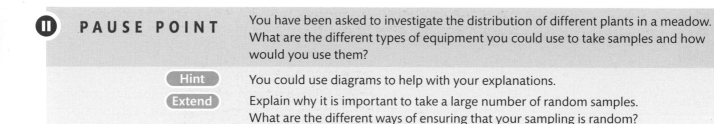

ⅠⅠ **PAUSE POINT**   You have been asked to investigate the distribution of different plants in a meadow. What are the different types of equipment you could use to take samples and how would you use them?

> **Hint**   You could use diagrams to help with your explanations.
>
> **Extend**   Explain why it is important to take a large number of random samples. What are the different ways of ensuring that your sampling is random?

Investigations for the plants and their environment topic

1   Investigating the effect of different amounts of water on the growth of mung bean shoots. This is an investigation that you could plan and carry out. You can then analyse and evaluate the results.

2   Investigation using quadrats to count the number of daisies in different areas of a field. This investigation will give you an opportunity to use statistics to calculate means and standard deviations. You can go on to use the t-test to compare the number of daisies growing in shaded and unshaded areas and use t-tables to decide whether the null hypothesis is rejected or accepted at a 5% significance level.

3   Use a line transect to investigate the abundance of a particular plant along a footpath or along the edge of a field. Use your results to draw conclusions as to why some areas may be more populated than others due to different environmental factors.

4   Plan an investigation to find the frequency of distribution of different plants in a field, lawn or meadow. You can then carry out the investigation and analyse the results. You can plot a bar chart to show the numbers of the different species of plant.

Assessment practice 3.6

Roisin and Adam were studying the distribution of dandelion plants on a lawn and in a vegetable patch.

They placed a quadrat at five different places on the lawn and each time counted the numbers of dandelion plants.

They then repeated this on the vegetable patch.

Here are their results.

Quadrat	1	2	3	4	5
Number of dandelions on lawn	4	3	1	5	4
Number of dandelions in vegetable patch	0	2	1	4	1

1   Calculate the means for the two sets of results.

2   Find the difference between the means.

3   Find the standard deviations for the two sets of data.

4   Calculate the standard error in the difference using the equation $\sqrt{\dfrac{s_1^{\ 2}}{n_1} + \dfrac{s_2^2}{n_2}}$

5   Calculate the value of t, where t = standard error in mean ÷ standard error in difference.

6   Use a t-test table to determine the critical value at a significance level of 5%.

7   Considering a 5% significance level, is there a significant difference between the numbers of dandelions in the lawn and the vegetable patch?

8   Explain what Roisin and Adam could do to improve the reliability of their data.

G Energy content of fuels

Key term

Fuel – a substance that undergoes combustion with oxygen to produce energy.

In this section, you will learn about different types of **fuel** and the hazards and risks associated with using fuels. You will also learn how to calculate the heat energy released by a fuel, so that you can investigate different fuels and compare their efficiency in terms of the amount of energy they release.

Discussion

Without looking at the book, how many different fuels can you think of?
Do you know what each of these fuels is used for?

What is a fuel?

Key terms

Organic – derived from living things.

Carbohydrate – a food source made up of the elements carbon, hydrogen and oxygen.

There are many types of fuel, ranging from the petrol used in cars to the food you eat. Most fuels are **organic** compounds containing carbon, hydrogen and, in some cases, oxygen. Normally, in order for combustion of a fuel to occur, you have to ignite it. However, this is not the case with the food you eat. The main energy-providing foods are **carbohydrates**. They are fuels as they are broken down in the human body to produce glucose, which is then broken down further by respiration to produce energy.

As most fuels contain carbon and hydrogen, a word equation for the combustion of a fuel can be written as:

$$\text{fuel} + \text{oxygen} \longrightarrow \text{carbon dioxide} + \text{water}$$

Fuels from crude oil

Key terms

Hydrocarbon – a compound made up of only hydrogen and carbon atoms.

Alkane – a hydrocarbon with the general formula C_nH_{2n+2}.

Homologous series – a group of organic compounds with similar chemical properties where one member of the series differs from the next by a CH_2 group.

Fractional distillation – separation of a mixture of liquids into fractions with different boiling point ranges.

Many of the fuels used in everyday life are obtained from crude oil. Crude oil is a mixture of substances made up mainly of **hydrocarbons**.

Most of the hydrocarbons in crude oil are **alkanes** (see Figure 3.29), which are saturated hydrocarbons. This means there are only single bonds between carbon atoms and they contain as many hydrogen atoms as possible.

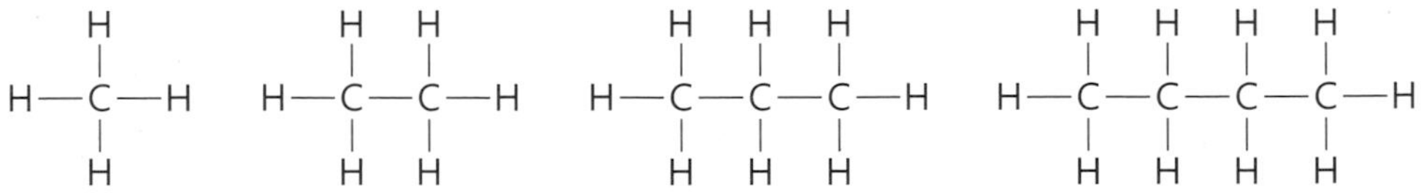

▶ **Figure 3.29:** The first four alkanes. These are all gases at room temperature.

These alkanes form part of a **homologous series**.

The alkanes in crude oil range from methane gas, CH_4, with only 1 carbon atom, up to solid alkanes with approximately 60 carbon atoms.

Crude oil in itself is not particularly useful, but it can be separated into many useful fractions by the process of **fractional distillation** (see Figure 3.30). This involves heating the crude oil in a furnace to turn it all into vapour and then passing it into a fractionating column, where it is separated into different fractions with different boiling point ranges. The gases with the lowest boiling points are collected at the top of the column and the fractions with the highest boiling points are collected at the bottom of the column.

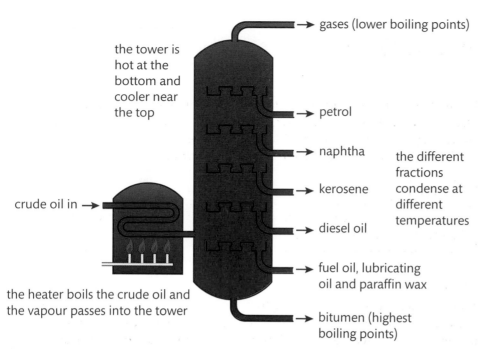

the tower is hot at the bottom and cooler near the top

gases (lower boiling points)

petrol

naphtha

kerosene

diesel oil

the different fractions condense at different temperatures

fuel oil, lubricating oil and paraffin wax

crude oil in →

the heater boils the crude oil and the vapour passes into the tower

bitumen (highest boiling points)

▶ **Figure 3.30:** The fractional distillation of crude oil

Each of the fractions produced is a mixture of alkanes in a particular boiling point range. The lower boiling point fractions are used as fuels. Table 3.12 shows the uses of the different fractions.

▶ **Table 3.12:** Fractions and their uses

Fraction	Approximate number of carbon atoms	Approximate boiling point range / °C	State at room temperature	Uses
Petroleum gas	1 to 4	< 20	Gas	Fuel for industry
Gasoline (petrol)	5 to 8	40 to 120	Liquid	Fuel for cars
Naphtha	9 to 10	100 to 180	Liquid	To make petrochemicals
Kerosene (paraffin)	11 to 12	160 to 250	Liquid	Fuel for aircraft and domestic heating
Diesel (gas oil)	13 to 20	220 to 350	Liquid	Fuel for diesel engines
Fuel oil	21 to 25	320 to 400	Liquid	Fuel for large ships
Lubricating oil	26 to 28	400	Liquid	For engine oil to lubricate moving parts
Paraffin wax	29 to 30	> 400	Solid	To make candles
Bitumen	> 30	> 400	Solid	For road surfacing and waterproofing

Properties of hydrocarbon fuels

As the length of the carbon chain increases, the fuels become darker in colour, more viscous, less flammable and therefore harder to ignite. Fuels with short carbon chains tend to burn more cleanly and produce less soot than those with longer carbon chains. Table 3.13 describes fractions with an increasing number of carbon atoms.

Fractions with an increasing number of carbon atoms	Colour	Viscosity	Ease of ignition	Sootiness of flame
Petroleum gases Petrol Kerosene Diesel Fuel oil Lubricating oil Wax Bitumen	**Darkens** as number of carbon atoms increases, e.g. petrol is colourless, fuel oil is dark orange, bitumen is black.	**Increases** as number of carbon atoms increases, e.g. petrol is runny and flows easily, fuel oil is thick and viscous.	**Decreases** as number of carbon atoms increases, e.g. petrol ignites easily, fuel oil is difficult to ignite and bitumen will not ignite.	**Increases** as number of carbon atoms increases, e.g. petroleum gases burn cleanly with a blue flame, kerosene produces a lot of smoke and soot when burnt.

> **Key term**
>
> **Viscosity** – a measure of how easily a liquid flows. The thicker and less runny the liquid, the more viscous it is.

Using alcohols as fuels

Alcohols are another homologous series of organic compounds made up of carbon, hydrogen and oxygen atoms. They have a general formula $C_nH_{2n+1}OH$.

The first five alcohols in the series are all liquids with the following names and formulae:

▶ methanol, CH_3OH

▶ ethanol, C_2H_5OH

▶ propan-1-ol, C_3H_7OH

▶ butan-1-ol, C_4H_9OH

▶ pentan-1-ol, $C_5H_{11}OH$.

Figure 3.31 shows models of the first three alcohols.

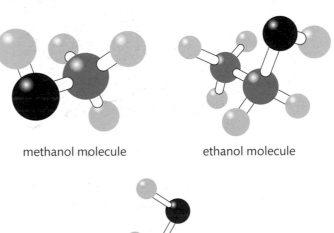

methanol molecule   ethanol molecule

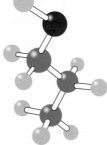

propan-1-ol molecule

▶ **Figure 3.31:** Models of the first three alcohols: methanol, ethanol and propan-1-ol

Alcohols such as these with a small number of carbon atoms make good fuels, as they burn more cleanly than liquid hydrocarbon fuels such as petrol. Methanol and ethanol can also be obtained from renewable energy sources, which is also an advantage. Using renewable energy sources helps to conserve fossil fuels such as crude oil and natural gas, which are in danger of running out.

Ethanol, for example, can be produced by fermentation and can be used as a fuel for cars, either on its own or mixed with petrol to make gasohol.

Burning ethanol produces carbon dioxide and water according to the following equation:

ethanol + oxygen $\rightarrow$ carbon dioxide + water

$C_2H_5OH + 3O_2 \rightarrow 2CO_2 + 3H_2O$

Even though the greenhouse gas carbon dioxide is produced, the use of ethanol as a fuel is considered to be carbon neutral, as carbon dioxide is used for photosynthesis by the plants which are grown to produce the ethanol.

Another fuel which can be used as an alternative fuel in diesel engines is biodiesel. This can be produced from the oilseed rape plant. In the countryside, you sometimes see fields of yellow oilseed rape plants. The vegetable oil can be extracted from the plants and used as biodiesel. As with ethanol, because it is a bio-fuel (a fuel from a plant), it is also carbon neutral.

▶ Filling station with a bio-fuel petrol pump

Other fuels that can be investigated are different types of cooking oil.

Ⅱ PAUSE POINT   Consider what you have learnt about the properties of fuels.
Write a list of all the things you can think of which make a good fuel.

    Hint   You should include physical properties, chemical reactivity and environmental considerations in your list.

    Extend   Write balanced chemical equations for the combustion of: (a) propane (b) methanol.

Hazards associated with fuels

Fuels are very useful substances for everyday life. However, you need to be aware of the hazards and risks associated with using them.

Toxicity

Some fuels are toxic to humans. An example of a toxic fuel is the alcohol methanol. As little as 10 cm³ of methanol can attack the central nervous system and may lead to blindness, coma or even death. Methylated spirit is mainly ethanol, but it contains a small amount of methanol. It is used in industry as a solvent or cleaning agent and as a fuel in some camping stoves. Because of the methanol content, it is toxic to humans.

Flammability

You need to be able to ignite fuels in order to burn them, so many fuels are flammable, and containers of these fuels need to display the appropriate hazard symbol, warning of their flammability. Careless use of flammable fuels could cause a fire.

Risk of explosion

If a large quantity of a gaseous fuel or vapour from a liquid fuel is released into the air, a spark is likely to cause an explosion.

Case study

The Buncefield fire

On Sunday 11 December 2005 at 6 am, there was a huge explosion at the Buncefield oil storage depot in Hertfordshire. Apparently there was a faulty gauge in a large storage tank containing petrol. Normally the tank would only fill to a safe level, but because the gauge was faulty, the tank continued to fill up until it started to overflow. Petrol vapour mixed with the air outside the tank and the fuel and air mixture ignited and caused a huge explosion.

This explosion triggered further explosions in other fuel tanks, affecting 20 tanks in total. The explosion was heard by people living up to 20 miles away and the smoke cloud produced was seen by people living 70 miles away from the site. Windows shook in houses up to 10 miles away and a window in St Albans Abbey, which was five miles from the site, was blown out completely.

It was very lucky that the explosion took place on a Sunday morning as the windows in offices nearby were completely blown out and, if it had happened on a weekday when people were at work, there could have been many deaths and serious injuries. Luckily no one died but around 45 people were injured.

Around 2000 people who lived nearby had to evacuate their homes to avoid smoke inhalation. Many schools in the surrounding area were closed on 12 and 13 December, and people were advised to stay indoors and keep their windows closed.

It took a crew of 180 fire fighters with 25 fire engines at their disposal over two days to completely extinguish the fire caused by the explosion.

Explosions on the scale of the Buncefield fire are rare, but they still happen from time to time in different parts of the world. This example illustrates how careful you need to be when storing and using fuels.

Check your knowledge

1. As well as explosions, what other risks are there when transporting and storing fuels?
2. Use the Internet to find out about other accidents that have occurred involving fuels.
3. Choose one of these accidents and write a short report about what happened and the consequences of the accident. (You could choose either an accident with a fossil fuel or a nuclear fuel for your report.)

Incomplete combustion

When fuels burn in pure oxygen or a plentiful supply of air, the products of combustion should be carbon dioxide and water. However, if the air supply is limited in any way, such as in a car engine, then incomplete combustion is likely to occur.

When this happens, carbon monoxide gas and carbon can form as well as carbon dioxide. Some of the fuel may not burn at all, leading to unburnt hydrocarbons being released into the atmosphere. The carbon and unburnt hydrocarbons formed by incomplete combustion are particulates.

Carbon monoxide

Carbon monoxide gas is toxic. It is a colourless and odourless gas so you would not be able to see it or smell it.

When you inhale carbon monoxide, the carbon monoxide molecules attach themselves to the haemoglobin molecules in the red blood cells more readily than oxygen molecules. This means that the blood is no longer able to carry oxygen around the body. The body cells become starved of oxygen, leading to asphyxiation and, if large amounts are inhaled, possibly death.

Safety tip

If you have a gas boiler, gas fire or use paraffin heaters in your home, you should have a carbon monoxide detector. This will set off an alarm if carbon monoxide levels become too high.

Particulates

The soot that forms when you burn hydrocarbon fuels is carbon. When large amounts of particulates are released into the atmosphere in a city, the soot causes buildings to become dirty. Particulates can also cause global dimming, meaning less sunlight can reach the Earth. It is also thought that these particulates can cause respiratory problems.

Most cars these days are fitted with catalytic converters. These convert the carbon monoxide and carbon in the exhaust gases into carbon dioxide before they are released into the atmosphere. Even though carbon dioxide is a greenhouse gas, which contributes to global warming, it is less dangerous than releasing the products of incomplete combustion into the atmosphere.

Safety tip

When using Bunsen burners or burning fuels in the laboratory, make sure that the room is well ventilated.

Pollution from sulfur impurities

Coal, crude oil and natural gas are called fossil fuels because they were formed from the remains of plants or animals which died millions of years ago. These fuels all contain some sulfur as an impurity. When these fuels or any substances, such as petrol, obtained from these fuels are burnt, the sulfur in them reacts with oxygen in the air to form the gas sulfur dioxide.

sulfur + oxygen $\longrightarrow$ sulfur dioxide

$S(s) + O_2(g) \longrightarrow SO_2(g)$

Getting to know your unit

Using a range of laboratory techniques is a regular part of a laboratory technician's role. They also have to ensure health and safety regulations are followed. It is important to communicate within an organisation as well as keep up to date on information management systems.

How you will be assessed

This unit will be assessed by a series of internally assessed tasks set by your tutor. Throughout this unit, you will find assessment activities that will help you work towards your assessment. Completing these activities will not mean that you have achieved a particular grade, but you will have carried out useful research or preparation that will be relevant when it comes to your final assignments.

In order for you to achieve the tasks in your assignments, it is important to check that you have met all of the Pass grading criteria. You can do this as you work your way through the assignments.

If you are hoping to gain a Merit or Distinction, you should also make sure that you present the information in your assignments in the style that is required by the relevant assessment criterion. For example, Merit criteria require you to demonstrate and compare, and Distinction criteria require you to analyse.

The assignments set by your tutor will consist of a number of tasks designed to meet the criteria in the table. This is likely to consist of written assignments, but may also include activities such as:

▶ creating a section for a laboratory's health and safety procedures file containing information about safety and security

▶ exploring different laboratory techniques and analysing possible methods to be used

▶ creating a database to store and communicate scientific information.

Assessment criteria

This table shows what you must do in order to achieve a **Pass**, **Merit** or **Distinction** grade, and where you can find activities to help you.

Pass	**Merit**	**Distinction**
Learning aim **A** Understand the importance of health and safety in scientific organisations		
A.P1 Explain how health and safety measures in a scientific organisation comply with legislation **Assessment practice 4.1**	**A.M1** Compare the health and safety measures taken in relation to legislation for different scientific working environments, referencing potential hazards **Assessment practice 4.1**	**A.D1** Evaluate the measures taken for different working environments to ensure high standards of health and safety that comply with legislation **Assessment practice 4.1**
A.P2 Describe the potential hazards relevant to different scientific working environments **Assessment practice 4.1**		
Learning aim **B** Explore the manufacturing techniques and testing methods for an organic liquid		
B.P3 Correctly prepare and test the purity of an organic liquid and draw conclusions **Assessment practice 4.2**	**B.M2** Demonstrate skilful application of techniques in preparing and testing the purity of an organic liquid and draw detailed conclusions **Assessment practice 4.2**	**B.D2** Analyse the factors affecting the yield and purity of an organic liquid in the laboratory and their relevance to its industrial manufacture **Assessment practice 4.2**
B.P4 Describe the industrial manufacture and testing of an organic liquid **Assessment practice 4.2**	**B.M3** Compare the laboratory and industrial manufacture and testing of an organic liquid **Assessment practice 4.2**	
Learning aim **C** Explore the manufacturing techniques and testing methods for an organic solid		
C.P5 Correctly prepare and test the purity of organic solids and draw conclusions **Assessment practice 4.3**	**C.M4** Demonstrate skilful application of techniques in preparing and testing the purity of an organic solid and draw detailed conclusions **Assessment practice 4.3**	**C.D3** Analyse the factors affecting the yield and purity of an organic solid in the laboratory and their relevance to its industrial manufacture **Assessment practice 4.3**
C.P6 Describe the industrial manufacture and testing of an organic solid **Assessment practice 4.3**	**C.M5** Compare the laboratory and industrial manufacture and testing of an organic solid **Assessment practice 4.3**	
Learning aim **D** Understand how scientific information may be stored and communicated in a workplace laboratory		
D.P7 Explain how scientific information in a workplace laboratory is recorded and processed to meet the needs of the customer and to ensure traceability **Assessment practice 4.4**	**D.M6** Analyse the differences in the storage and communication of scientific information in different work place laboratories **Assessment practice 4.4**	**D.D4** Evaluate the challenges to organisations in making available large volumes of scientific information **Assessment practice 4.4**
D.P8 Explain how useful scientific information is obtained from large data sets and the potential issues and benefits **Assessment practice 4.4**		

Getting started

A good laboratory technician knows a range of techniques to make and test products. Write down a list of techniques that you have used and what you have produced or tested when using them. When you have done this, write down ways you made sure you followed health and safety guidelines while using these techniques. Suggest one way you might store and communicate this information to others in your group.

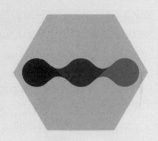

 A

Understand the importance of health and safety in scientific organisations

Key terms

Hazard – something that has the potential to cause harm.

Risk – the harm that could be caused by a hazard and the chances of it happening.

Manufacturing and testing products uses techniques that will always include some **hazards** with the associated **risks**. It is important that you understand the relevant health and safety legislation to ensure you minimise the risks to yourself and to those around you. You need to know how to prevent accidents and injury, as well as knowing what to do if an accident does happen.

▸ **Figure 4.1:** Health and safety responsibilities

Scientific organisations will have their own policies to ensure that they comply with a range of legislations. Ultimately we are all responsible for our own health and safety, following legislation as well as company procedures. Figure 4.1 shows various ways in which health and safety issues are dealt with in organisations.

Application of health and safety legislation in scientific organisations

Health and safety at work legislation

You will know from your own work in a laboratory that a risk assessment is produced before every activity is carried out. A risk assessment is produced even if the procedure is one that is carried out regularly, in order to ensure that the procedure is safe for the person carrying it out. The person producing the risk assessment also has to consider the level of knowledge and experience of the person carrying out the activity. They will include all possible hazards within the activity.

Sometimes you will have produced your own, but, even if you have, a member of the technical team in the centre will also have produced a risk assessment. The aim of the risk assessment is to minimise risks. You may produce your own as part of the evidence for the criteria. However, it is the centre's responsibility to ensure that risk assessments are appropriate and followed, which is why they will always produce one.

There will be a designated person or team in charge of managing health and safety in an organisation. They will ensure that everyone knows the latest legislation and carries out all necessary procedures. They will ensure that all necessary safety equipment is available.

Personal protective equipment (PPE) is equipment that protects the user from health and safety risks. These include safety goggles, protective clothing, face shields and helmets. Other equipment includes fume cupboards or laminar flow cabinets.

The use of hazardous substances is controlled by legislation: the Control of Substances Hazardous to Health regulations – 2002 (**COSHH**). One part of this legislation means that all hazardous substances must be correctly labelled.

Tankers carrying chemical substances have large labels on the side describing the substance being carried and the procedures to deal with it in an accident or spillage. When these chemical substances are transferred to smaller containers for laboratory use, the laboratory technician will put the correct labels on.

> **Key term**
>
> **COSHH** – Control of Substances Hazardous to Health (legislation).

Symbol	Meaning	
	Health hazard	Can cause eye damage, skin rashes and can be dangerous if ingested
	Corrosive	Can cause skin burns and permanent eye damage
	Flammable	Can catch fire if heated or comes into contact with a flame
	Acute toxicity	Can cause life-threatening effects, even in small quantities

▶ **Figure 4.2:** Hazard symbols

You may see some containers with old square orange labels. These are now out of date. The symbols above in the red diamond shape are the correct ones to use.

You may have used **CLEAPSS** Hazcards when producing a risk assessment. These give the potential risks for all chemicals and biological substances as well as storage and disposal information. Hazcards also provide information to help technical staff when they are preparing dilute **solutions** of concentrated chemicals, such as HCl.

> **Key terms**
>
> **CLEAPSS** – Consortium of Local Education Authorities for the Provision of Science Services.
>
> **Solution** – a mixture where one substance is dissolved in another.

Case study

Following health and safety legislation

Jo is a laboratory technician working in a college prep room. One of her roles is to ensure health and safety legislation is followed.

Jo has noticed that there are a lot of chemicals stored in the prep room that have square orange hazard labels on them.

Check your knowledge

1 Why is it a problem that the chemicals have the square orange labels on?

2 What do you think Jo should do to solve this problem?

3 Where can Jo find out how to label the chemical bottles correctly?

Chemical manufacturers produce data sheets for the technicians in the workplace. These give all relevant information for the products' uses. You can find these on manufacturers' websites. Hazard data sheets give details such as what to avoid when using the substance, how to store it, exposure limits and any specific risks associated with the hazards such as the substance being a **carcinogen**, a **teratogen** or a **mutagen**.

Figure 4.3 shows an example of a Hazcard. The information it contains must not be used directly to inform any risk assessment, as users looking for this information must use the most up-to-date version available via the CLEAPSS website: www.cleapss.org.uk.

98A Risk Assessment Guidance **Sulfuric(VI) acid, H_2SO_4**

Sulfuric(VI) acid	H_2SO_4 (98.07)
DANGER	Causes severe skin burns and eye damage [H314]. **This substance (concentrated acid) is dangerous in contact with:** • WATER. A vigorous reaction occurs. **When diluting, add the concentrated acid slowly *to* cold water (or ice) never the reverse. Avoid creating a spray or mist. Stir frequently to mix and minimise temperature rise.** For full details – *Recipe Book* 98. Seek additional guidance or training before attempting this procedure for the first time. • HYDROCHLORIC ACID (concentrated), CHLORIDES. Hydrogen chloride gas is given off. • CHLORATE(V), MANGANATE(VII) compounds. Spontaneously explosive products form. • SODIUM, POTASSIUM and many other metals. Dangerous reactions can occur. • PHOSPHORUS (WHITE). Ignition can occur. **Note also**: • WEL (mg m^{-3}): 0.05 (LTEL), 0.15 (STEL); as a mist • Do **not** use concentrated sulfuric acid for drying gases (especially hydrogen). • Fuming sulfuric(VI) acid (oleum) is more dangerous. It is NOT recommended for school use.
Storage	CORROSIVE LIQUID – acid (CLa) [Colourless 'oily' liquid] • Ventilated chemical store/cupboard, at floor level. Protect bottles from being knocked over. Provide a means of containing spills (eg, stand them in a tray filled with mineral absorbent). • Plastic bottles can become brittle and the acid discoloured. If *newly-purchased* acid is discoloured - return to supplier. Keep containers tightly closed; once opened, concentrated acid absorbs water from the atmosphere. • Full bottles of concentrated acid are *very* heavy. Use a bottle carrier when moving bottles from one area to another. Avoid times of the day when corridors are busy.
Emergencies	**Follow standard procedures in Section E, *About Hazcards* (GL 120), BUT NOTE for concentrated sulfuric(VI) acid**: • **If splashed in the eye:** immediately irrigate the eye with gently-running water and call for a first-aider to assist. Remove contact lenses if present and easy to do, and continue irrigating. Call the emergency services, tell them the quantity of chemical(s) involved and ensure that irrigation is continued until the patient is handed over to qualified medical staff. • **If spilt on skin (or clothes)**: remove contaminated clothing and quickly wipe as much liquid as possible off the skin with a **dry** cloth then immediately drench the affected area with a large volume of cool water. If a large area is affected or blistering occurs (or any other concerns) – call the emergency services and tell them the quantity of chemical(s) involved. • **General spills**: Neutralise contaminated mineral absorbent with solid sodium carbonate.

This *Hazcard* should be read in conjunction with guidance leaflet *About Hazcards* (GL 120), which provides additional important information. ©CLEAPSS Aug 2014

98A Risk Assessment Guidance **Sulfuric(VI) acid, H_2SO_4**

Detailed guidance on specific activities and techniques involving this substance can be found in the *Practical Procedures* section of the CLEAPSS website: www.cleapss.org.uk

General use of:	**Hazard information**	**User***	**Suggested general control measures and guidance**
Concentrated acid and solutions ≥ 1.5 M	DANGER Causes severe skin burns and eye damage.	TT (Y9)	• Wear splash-proof goggles or a face shield. • Wear chemical-resistant gloves if transferring/dispensing large volumes. • Gloves may be also advised for other practical procedures or for users with wounds or skin conditions. See activity-specific guidance and/or GL 120. **Note**: Student use of small volumes of the acid at these higher concentrations is acceptable only if the teacher is confident that the risks can be adequately controlled. Design activities to minimise the need for students to directly use or transfer concentrated acid solutions. Large bottles of concentrated acid should not be handled by students or left where they are accessible to them. Bottles have been stolen and accidents have occurred. **Disposal**: W7 → 0.1 M; or W4. For concentrated acid see Wspec below.
Solutions < 1.5 M and ≥ 0.5 M	WARNING Causes skin irritation and serious eye irritation.	Y7	• Wear eye protection even when dilute solutions are used. • Gloves may be advised for some practical procedures or for users with wounds or skin conditions. See activity-specific guidance and/or GL 120. **Note**: For many pre-16 activities, 0.4 M is adequate.
Solutions < 0.5 M	—		**Disposal**: W7 → 0.1 M; or W4

* Provides an *indication* of the level of practical skill/competence *typically* required for using the chemical in this form or at this concentration. This guidance should be taken into account when checking, updating or customising risk assessments.

Disposal	**Follow general guidance in Section F, *About Hazcards* (GL 120), BUT NOTE for the concentrated acid**: • Wear goggles or a face shield and chemical-resistant gloves. • Add acid in small portions (~ 10 cm^3) to 1 M sodium carbonate solution (1 dm^3 of 1 M sodium carbonate will neutralise ~ 50 cm^3 of concentrated acid). Maintain constant stirring and allow cooling between additions of acid (or add ice). Try to avoid creating a spray or mist. Use an indicator (eg, litmus) to check solution is *just* alkaline and then rinse away down a foul-water drain [**Wspec**]. • Immerse glassware contaminated with concentrated acid in a large volume of cold water (or 1 M sodium carbonate) then rinse away down a foul-water drain.

This *Hazcard* should be read in conjunction with guidance leaflet *About Hazcards* (GL 120), which provides additional important information. ©CLEAPSS Aug 2014

Figure 4.3: Hazcard

The Classification, Labelling and Packaging (CLP) regulation must be followed by technicians in charge of storing and using chemical substances. These give guidelines on how to classify, label and package substances used in the laboratory.

Key terms

Carcinogen – an agent or substance that has been suspected of causing or increasing the risk of cancer.

Teratogen – a drug or other substance capable of interfering with the development of the fetus (unborn child), therefore causing birth defects.

Mutagen – an agent such as a chemical, ultraviolet light or a radioactive element, that can induce or increase the risk of genetic mutation in an organism.

Step by step: Risk assessment for practical work   `7 Steps`

1 You should be provided with a template by the organisation/centre, such as the following.

Equipment/chemical/hazard	Risk	Existing controls	Likelihood (high, medium, low)	Severity (high, medium, low)	Procedure in case of accident

▼

2 List all hazards – all equipment and substances to be used.

▼

3 Research the equipment and substances using a health and safety website, e.g. HSE, CLEAPSS, COSHH.

▼

4 Complete risks column for all hazards. What is the risk? How likely is the risk?

▼

5 Complete control measures for each hazard.

▼

6 Complete the likelihood and severity columns for each hazard, where H = high, M = medium and L = low.

▼

7 Give the procedure to be carried out in case of accidents.

Concentrated sulfuric acid has the hazard symbol shown in Figure 4.4. This shows that concentrated sulfuric acid is corrosive. It will burn tissue and other materials. This means that when you use concentrated sulfuric acid you must wear safety glasses, gloves and a lab coat. Any spills must be reported to your supervisor. If it is a large spill it may need to be cleaned up professionally. Windows must be opened. You must wash it off skin or out of eyes immediately and both should be rinsed for at least 15 minutes.

▶ **Figure 4.4:** Corrosive hazard symbol

(II) **PAUSE POINT**   List all the health and safety rules you need to use when working in the school or college laboratory.

Hint   Think about your behaviour, the equipment and chemicals you use and any personal protective equipment (PPE).

Extend   Can you explain why you have to follow each rule?

As well as laboratories' own safety standards, there are many organisations that keep a check on laboratories to make sure they are maintaining the standards required for their particular scientific area and that the staff that work there are not put at risk.

The people who use the products or services supplied by the scientific workplace are also protected by other organisations, as is the environment around the workplace.

▶ **Figure 4.5:** A poster for health and safety at work

Health and Safety at Work Act

This is the law that covers all aspects and areas of the workplace. It is important when researching or using this law that you look up the most up-to-date version, as parts of it are regularly updated to meet current standards.

The poster shown in Figure 4.5 should be displayed in all workplaces, and it makes workers aware of the basics of the law.

There are many regulations and laws that must be obeyed in the workplace.

Some of these are made by the Health and Safety Executive (HSE). Their purpose is to prevent death, injury and ill health to those at work and those affected by work activities.

Organisations with more than five employees have to produce a health and safety policy. The HSE oversees this document which sets out the general approach, objectives and the management of health and safety in the business. To do this, the HSE carries out inspections to ensure that the workplace:

▶ writes and implements the Health and Safety policy
▶ adequately assesses the risks in the workplace
▶ provides the facilities for workers to work safely, including the provision of PPE
▶ trains the workers
▶ consults the workers on Health and Safety issues
▶ displays the posters for Health and Safety at Work.

Failure to comply with these requirements can have serious consequences for both organisations and individuals. Sanctions include:

▶ fines
▶ imprisonment.

In 2010, Auto-Plas (International) Limited, a company that makes plastics, was prosecuted for not following the Work at Height Regulations 2005. They were prosecuted by the HSE and had to pay costs of £2502.45.

The HSE works in conjunction with other agencies to ensure all aspects of health and safety are covered in a common way. They have their own laboratories with scientists and technicians researching the problems seen in different types of workplaces.

Case study

HSE

Rani works for the HSE in the healthcare industry. She is an HSE inspector. She carries out visits to different organisations in the health care service, e.g. hospitals, doctors' surgeries and care homes. It is her role to check if organisations are following health and safety guidelines. She makes sure they have procedures in place to prevent accidents and policies on how to deal with accidents if they do happen.

Check your knowledge

1 Why is it important to have a health and safety policy in health care organisations?

2 What do you think Rani does during and after her visits to healthcare organisations to make sure that they are following relevant health and safety policies and legislations?

As well as the safe use of chemical or biological substances, technicians in a laboratory also have to consider other hazards, such as the manual handling of heavy objects. Employers and employees need to follow the requirements set out in the Management of Health and Safety at Work Regulations 1999 to produce and follow risk assessments. They must also ensure that they comply with requirements in the Manual Handling Operations Regulations (MHOR) (amended in 1992) and implement procedures to minimise the risk of injury from manual handling tasks.

Wherever possible, you must:

▸ avoid hazardous manual handling operations so far as reasonably practicable

▸ assess the risk of injury in any hazardous manual handling operations that cannot be avoided

▸ reduce the risk of injury so far as reasonably practicable.

Other hazards can be due to using computer and laptop screens regularly. The Health and Safety (Display Screen Equipment (DSE)) Regulations 1992 must be followed to protect the health of people who work with DSE. The regulations were introduced because DSE has become one of the most common kinds of work equipment. It is important that employers and employees understand and follow the DSE regulations in order to avoid any long-term injury.

Display Screen Equipment (DSE) includes a device or equipment that has an alphanumeric or graphic display screen, regardless of the display process involved. It includes both conventional display screens and those used in technologies such as laptops, touch-screen tablets and smartphones.

This sort of equipment can be associated with neck, shoulder, back and arm pain, as well as with fatigue and eyestrain. Most issues do not cause serious health problems, but medical conditions such as repetitive strain injury can cause a lot of discomfort and make it hard to work, and so should be avoided if possible.

The HSE has given guidelines for employers and employees. These discuss the types of equipment that should be used, from screens to office furniture, and about how many hours should be spent working at a screen. There is also guidance on how to best care for your vision whilst viewing screens.

If you want to find out more, go to the HSE website.

However well an organisation is run, illness and accidents do sometimes still occur. This is where Reporting of Injuries, Diseases and Dangerous Occurrences Regulations 2013 (**RIDDOR**) are used. Employers have a legal duty to report ill health, accidents and accidents that did not quite happen ('near misses').

As part of this procedure, the employer must keep a register of all accidents and near misses so that they can see if changes can be made to make the workplace safer. HSE inspectors will also need to see the documentation if an incident occurs. In the event of an accident, details must be logged in the organisation's accident book. On the government-run HSE website, there is information on what types of incidents and accidents must be reported to the HSE. There are also forms on the website that allow you to report these incidents.

> **Key term**
>
> **RIDDOR** – Reporting of Injuries, Diseases and Dangerous Occurrences Regulations 2013.

Case study

Accident report form

Julie was involved in an accident in the lab. She spilt some sulfuric acid on her hand which caused a bad burn. You are the health and safety officer for the lab so you need to fill out an accident report form.

Check your knowledge

1 Using the HSE website, access the form for reporting accidents.

2 Complete the details required (you can make up dates, times and names for this imaginary scenario;a use your knowledge of laboratory work to help you).

As well as laws, there are regulations and standards. Laws must be enforced in the workplace; however, each type of laboratory will have laws, regulations and standards that are specific to them.

The United Kingdom Accreditation Service (UKAS) is the only national accreditation body which assesses organisations that provide certification, testing and inspection services. UKAS makes assessments against internationally-agreed standards, and gaining accreditation can show that the organisation is competent, impartial and capable of providing a high quality service. UKAS is a non-profit company that assesses and accredits testing and calibration laboratories, certification bodies, proficiency testing schemes and medical laboratories.

 PAUSE POINT   List the health and safety regulations and laws discussed in this unit.

> Hint
>
> Think about the different types of scientific organisations and the regulations they will need to follow.

> Extend
>
> How do these regulations and laws affect how you behave in the laboratory?

Hazards in a scientific organisation

> **Key term**
>
> **COMAH** – Control of Major Accident Hazards.

Major industrial accidents involving dangerous substances can be a significant threat to humans and the environment. They can cause serious injury to human health or serious damage to the environment, both at and away from the site of the accident. Legislation has been put in place by the HSE to deal with these substances. Control of Major Accident Hazards (**COMAH**) 2015 is legislation that supports and guides science organisations when dealing with dangerous substances.

Case study

Accident in Seveso

The Seveso Disaster was an industrial accident that occurred just after midday on 10 July 1976, in a small chemical plant in Italy. A chemical called 2,3,7,8-tetrachlorodibenzo-p-dioxin was released and the local towns were exposed to high levels of it. Over 3000 animals were found dead and around 80 000 had to be killed to prevent the chemical entering the food chain. It also led to birth defects and increased cancer in humans in the area. This accident was one of the events that led to the introduction of COMAH.

Check your knowledge

1 Explain the importance of COMAH to your group.

There are a variety of specific dangers related to working in scientific workplaces. Not all dangers are present in all workplaces. There are regulations to deal with each of the specific dangers possible.

The Dangerous Substances and Explosive Atmospheres Regulations 2002 (**DSEAR**) require employers to control the risks to safety from fire, explosions and substances corrosive to metals. This would be particularly relevant in a firework factory or on an oilrig.

> **Key term**
>
> **DSEAR** – Dangerous Substances and Explosive Atmospheres Regulations 2002.

Skin and respiratory sensitisers are covered by the COSHH regulations 2002. These are substances that may cause you to get skin problems such as eczema, or lung problems such as asthma. Not all organisations use these types of substances but, for example, they may be present in labs or hospitals.

Examples of occupations where you may use sensitisers are shown in Table 4.1.

▶ **Table 4.1:** Sensitisers used in different occupations

Occupation	Sensitiser
Engineer	Cobalt, chromium
Beautician	Cosmetics and fragrances
Builder	Epoxy resins
Farmer	Plant pollen
Food technician	Preservatives
Health care	Resins

Not all hazards are related to chemical or biological substances. Other hazards include electrical hazards (use of circuits and electrical equipment such as lighting), working at heights (on ladders or scaffolding), lone working (sometimes workers are alone in a lab due to an experiment that is running for 24 hours), working with vehicles (cars, lorries, plant equipment), and noise (machinery in factories can produce significantly dangerous noise levels). These hazards are also covered in the Health and Safety at Work Act and employers and employees must know their rights and responsibilities when working with these hazards.

The Health and Safety at Work Act covers both school and college labs as well as workplace laboratories. The law is the same but there will be differences in how it is implemented due to the differences in equipment, method and substances used.

Reflect

How do you behave in the laboratory? Are you aware of health and safety practices?

Do you consider yourself responsible for the safety of others?

Who is responsible for safety in your laboratory?

Chemical substances can be analysed by titration in school laboratories and in industrial research laboratories. The risk assessments for this technique will be very similar in each type of laboratory but may have some differences. PPE should be worn, e.g. safety glasses and lab coats. Glassware and chemicals should be handled with care. The main differences will be due to the types and concentrations of the chemicals used in each type of lab. Concentrated solutions are more likely to be used in industry, as are more dangerous chemicals such as lachrymators (these can cause respiratory problems as well as burn eyes and skin) or bacteria that cause diseases. In these cases, the risk assessment will have more detail, for example, where the experiment can be carried out (in a fume cupboard or a clean room), or special precautions, such as having a negative air pressure inside the laboratory to prevent bacteria spread. There may be extra training necessary and this will be part of the risk assessment. Both risk assessments will be written on a template that is produced by the organisation and will follow health and safety regulations and the organisation's policies.

You are an HSE inspector. You are visiting two different scientific organisations. One is a research and development pharmaceutical lab and the other manufactures chemicals for the plastics industry.

Each organisation needs to understand what health and safety legislation is relevant to their working practice. They also want advice on how to implement health and safety procedures effectively within their organisation.

Produce a report for each organisation, to include:

- a description of relevant health and safety legislation
- a description of potential hazards relevant to the organisation
- an explanation of the measures taken by each organisation to ensure high standards of health and safety that comply with the legislation
- a comparison and evaluation of the measures taken in each organisation to ensure high standards of health and safety.

Plan

- What is the task? What am I being asked to do?
- How confident do I feel in my own abilities to complete this task? Are there any areas I think I may struggle with?

Do

- I know what it is I am doing and what I want to achieve.

Review

- I can explain what the task was and how I approached the task.
- I can explain how I would approach the hard elements differently next time (i.e. what I would do differently).

B   Explore manufacturing techniques and testing methods for an organic liquid

Chemists work in a range of industries from forensic science services to medical research. They may use laboratory techniques to examine evidence from a crime scene or similar techniques to produce and test a new life-saving drug. To be involved in practical chemistry, you need to be familiar with lots of techniques.

Chemists are involved at every stage of the process, from the initial discovery to getting the reactions to work on the chemical plant, to testing the products. They often use high-tech, expensive apparatus. They may also use dangerous chemicals and carry out potentially risky procedures and techniques. It is important that any chemist is trained and experienced at using this apparatus and carrying out these reactions safely.

Manufacturing techniques

In this unit, you will be asked to prepare and purify an organic liquid. To do this, you will need to carry out organic reactions using a range of techniques. Many organic reactions are very slow and need heat to take place at an appropriate rate. Unfortunately, the chemicals involved in the reactions are often also very volatile, which means they evaporate if they are heated. This means that the reactants and products are lost to the atmosphere, giving a very small yield.

Reflux is a technique that allows organic substances to be heated for a long time whilst minimising the loss of substances to the atmosphere. You have to heat the reaction mixture in a flask fitted with a reflux condenser (also called a Liebig condenser), as shown in Figure 4.6. A Liebig condenser consists of two tubes, one inside the other. The space between the tubes allows water to flow through, cooling down any gases present, to cause condensation.

All the vapours rising from the reaction mixture during heating enter the condenser and change back into liquids and return to the flask so that the unreacted compounds can react.

The flask can be heated using a hot water bath or if higher reaction temperatures are required, oil can be heated up to raise the temperature of the reaction flask. This can be dangerous, so an electric mantle or hotplate should be used if available. The use of anti-bumping granules in the reaction mixture prevents formation of large gas bubbles by providing **nucleation sites** for small bubbles to develop, which provides a safer and more stable reaction mixture that helps to make the boiling smooth so that it does not bubble over. A gentle flow of cold water enters at the bottom of the condenser. This cools and condenses the vapours so that they return to the flask.

> **Key term**
>
> **Reflux** – a method involving heating a reaction mixture to the boiling point temperature of the reaction solvent and using a condenser to recondense the vapours back into the reaction flask. This allows a longer reaction time so that the reaction can complete.

> **Key term**
>
> **Nucleation sites** – site on anti-bumping granules where small bubbles can form, preventing rapid boiling of a liquid during a reaction.

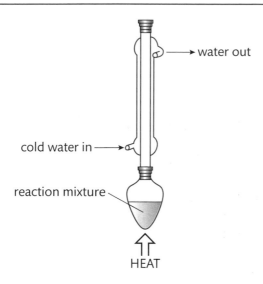

water out

cold water in

reaction mixture

HEAT

▶ **Figure 4.6:** Reflux

Reflux is used in a variety of industries, such as the petrochemical and beverages industry, in order to improve the efficiency of distillation. Large scale distillation can be used to separate mixtures of two or more organic liquids. However, this process may not produce a pure product (distillate). Reflux can be used to further evaporate and condense the distillate in order to increase the purity. The distillate is heated and the vapours cooled and collected in a reflux drum. As the liquid falls back into the drum, it cools down any vapours that are traveling upwards, causing them to also condense. This leads to a more efficient separation of materials with different boiling points.

A reflux still is used in the production of alcoholic drinks. By controlling the temperature at the condenser's outlet, the reflux still ensures that components with higher boiling points are returned to the still, while components with lower boiling points are routed to a secondary condenser. This produces high-quality alcoholic drinks, while making sure that impurities and unreacted substances are returned to the reflux still.

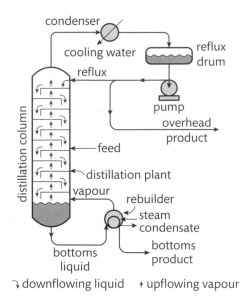

▶ **Figure 4.7:** Industrial distillation and reflux

Key term

Distillation – the action of purifying a liquid by a process of evaporation and condensation.

Distillation is the process of separating out compounds within a liquid state because of the differences in their boiling points – a physical property. It relies on the principal that all liquids have a specific temperature at which they boil. As liquids are heated, their vapour pressure increases. When this pressure reaches the point at which it is equal to atmospheric pressure, the liquid starts to boil. A liquid with a low vapour pressure has a higher boiling point than one with a higher vapour pressure. This method has been used for thousands of years in the separation of perfumes. It also provides the basis for crude oil processing, industrial dry cleaning and the production of alcoholic drinks like vodka (see Figure 4.7).

In distillation the flask containing the mixture to be separated is heated up. The liquid with the lower boiling point starts to boil first and turn into vapour. The vapour travels into the Liebig condenser and is cooled down. The vapour condenses in a liquid and drips into the second flask, producing the distillate. (see Figure 4.8).

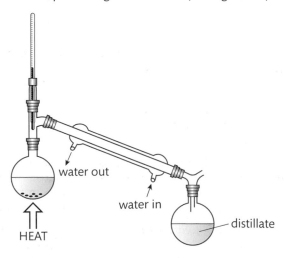

▶ **Figure 4.8:** Distillation equipment

Key term

Fractional distillation – distillation where components in a chemical mixture are separated according to their different boiling points.

Fractional distillation is usually used to separate several liquids from a mixture, or where the differences in boiling points are small (see Figure 4.9). It uses the same apparatus as simple distillation, but with a fractionating column between the heating flask and the still head. The column is usually filled with glass beads or pieces of broken glass, which act as surfaces on which the vapour leaving the column can condense, and then be evaporated again as more hot vapour passes up the column. The vapour undergoes several repeated distillations as it passes up the column. This results in a better separation. Fractional distillation takes longer than simple distillation.

Fractional distillation is a common method in industry to separate mixtures. You may have studied the separation of fractions in crude oil at level 2.

▶ Crude oil fractions with higher boiling points, which consist of larger molecules, separate at the bottom of the distillation tower.

▶ The smaller fractions, with lower boiling points, separate and are collected at the top of the tower.

It is a **continuous process** with crude oil being added at the bottom at the same time as products are removed. Unless there is a change in conditions, the amount of feed being added and the amount of product being removed is usually the same. This is continuous steady state fractional distillation.

> **Key term**
>
> **Continuous process** – production that occurs 24 hours a day, seven days a week. It is rarely shut down. Reactants are continually being added and products are continually being removed.

Large volumes of crude oil are separated at any time and so the towers are very large. They can be 0.5–6 metres in diameter, and as tall as 60 metres (or even taller).

Reflux is often also used alongside the distillation towers. As we have seen, reflux allows the chemicals to boil and condense for as long as necessary to get as complete separation of compounds as possible. The reflux liquid flowing downwards also condenses the vapours flowing upwards, giving them longer to separate.

Table 4.2 shows some of the differences between fractional distillation in the laboratory and in distillation towers used in industry.

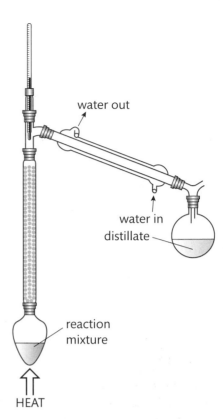

▶ **Figure 4.9:** Fractional distillation in a laboratory

▶ **Table 4.2:** Distillation in college laboratories compared with distillation in industry

Similarities between procedure in collge lab and industry	Differences in procedure	
	College	**Industry**
Crude oil is heated.	Oil is heated with Bunsen burner or heating mantle (see Figure 4.8).	Large-scale burners are used (see Figure 4.7).
Fractions are removed at their boiling point.	Crude oil is heated in a boiling tube or round bottleneck flask.	Crude oil is heated and then pumped into a fractionating tower that can be several storeys high. It is separated in the tower.
	Smaller fractions are removed first.	Continuous process so once started all fractions are removed at the same time.
	Fractions are removed through a delivery tube one at a time.	Fractions are removed through own pipe.
	Small volumes are used.	Extremely large volumes are used continuously.

Solvent extraction is a method whereby compounds can be separated based on their differing solubility in two **immiscible** liquids. Immiscible liquids are liquids that do not mix, like water and petrol. For example, take a compound that is dissolved in water, but can dissolve more readily in petrol. If you add petrol and shake the mixture, the compound will move out of the water into the petrol. If left to stand, the water will separate from the petrol-compound mixture. This is because the water and the petrol are immiscible liquids. The compound is then separated from the water.

Investigation 4.1

Solvent extraction of iodine from seaweed (tutor demonstration)

Steps in the investigation	Pay particular attention to...	Think about this...
1. In a fume cupboard, burn a quantity of seaweed (kelp) to ash.	**Safety tip**: make sure that the fume cupboard is used correctly so no dangerous fumes are released into the lab.	
2. Gently heat the ash in water to dissolve the iodine that is present in the seaweed ash.	Take care not to heat the water too quickly.	
3. Filter away any iodine ions using a funnel and filter paper. Collect the filtrate.	Make sure the filter paper is the right size for the funnel.	
4. Add a few drops of dilute sulfuric acid and approximately 20 cm^3 of hydrogen peroxide (1.5 M) to the filtrate. It will turn yellow.	The problem here is deciding when the colour change has completed.	Peroxide oxidises iodide to iodine.
5. Pour the filtrate into a separating funnel containing 20 cm^3 of cyclohexane.	Take care that all the filtrate is transferred.	
6. Put a stopper into the separating funnel and shake. Keep your finger on the bung.	**Safety tip**: keeping the bung in place stops the liquid escaping.	
7. Unscrew the stopper carefully to release pressure from any gaseous cyclohexane in between each shaking.	**Safety tip**: do this regulary so that pressure does not build up.	
8. Repeat until the water becomes colourless and the cyclohexane layer is purple.	**Safety tip**: take care to release the pressure between each shaking.	The two obvious colours means the iodine has all been dissolved in the cyclohexane layer.
9. Allow the layers to settle out.		The layers separate because they are immiscible.
10. Release the stopper then run off the water layer and discard it.		
11. Collect the iodine-rich cyclohexane layer into a second flask.	.	
12. Evaporate the cyclohexane in a fume cupboard to produce crystals of pure iodine.	**Safety tip**: do this so dangerous gases are not released into the lab.	

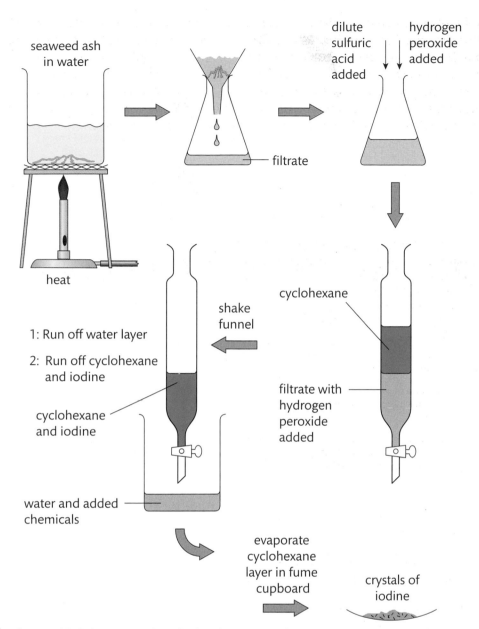

▶ **Figure 4.10:** Solvent extraction of iodine from seaweed

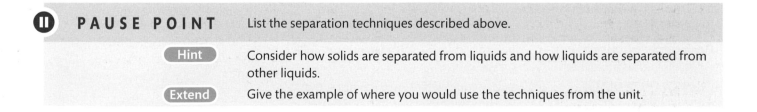

❚❚ PAUSE POINT   List the separation techniques described above.

Hint   Consider how solids are separated from liquids and how liquids are separated from other liquids.

Extend   Give the example of where you would use the techniques from the unit.

When products are made in the lab or in industry, it is important for them to be as pure as possible. This is because impurities may make the product behave differently from what is expected. Impurities in medicines, for example, could make a patient sicker.

A range of techniques and chemicals can be used to remove impurities from manufactured chemical products.

Purification of impure ester made in a lab

Steps in the investigation	Pay particular attention to...	Think about this...
1. Pour the impure ester mixture into a separating funnel containing 20 cm³ of water.	Make sure that you keep the stopcock closed.	This will remove any impurities that are in the ester and are water-soluble, e.g. alcohol, carboxylic acid and sulfuric acid.
2. Put a bung on top of the stopcock and, holding the bung, shake the funnel.	**Safety tip**: remember to keep hold of the bung to stop the liquid escaping while shaking. Vent the separating funnel whenever it has been shaken to release vapour pressure.	
3. Allow the layers to separate then pour out the denser aqueous layer from the bottom.	Remember to remove the bung before opening the stopcock.	
4. Repeat steps 1–3 to leave the organic layer in the separation funnel.		This will ensure most of the water soluble impurities are removed.
5. To the organic layer, add 5 cm³ of sodium carbonate solution and swirl until no more gas bubbles are seen.	**Safety tip**: swirl carefully so that the liquid does not escape the funnel. Vent the separating funnel whenever it has been shaken to release pressure, particularly as carbon dioxide gas is being released.	This will neutralise any acid still present and produce carbon dioxide gas.
6. Allow the layers to separate then once more pour out the denser aqueous layer from the bottom.		
7. Wash the remaining organic layer with about 20 cm³ of water, allow to stand and then run off the aqueous layer.	Make sure you give the mixture time to separate fully.	
8. Add a spatula full of anhydrous magnesium sulfate to a clean dry conical flask.	It is important that the flask is dry.	
9. Pour the contents of the funnel into the flask and allow to stand for 15 minutes.	The anhydrous magnesium sulfate will remove any water.	
10. Decant the liquid into a clean flask.	This will give you your pure ester.	

You can also use other chemicals. Anhydrous calcium chloride can be used instead of magnesium sulfate to remove water. Anhydrous calcium chloride is hygroscopic, which means it attracts and holds water molecules. It is often used in a **desiccator** with a wet product. It will remove the water from the wet product, allowing the product to dry out. It is a good drying agent for many solvents. (Remember if water is present, the product is described as wet, even if the product is itself a liquid.)

▶ A desiccator

Molecular sieves can be used to remove water and other impurities. A molecular sieve is a material containing pores of uniform size. The material has pore diameters that allow small molecules, such as water, to fit through. Large molecules cannot fit through the pores and cannot be absorbed, while small molecules can. Molecular sieves can be made with a pore size that is designed to separate a specific impurity from a product. Many molecular sieves are used as desiccates to remove water. Examples include activated charcoal, silica gel and aluminasilicates.

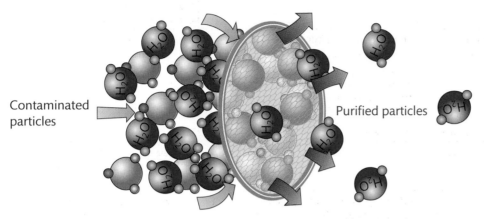

Contaminated particles

Purified particles

▶ **Figure 4.11:** A molecular sieve.

You can use water in organic preparations to remove impurities that are soluble in water. In this case the water acts as the solvent in a solvent extraction.

Ⅱ PAUSE POINT List four ways you can remove impurities from organic preparations.

Hint Think first what the impurities might be.

Extend Explain how using anhydrous calcium chloride in a desiccator can be used to remove impurities.

Esters are very useful compounds. They are used in industry as solvents and as a reactant when making polyesters. Esters have a sweet smell, like pear drops, and are very common in nature. The way fruits smell is due to esters, and most animal fats and vegetable oils are esters.

You may have heard of ethyl ethanoate. It is an ester that is used in glues and nail polish removers. It is also used to remove caffeine from tea and coffee.

It is made by a reaction between ethanol and ethanoic acid. Concentrated sulfuric acid is used as a catalyst as the reaction is slow and reversible.

The equation for the reaction between ethanoic acid and ethanol is:

Key term

Ester – an organic compound made by replacing the hydrogen of an acid by an alkyl or other organic group. It is the product of the condensation reaction between an alcohol and carboxylic acid.

sulfuric
acid

ethanoic acid + ethanol $\rightleftharpoons$ ethyl ethanoate + water

$$CH_3COOH + C_2H_5OH \rightleftharpoons CH_3COOC_2H_5 + H_2O$$

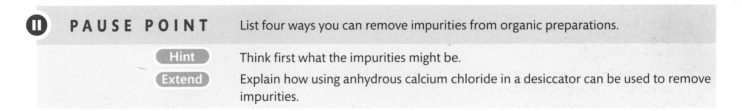

Making ethyl ethanoate

Steps in the investigation	Pay particular attention to...	Think about this...
1. Add 15 cm³ of ethanol and 10 cm³ glacial ethanoic acid to a 50 cm³ round-bottomed flask.		
2. Add a few anti-bumping granules.	**Safety tip**: the anti bumping granules allow the reaction to bubble slowly and not get too vigorous.	
3. Slowly add about 1 cm³ of concentrated sulfuric acid.		The reaction can get too vigorous if the acid is added quickly.
4. Attach a condenser to the round-bottomed flask.		
5. Attach a condenser to the cold water tap with rubber tubing and allow water to flow.	Ensure you attach the tap to the bottom connection on the condenser and turn the tap on before heating. Ensure that the other end of the rubber tubing is in a sink.	This cools the vapour so that it returns to the flask to allow more time for the reaction to complete.
6. Heat using an electric mantle or hotplate until the mixture is boiling.	**Safety tip**: alcohols are flammable – so make sure there are no naked flames.	
7. Reflux the mixture for 30 minutes.		This allows time for the reaction to complete.
8. Turn off the heat and allow to cool.		This will give an impure ester so you will need to use appropriate separation techniques to purify it (see Investigation 4.2).

Safety tips

- Eye protection must be worn.
- Concentrated sulfuric acid is corrosive.
- Carboxylic acids are corrosive.
- Disposable gloves should be worn.

Key term

Disproportionation reaction – a type of redox reaction in which a reactant is simultaneously reduced and oxidised to form products.

Reflect

How easy did you find it to follow the standard operating procedure to make ethyl ethanoate?

How pure was the product you made?

What did you do well?

What could you do differently next time?

Make a list of the areas you think you need some help on and practical techniques that you may want to practise again before carrying out an assessment activity.

Because ethyl ethanoate is such a useful product, large amounts of it are made in industry. There are several different processes. Some are given below, and you may want to research others.

Tishchenko reaction

This is a **disproportionation reaction** of ethanal (the aldehyde is reduced and oxidised simultaneously. (You will have covered redox reactions in Unit 1.) The conditions for the reaction are: temperature 0–5 °C in the presence of aluminium alcoholate as a catalyst.

There is high conversion of ethanal (up to 98%) into the ester ethyl ethanoate, therefore giving a high yield.

ethanal → ethyl ethanoate

Esterification reaction: esterification of ethanoic acid with ethanol in the presence of acid catalysts (e.g. sulfuric acid)

This process may be a **batch process** or a continuous process. Water formed in the reaction is removed by distillation. It is important to achieve maximum conversion of ethanoic acid, which is costlier than ethanol. The conversion of reactants in this process is about 95%.

The batch procedure involves a single reactor that is filled with the ethanoic acid and ethanol. Sulfuric acid catalyst is added and the water is removed as the reaction proceeds. This method can be used to make large quantities of esters. The batch process requires reactors that hold extremely large volumes of reactants. Heating coils are used to heat the reactants. The continuous process for making esters is often used to manufacture large quantities of esters. This procedure involves the mixing of streams of the reactants into a reaction chamber while the product is removed at the same time. Continuous esterification has the advantage that larger quantities of products can be prepared in shorter periods of time. This procedure can be run for days or weeks without interruption.

> **Key term**
>
> **Batch process** – the production of materials in a small or limited number. The production does not go on all the time.

Liquid-phase oxidation of n-butane

This is a process used in ethanoic acid manufacture. Ethyl ethanoate is a by-product of the oxidation reaction of n-butane by oxygen from air.

Alkylation of ethanoic acid

Ethanoic acid reacts with ethene. (This is a type of hydrocarbon, an alkene. You will study these in Unit 5.) Conditions: temperature 150 °C and pressure 7.7 MPa in the presence of sulfuric acid or solid acid catalyst (*Avada process*).

Preparation of another ester: 3-methyl but-1-yl ethanoate

Banana oil is another ester. It is found naturally in bananas, but can also be made synthetically. The chemical name for banana oil is 3-methylbut-1-yl ethanoate. Bananas tend to go off very quickly and so are not used to flavour foods. This is because the food, such as banana ice cream, would start to go brown or black, the way a banana does when it is over-ripe. The synthetic banana oil is used instead.

The 3-methylbut-1-yl ethanoate can be made by an esterification reaction similar to the one used to make ethyl ethanoate. You should be familiar with at least one of these preparations and be able to make one of these products in order to pass this unit.

Preparation of 3-methylbut-1-yl ethanoate

Steps in the investigation	Pay particular attention to...	Think about this...
1. In a 25 cm³ round-bottomed flask, add 5 cm³ 3-methyl-1-butanol, 7 cm³ glacial ethanoic acid and a couple of anti-bumping granules.	**Safety tip**: the anti-bumping granules allow the reaction to bubble slowly and not get too vigorous.	
2. Add 0.5 cm³ of concentrated sulfuric acid and swirl to mix the solution.	**Safety tip**: take care to swirl gently, and wear safety glasses.	
3. Attach the flask to a reflux condenser and heat the mixture to reflux for 60 minutes.	The following shows how you should set up the equipment. Make sure all the glassware fits snugly.	
4. Allow to cool to room temperature. Then transfer the contents to a large separating funnel.		
5. Add 15 cm³ of distilled water to the solution and stir. Then let the layers separate.		This removes water-soluble impurities.
6. Remove the aqueous layer and discard.		
7. Wash the organic layer with 9 cm³ portions of a saturated sodium bicarbonate solution until it tests basic after removing from the tap funnel.	You can test this with litmus paper.	Sodium bicarbonate solution reacts with excess acid.
8. Wash the organic layer with 6 cm³ of a saturated sodium chloride solution.		Sodium chloride solution removes excess water.
9. Dry the organic layer with anhydrous sodium sulfate for 10 to 15 minutes.		The anhydrous sodium sulfate dries the product.
10. Transfer the organic layer to a 10 cm³ flask, filtering it through a cotton plug.		This will remove any solid impurities.

PAUSE POINT   What techniques do you need to make an ester?

Hint   Think about the different steps in the methods above.

Extend   Explain why the methods you would use in the lab are batch processes.

Testing methods and techniques

Estimating the **purity** of a substance in chemical terms is very important in many industrial applications. For example, in the pharmaceutical industry, if a medicine is impure it may cause unwanted side effects or reactions.

Boiling point determination

Boiling points are known very accurately for most elements and compounds and are listed in data books.

> **Key terms**
>
> **Purity** – freedom of a substance from other matter of different chemical composition. In chemistry, elements and compounds are pure, a mixture is not.
>
> **Boiling point** – the temperature at which a liquid turns into a gas.

You can identify a substance as high purity or low purity by comparing the experimental values of its boiling point with those from a data book.

Water, for example, boils at 100 °C at 1 bar atmospheric pressure, but the boiling point is increased if salt, NaCl, is added. In the manufacture of sugar, the liquid that is boiling is a solution of sugar, or syrup. If the concentration of sugar rises, then the boiling point will rise as long as the pressure is constant. This tells the technicians how much sugar is present.

Any increase in concentration of a solution increases boiling point.

The boiling point of a substance is dependent on the intermolecular forces, or the strength of the bonds within the substance. The stronger the intermolecular forces, or bonds, the higher the boiling point will be.

> **Link**
>
> Look back to *Unit 1: Principles and Applications of Science 1* for more information on intermolecular forces and bonds.

 PAUSE POINT   Explain how different intermolecular forces and bond strengths affect boiling point.

Hint   Review your work on bonds and forces for Unit 1.

Extend   Can you explain why NaCl has such a high boiling point?

You can use distillation apparatus to determine the boiling point of a liquid. When a thermometer is added to the distillation apparatus, as in Figure 4.8, the vapour of a boiling liquid will condense on the thermometer bulb. Since a substance condenses at the same temperature that it boils, the temperature at which the vapour condenses on the thermometer will be the boiling point.

Determining boiling point

Steps in the investigation	Pay particular attention to...	Think about this...
1. Set up the apparatus as in Figure 4.8, but do not insert the thermometer.	Ensure the glassware is snugly fitted.	This is so the vapours cannot leave the apparatus before the temperature is measured by the thermometer.
2. Add 20 cm³ of ethanol along with a few anti-bumping granules to the round-bottomed flask.	**Safety tip**: the anti-bumping granules allow the reaction to bubble slowly and not get too vigorous.	
3. Insert the thermometer as in Figure 4.8.	The thermometer bulb must sit by the entrance to the condenser, not higher, not lower.	This is where the boiling point is measured, as this is where the vapours start to turn to gas.
4. Heat the liquid slowly until it gently boils.	If you do this too quickly you may not notice when the boiling point is reached. if it is heated slowly, there is no bumping/shaking of the apparatus.	
5. Note the temperature on the thermometer once it remains constant.		The vapour is condensing, i.e. the temperature is the boiling point.
6. Compare your findings to published data.		The closer your experimental value is to the published value, the more pure the substance is.

Theory into practice

A manufacturer of pharmaceuticals wants to use ethanol as a solvent. They need to know how pure the ethanol is.

Explain how way the manufacturer could test the purity of the ethanol they are going to use.

Often samples to be tested are very small volumes. Consider trace samples found by forensic scientists at crime scenes. Liquids boil when the vapour pressure of the liquid is equal to atmospheric pressure. This means the boiling point of a small sample of liquid can be determined using the Siwoloboff method.

Investigation 4.6

Siwoloboff method of determining boiling point

Steps in the investigation	Pay particular attention to...	Think about this...
1. Fill a boiling tube two-thirds full of water (or water-glycerol mixture if the boiling point to be determined is above 100 °C).		Water does not boil above 100 °C but a water – glycerol mixture does.
2. Add a stirrer to the water.		
3. Half-fill a dry sample tube with the liquid to be tested.	The test tube must be dry as any water in the test tube will mean the substance is less pure.	
4. Insert a capillary tube sealed at one end into sample tube with open end down.		
5. Attach the thermometer to a sample tube using a rubber band so that the bulb of thermometer is at the same level as the bulb of the sample tube.	This is so that the temperature measured is the same as the boiling point that is being recorded.	
6. Place the sample tube in a water bath and heat, stirring constantly.	Stirring the liquid means the heat is evenly distributed throughout. If the liquid is not stirred, an accurate reading may not be obtained.	
7. When a rapid stream of bubbles or vapour from the capillary tube is observed, withdraw the heat source and allow to cool. Stir constantly.		
8. When bubbles no longer issue from the capillary, and the liquid starts to suck back, read and record the boiling point.	You will have to watch this carefully to make your decision on when the liquid is being sucked back.	
9. Repeat this investigation a number of times.	Scientific investigations produce more accurate results if they are carried out a number of times.	

The Siwoloboff method gives a boiling point at a particular atmospheric pressure. In a school lab this is usually fairly constant, but when comparing to published data you should take care to check with data gathered at the same atmospheric pressure. There are sources of error in both these determination of boiling point methods. The precision of the thermometer will affect how accurate the results are. In both distillation and the Siwoloboff methods you need to make judgements. For the first you need to decide when the temperature is staying constant and in the second you need to decide when there are no longer any bubbles produced. This is why repeating the procedures helps get reliable results.

Spectroscopy

One use of spectroscopy is to determine the purity of a chemical substance (**spectroscopic analysis**). This method is very reliable and extremely accurate. Spectroscopy uses the principle that substances absorb, emit or scatter electromagnetic radiation. Electromagnetic radiation has a range of wavelengths which are shown on the **electromagnetic spectrum**. Each chemical substance interacts with electromagnetic radiation to a different extent, depending on the wavelength of the radiation. Shorter wavelengths have greater energies. The spectrum produced by the sample indicates which substances are present in a sample.

Link

You may have used spectroscopy in *Unit 2: Practical Scientific Procedures and Techniques* to determine concentration of substances.

Absorption

Different wavelengths of electromagnetic radiation are absorbed by different atoms. A spectrum is produced with dark lines corresponding to the wavelength of the energy absorbed. When a sample is tested, the purity can be measured by comparing the spectrum produced by the sample to a spectrum produced by a pure sample of the same substance.

Link

You have covered energy levels in *Unit 1: Principles and Applications of Science 1.*

Emission

Atoms also emit energy as electromagnetic radiation. When electrons in the atoms move to a higher energy level and then return back again, energy is released as electromagnetic radiation. This radiation can be detected and produces bright lines on a spectrum. The lines on the spectrum correspond to the wavelength of the energy given off. When a sample is tested, the purity is measured by comparing the spectrum produced by the sample to a spectrum produced by a pure sample of the same substance.

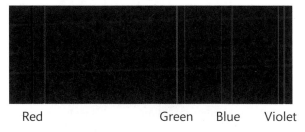

Red   Green   Blue   Violet

▶ Emission spectrum

Link

You have covered how electromagnetic waves are deflected in *Unit 1: Principles and Applications of Science 1.*

Scattering

Information about a chemical substance can also be determined from the way it deflects electromagnetic waves. The pattern produced on the spectrum is specific to the chemical substances being tested.

Infrared spectroscopy

Spectroscopy is used in a vast number of applications for physics, biology and chemistry. The various types of spectroscopic analysis methods have specific names related to the part of the electromagnetic spectrum used; for example, ultraviolet spectroscopy and infrared spectroscopy.

In infrared (IR)spectroscopy the amount of energy absorbed corresponds to the increased vibration in the bonds that join the atoms in each molecule. The wavelengths used in IR spectroscopy are in three distinct groups: near infrared (NIR), mid-infrared (MIR) and far infrared (FIR). These descriptions correspond to the position of the wavelengths relative to the red end of the visible wavelengths in the electromagnetic spectrum.

The atoms in a molecule vibrate with a set frequency. These can be stretching bonds or bending bonds. The molecule absorbs IR energy that is at the same frequency as the vibrations. When IR radiation is passed through the molecule, the amounts of energy absorbed at different frequencies can be measured and recorded as a spectrum. Each frequency is proportional to 1/wavelength and can be expressed as its wavenumber, which is what is recorded on the spectrum.

Scientists call this spectrum the infrared fingerprint as it is unique for each molecule. You can identify unknown molecules by comparing them to known infrared fingerprints, for example:

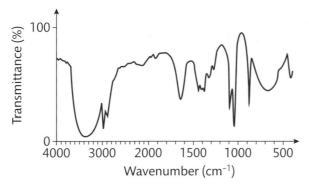

▶ **Figure 4.12:** Infrared spectrum for a sample of ethanol CH_3CH_2OH

If you look at the two fingerprints below you can see that A is the same as the one for ethanol and so A must be ethanol.

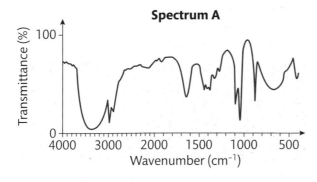

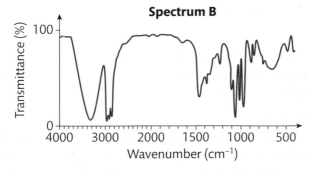

▶ **Figure 4.13:** Infrared spectra

The fingerprint for B is different. This is for a sample of propan-1-ol.

Infrared spectroscopy is used extensively in research and in industry. It is used in areas such as forensic science and the manufacture of polymers. It also provides a quick way of analysing exhaust gases in cars. This is because the carbon–oxygen bond in carbon monoxide and the bonds in nitrogen oxides and unburnt fuel all have distinct absorption characteristics.

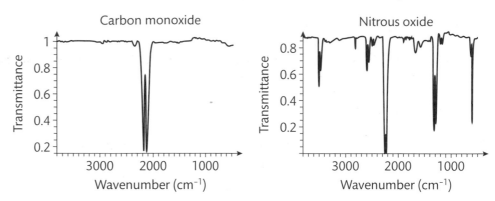

▶ **Figure 4.14:** Infrared spectrum for two exhaust gases: carbon monoxide and nitrous oxide

In order to use spectroscopy to analyse a substance for purity you must have a pure substance to compare. In all cases you would either carry out spectroscopy on a known pure substance or you would compare your spectrum to published data.

Thin-layer Chromatography (TLC)

> **Link**
>
> You will have carried out TLC in *Unit 2: Practical Scientific Procedures and Techniques*.

One application of TLC is to assess the purity of a substance.

There are many forms of chromatography used in industry. These include high performance liquid chromatography (HPLC) and gas chromatography (GC).

The basic versions of paper chromatography and thin-layer chromatography can be carried out using simple apparatus in the laboratory.

The more soluble the compound is in the solvent, the faster it will move through the solvent. So a very soluble compound will move higher up the TLC plate than a less soluble one and this will separate the compounds out.

Pure pigments can also be identified as they will only have one component present, so this is also a way to test if a substance is pure.

Key term

Chromatogram – the pattern of separated substances produced by chromatography (e.g. as seen on a TLC plate).

You can identify compounds by comparing them to the **chromatogram** of a pure known compound. They can also be identified using their retention factors R_f. These are worked out by measuring the distance the solvent has moved up the TLC plate and the distance each spot has moved up the TLC plate.

$$R_f = \frac{\text{distance travelled by compound}}{\text{distance travelled by solvent}}$$

You can compare these R_f values to known values for an identical solvent in order to identify the compounds.

Worked example

A scientist need to work out the R_f value of one component of an ink.

Step 1: He measures the distance between the base line and the solvent front. This is the distance travelled by the solvent.

Step 2: He records this as 108 mm.

Step 3: He measures the distance between the base line and the centre of the spot for the ink component. This is the distance travelled by the ink component.

Step 4: He records this as 90.5 mm.

Step 5: He puts these figures into the calculation.

$$R_f = \frac{\text{distance travelled by compound}}{\text{distance travelled by solvent}}$$

$$R_f = \frac{90.5}{108}$$

The R_f value for the ink component is 0.84.

Chromatography can also be used to estimate how pure a substance is. For example, the number and size of the spots on the chromatogram can give you an idea of how many impurities are present and how much impurity in relation to the wanted compound there is. However, this is not an accurate estimation of how many impurities or how much they are.

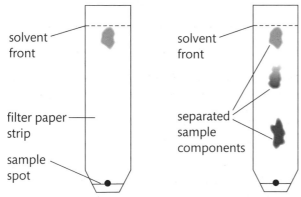

▶ **Figure 4.15:** Chromatograms of a pure and an impure substance

The chromatogram on the left shows a pure substance as it only has one dot. The one on the right shows that the substance had three components as it has three separate dots.

In all types of chromatography, there is a mobile phase, and in TLC this is the solvent. There is also a stationary phase, and in TLC this is the silica (SiO_2) gel on the TLC plate. Silica gel is polar. If a compound is strongly adsorbed to the gel, it will need a highly polar solvent to move it up the plate. If the compound is weakly adsorbed to the gel it will move easily up the plate. If the solvent dissolves all the compounds in the mixture easily, all the components of the mixture will move up the plate.

The solvent chosen for the mobile phase must be suitable to move all the components to some extent.

For example, pure propanone is not suitable to use to separate plant pigments as it moves all the pigments to the top of the plate. Petroleum ether is not suitable either as it only moves adsorbed carotene (a plant pigment) up the plate slightly. So a mixture of 70% (by volume) petroleum ether and 30% propanone is usually used to separate plant pigments in TLC. This gives the optimum separation of components. If components are not fully separated by the solvent, then it is difficult to state exactly when components are present.

Note that the R_f is always the same for a particular chemical substance if the chromatography procedure is kept constant (i.e. same solvent and stationary phase), and so specific chemical substances can be identified.

Sometimes it is not possible to see the chemical substances that have separated out because they are colourless. In this case, a dye or ultraviolet light is used.

For amino acids, the chromatogram can be sprayed with ninhydrin, which turns amino acids coloured so the spots can be seen. The reaction between ninhydrin and the coloured acids is quite slow and so can take hours or even weeks to develop fully. This process can be speeded up by warming the chromatogram in an incubator.

Safety tips

- You must wear eye protection when using TLC, as ultraviolet light is hazardous and can damage the eyes.
- Disposable gloves should be worn and skin contact avoided when using dyes.
- Ninhydrin is harmful if swallowed and irritating to the skin, eyes and respiratory system. The spray is particularly hazardous and should be used in a fume cupboard.

Case study

Analysing amino acids

Rebecca is a trainee laboratory technician working in the food industry. To improve her laboratory technique, she is analysing a mixture of amino acids. She will be able to use this technique to identify compounds in substances such as food or drinks.

Many additives and colourants used in foodstuffs are harmful but can be recognised quickly and easily using TLC. Other compounds may also have been added to the food inadvertently, such as pesticides and insecticides, and these may also be identified using TLC.

Check your knowledge

1 What measurements does Rebecca need to take in order to work out which amino acids are present in the mixture she is analysing?

2 Why is TLC a good technique to use in order to identify unknown substances?

High Performance Liquid Chromatography (HPLC)

This is a technique that you can use for numerous applications. For example, you can use it to analyse proteins, water quality, additives and contaminants in food. You can use it in quality control and to assess the purity of raw materials, to monitor how substances degrade over time, etc.

- Liquid solvent is forced through an HPLC column at high pressure.
- The column contains small silica particles which provide a large surface area. The substance to be separated is dissolved in the solvent and passes through the column. The components being separated interact with the packing material and the large surface area enhances the separation of the components.
- Detection of the separated components is automated and very sensitive; ultraviolet, UV, absorption is often used. This can give you the type and quantities of substances present.

- Many **organic compounds** can be identified by the amount of UV radiation they absorb at particular wavelengths. A beam of UV light is shone through the stream of liquid coming out of the column. The amount of UV light absorbed is detected and, allowing for absorption of UV by the solvent itself, the display on the processor will indicate which organic compounds are present.
- This will indicate the purity of the sample being tested.
- HPLC is a very sensitive technique and can give precise results as well as quantitative results. This can make it more useful than TLC carried out in a school laboratory.
- HPLC is not sensitive to all compounds. This is a limitation. HPLC is often preferred over gas chromatography (GC) as it can handle involatile as well as volatile substances. GC is only good if the substance is volatile. Where it cannot detect a substance, gas chromatography should be used. Using HPLC will not always show if impurities are present.

Gas chromatography

Gas chromatography (GC) is another type of column chromatography. You can use it for many tests. It is used to analyse samples from athletes to test for banned substances in sports competitions, animal fat contamination in vegetable oils, and alcohol concentrations in a motorist's blood sample.

- The liquid sample, or the sample dissolved in a solvent, being analysed is injected into the column. The column comprises a coiled steel tube packed with porous rock on which is adsorbed a liquid solvent.
- The coiled tube is heated up inside a thermostatically controlled oven. This turns the solvent and the sample to be separated into a gas (vapour).
- An inert gas, such as helium, also passes into the column.
- As the components of the sample move through the column, some will be carried with the inert gas (mobile phase) while others will dissolve into the liquid solvent (stationary phase). (See Figure 4.17.)
- The molecules travelling in the mobile phase will take less time to travel through the column than those in the stationary phase.
- The time taken for each component in the sample to pass through the column to the detector is the retention time and depends on the solubility of each component in the inert gas or liquid solvent. The gas is burnt and an electrical current is produced. This current can then be detected.
- The display on the processor shows a series of peaks. Each peak corresponds to the retention time of the sample. This enables you to identify each component present in the sample. This gives you a measure of the purity of the sample. See Figure 4.16.

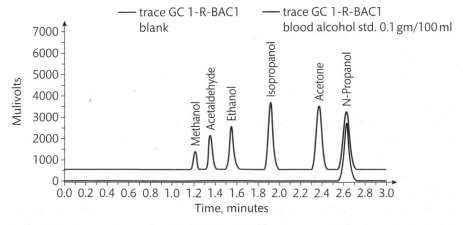

▶ **Figure 4.16:** Chromatogram showing blood alcohol levels

▶ If the sample is pure, i.e. it only contains one chemical, then there will be a single peak on the recorder.

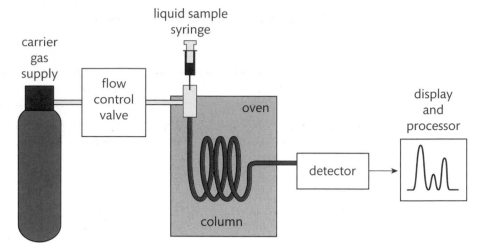

▶ **Figure 4.17:** Gas chromatography system

The relative concentrations of the different components are often displayed on a graph like Figure 4.18. The greater the reading, the more substance is present.

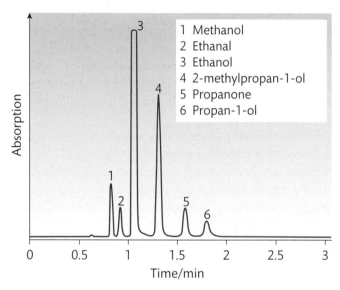

▶ **Figure 4.18:** Concentrations of components

Gas chromatography can show how pure a substance is, as each component that leaves the gas chromatography column is detected and measured. This gives a good estimation of the purity of the substance. The results can be compared to known data as long as the conditions used are the same. One problem with gas chromatography is that the substance to be analysed must be volatile, i.e. it must evaporate easily. If the substance is not volatile, then another method of analysis will be needed.

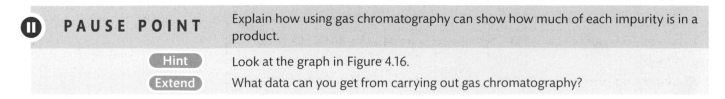

Ⅱ PAUSE POINT   Explain how using gas chromatography can show how much of each impurity is in a product.

        Hint   Look at the graph in Figure 4.16.

        Extend   What data can you get from carrying out gas chromatography?

Table 4.3 compares the different types of chromatography.

▶ **Table 4.3:** Types of chromatography

Types of chromatography	Types of sample separated	Mobile phase	Stationary phase	Uses	Limitations
Paper	Dried liquid samples.	Liquid/solid solvent.	Filter paper strip.	One of the most common types of chromatography: to analyse pen inks, lipsticks, food and fabric dyes, etc.	Can only show number of impurities and relative size of impurities.
TLC	Dried liquid samples.	Liquid/solid solvent.	TLC sheet – glass/plastic plate covered with a thin layer of silica gel.	To analyse dye composition of fibres, inks and paints; to detect pesticide or insecticide residues in food.	Can only show number of impurities and relative size of impurities.
HPLC (liquid)	Liquid samples that may incorporate insoluble molecules.	Liquid solvent or solution.	Column composed of silica or alumina gel powder or suspension of solid beads in a liquid and absorbed liquid.	To test water samples to look for pollution in lakes and rivers; to analyse metal ions and organic compounds in solutions; to analyse blood found at a crime scene.	Not sensitive to all substances.
GC (gas)	Vaporised samples and gas mixtures.	Carrier gas, e.g. nitrogen, hydrogen or helium, is used to move gaseous samples.	Column composed of a liquid or of absorbent solid beads.	To detect bombs in airports; to analyse fibres on a person's body; to test for the presence of accelerants in arson cases and residue from explosives; to analyse body fluids for the presence and level of alcohol and illegal substances.	Substance to be analysed must be relatively volatile.

PAUSE POINT   Explain how gas chromatography can be used to test a sample of body fluid from an athlete.

> **Hint**   Remember that the test is looking for banned substances.

> **Extend**   Why would you use gas chromatography rather than HPLC? Why would you use TLC rather than HPLC?

Theory into practice

In large developed cities around the world, like London or Manchester, drinking water is recycled. It is estimated that water consumed in these cities has already 'passed through' dozens of other people. There are several methods possible to ensure clean drinking water.

▶ Water can be distilled to remove dissolved impurities.

▶ Solid waste can be separated using filters and sieves.

▶ Molecular sieves can be used to remove very small particles that make water cloudy.

▶ Chromatography can be used to analyse water quality and test for impurities.

Quality control

Shelley works as a forensic scientist for a large drinks manufacturer. The company sells a successful brand of orange fizzy drink all around the world. One of the sales representatives for the company in Malaysia bought a bottle of the drink in a local shop. When the sales representative tried the drink, he did not think it tasted like it should. He bought some more bottles from the same shop and sent them to Shelley to be tested.

Check your knowledge

1 Why does Shelley test the product using TLC? What information might she find out?

2 What other methods might Shelley use to test the products made?

Reference data

Reference data is extremely important when testing chemical substances for purity.

The term 'reference data' refers to a comprehensive listing of components and particulars associated with them. In science this usually means: names and formulae of chemical elements and compounds, crystal structure, R_f values, melting and boiling point values, other physical properties, chemical properties, etc.

In research and education, it is essential to refer to other literature. In science in particular, it is very important to find information in the form of explanatory documents or data tables. It is vital that you use different sources in order to be sure that there is at least a general consensus of opinion.

Assessment practice 4.2   B.P3   B.P4   B.M2   B.M3   B.D2

You work for a perfume company that uses esters in its products. It is important that the esters used do not contain impurities, otherwise the fragrances produced may be of poor quality.

You must prepare an ester and test it for purity. You also need to research how the ester would be manufactured and tested in industry.

You then need to produce a report containing:

- notes and results from preparing the ester
- a description of the principles behind the preparative methods and tests used
- an analysis of ways to improve yield and purity and the reliability of testing methods as a guide to purity and their relevance to industry
- an explanation of the principles behind the industrial manufacture and testing of the ester comparing it to the methods you have used.

Plan

- What is the task? What am I being asked to do?
- How confident do I feel in my own abilities to complete this task? Are there any areas I think I may struggle with?
- Do I have all the information I need? Do I need to do more research?

Do

- I know what it is I am doing and what I want to achieve.

Review

- I can explain what the task was and how I approached the task.
- I can explain how I would approach the hard elements differently next time (i.e. what I would do differently).

C Explore manufacturing techniques and testing methods for an organic solid

Manufacturing techniques

Precipitation, crystallisation and recrystallisation

Solutions

Many reactions take place in liquid solutions. This allows for the reacting particles to move, collide and react. A solution contains a **solute** dissolved in a solvent. Where the solution is a liquid, the solvent is a liquid. Water and ethanol are common solvents. Solutes can be solid, liquids or gases. For example, carbon dioxide gas in a fizzy drink is the solute.

A solution in which the maximum amount of solute has been dissolved is called a **saturated solution**. Any more solute added will not dissolve and will sit on the bottom of the container. A **supersaturated** solution is when a solution contains more of the dissolved solute that can be dissolved under normal circumstances.

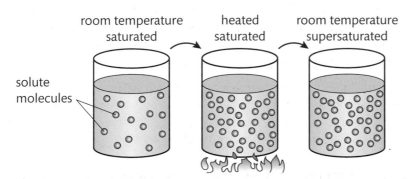

▶ **Figure 4.19:** Making a supersaturated solution

Key terms

Solute – the substance that dissolves in a solvent to form a solution.

Saturated solution – a solution in which the maximum amount of solute has been dissolved.

Supersaturation – the difference between the actual concentration and the solubility concentration at a given temperature.

Influence of temperature

When the temperature of a solvent is increased, the amount of solute that can dissolve in it can increase. This is because increased temperature means the the particles have more kinetic energy and so move faster. This allows them to move from one position to another more easily. The greater freedom of movement allows the system to change its state more easily, and in keeping with the Second Law of Thermodynamics (simply put, this law states that energy will disperse as much as possible). Solubility is temperature dependent, and solids are normally more soluble at high temperatures.

Solid particles are packed close together. Dissolving a solid means its particles are further apart as they have more energy to move. Therefore, an increase in temperature generally leads to an increase in the solubility of the solid. If this solution is then cooled, it can take a while for the excess solute to precipitate or crystallise from the solution. So there is more solute than normal dissolved in the solvent, meaning that the solution is supersaturated.

Inflence of polarity of solvents

In Unit 1, you looked at chemical bonding and the effect this has on the behaviour of chemical substances.

One way to think about how substances behave is in terms of **polarity**. Wax and hexane have covalent bonding and are non-polar. Molecules in wax and hexane have only weak forces of attraction between them. These are van der Waals forces.

Key term

Polarity – the property of molecules having an uneven distribution of electrons, so that one part is positive and the other part is negative.

Compounds that contain atoms with a large difference in electronegativity are usually polar. For example, water, a polar molecule, contains an oxygen atom and hydrogen, and oxygen is highly electronegative. The oxygen has a bigger share of the paired electrons in the water molecule than the two hydrogen atoms have. As a result, the oxygen end of the molecule is slightly negative and will attract a positive charge, while the hydrogen ends are slightly positive. In addition to van der Waals forces, there is also an electrostatic attraction between the oxygen end of one water molecule and the hydrogen ends of another water molecule. This attraction is called hydrogen bonding.

There is a spectrum of polarity in solvents, as shown in Figure 4.20. Some molecules are not polar, e.g. hydrogen molecules, as the hydrogens are identical in electronegativity and the electron distribution is even. Water is very polar.

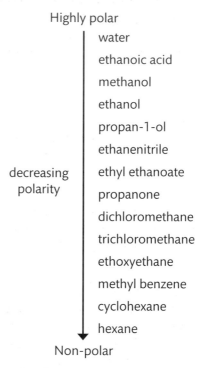

▶ **Figure 4.20:** Polarity of a molecule

Substances tend to dissolve in solvents whose polarity is similar.
▶ Non-polar dissolves non-polar. For example, hexane dissolves wax.
▶ Sucrose is a polar compound. It dissolves easily in water which is also highly polar.

For something to dissolve in water, the very strong forces of attraction between adjacent water molecules must be overcome. When sucrose dissolves in water, the strong attractions between the water molecules in water, and the strong attractions between the sucrose molecules in the sucrose, have to be overcome. The attraction between the sucrose molecules and the water molecules must be even stronger.

Precipitation

Precipitation from an organic solvent is an important method for purifying proteins and nucleic acids.

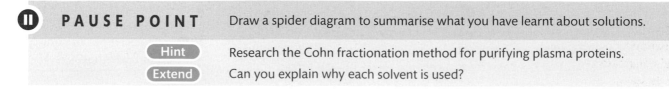

PAUSE POINT   Draw a spider diagram to summarise what you have learnt about solutions.

Hint   Research the Cohn fractionation method for purifying plasma proteins.

Extend   Can you explain why each solvent is used?

Crystallisation

Crystallisation is a process where a previously dissolved substance comes out of solution in a controlled way.

Supersaturation is important in order for crystallisation to occur. It causes crystal **nucleation** and growth. When there are more solids dissolved than normal for a saturated solution, crystals start to form. The initial formation of the crystals is called nucleation.

Key terms

Precipitation reaction – a chemical reaction where a suspension of small solid particles, a precipitate, is produced from a liquid or gas state.

Crystallisation – the process of forming crystals from a liquid or gas.

Nucleation – the initial process that occurs in the formation of a crystal when the dissolved substance starts to come out of the solution from a solution, a liquid, or a vapour, in which a small number of ions, atoms or molecules become arranged in a crystalline solid, forming a site upon which additional particles are deposited as the crystal grows.

Nucleation can occur spontaneously from within the solution (primary nucleation) or in the presence of existing crystals (secondary nucleation). Crystal growth is the increase in size of crystals which are seeded/placed into the solution as solute is deposited from solution.

At low supersaturation, crystals can grow faster than they nucleate, or initially form. The result is a larger crystal size distribution. However, at higher supersaturation, crystal nucleation dominates crystal growth, ultimately resulting in smaller crystals. More crystals are formed but they do not grow as large.

Recrystallisation

It is important to ensure that products are pure. When an organic solid has been prepared, it is likely to need purification. One way to remove impurities is to use the technique of **recrystallisation**. The principle is that different substances have different solubilities in a solvent at different temperature. Separation is possible due to the differing solubility of the product and impurity in hot and cold solvent.

Key term

Recrystallisation – a technique used to purify a chemical by dissolving both the chemical and the impurity in a solvent and warming the solution. Separation is possible due to the product and the impurity having different solubilities in hot and cold solvent.

Selection of the solvent for the recrystallisation process is very important. If a substance is soluble in a solvent when it is cold, the crystals will not come back out of the solution. You need to select a solvent in which the substance is soluble only when the solvent is hot. This also explains how insoluble impurities are removed by hot filtration.

It is preferable to cool the solution slowly as this encourages larger, more regular crystals that are less likely to have impurities trapped in them.

Some commonly used solvents are:
- water
- ethanol
- propanone
- ethyl ethanoate
- cyclohexane.

Recrystallisation

Steps in the investigation	Pay particular attention to...	Think about this...
1. Add the impure solid to a conical flask.		
2. Heat the solvent separately and then add it to the impure solid with a dropping pipette.	Warm it gently so no product is lost through spitting.	How well the product and the impurities will dissolve in the solvent will affect your choice of solvent. This should be tested first to establish solubility at different temperatures.
3. If there is still some undissolved solid, add further solvent and warm until the mixture boils again.		
4. The minimum amount of solvent to dissolve all the product should be used.		The more solvent used, the more of your product will stay in the solution when the solution is cooled. The more solvent used, the lower the yield.
5. Continue adding further solvent and heating until all the solid is dissolved.		
6. Filter solution whilst hot to remove insoluble impurities.		
7. Allow the solvent to cool.	Cool the solvent slowly as this allow for bigger crystals to form. You can obtain more crystals by cooling the solution below room temperature in an ice bath.	As the solution cools the solubility of the product and impurities drops. If the correct solvent has been used the impurities will stay in solution and the product will recrystallise.
8. Use filtration to separate the solid crystals and allow to dry.	You can use a Büchner or Hirsch funnel for filtration. (See Table 4.4.) Put crystals in a desiccator to dry.	The impurities will remain dissolved in the filtrate.
9. This method can be repeated to obtain even purer crystals.		

Filtration

Solids are separated from liquids by **filtration**. There are a range of different types of filtration, as shown in Table 4.4. It is important to know which one is appropriate to use.

Key term

Filtration – technique to separate solids from the liquid in which they are suspended.

▶ **Table 4.4:** Types of filtration

Type of filtration	Description	Use	Example
Gravity filtration: fluted filter paper	Simple filtration, using fluted paper in a filter funnel. The folding speeds up the rate of filtration as the surface centre is greater.	Remove solid impurities from a liquid.	Removes insoluble rock impurities from solution when rock salt is dissolved in water.
Gravity filtration: non-fluted paper	Similar to above. The paper is folded more simply.	Remove solid impurities from a liquid.	Removes sand from a mixture of sand and water.
Hot filtration	Similar to above but solution/liquid is kept warm.	Remove solid impurities from a liquid – prevents crystals of desired solute from forming. This is necessary because if the wanted solute recrystallises it will also be filtered out and so will still be mixed with the impurities.	Obtaining pure paracetamol from impurities present after synthesis.
Vacuum filtration: Büchner funnel	Uses vacuum to increase speed and efficiency of filtration. The reduced pressure also helps to dry the product.	Quickly removes solid impurities from a liquid. Only use with a cold solution.	Separating recrystallised antifebrin after synthesis. (antifebrin is a compound used to make paracetamol).
Vacuum filtration: Hirsch funnel	Similar to Büchner funnel but much smaller.	Used for small quantities.	Separating small amounts of antifebrin.
Vacuum filtration: sintered glass crucible	Glass crucibles with fitted glass disks sealed permanently into the bottom end.	They can have different size pores so can be used for very small crystals and avoids paper fibres contaminating the crystals.	Filter precipitates such as silver chloride.

PAUSE POINT   Draw a concept map to summarise what you have learnt about solutions.

Hint   Consider the terms solvent, solute, saturation and supersaturation.

Extend   Add information to your concept map about separating techniques and filtration.

Key terms

Evaporation – the process whereby a liquid turns into a vapour at a temperature below or at the boiling point of the liquid. It occurs at the surface of the liquid, where molecules with enough energy escape into the gas phase.

Anhydrous – a compound that contains no water, e.g. anhydrous copper sulfate, which is white and contains no water compared to blue copper sulfate, which contains water of crystallisation.

Evaporation and drying

Once you have crystallised and filtered your desired product, it is important to dry the product to remove any excess solvent.

You can carry this out in a variety of ways.

Once you have removed the product from the filter funnel, you can place it in an evaporating dish. This can then be left in a warm place to dry. Some solvents may take longer than others to dry, so you may want to place the evaporating dish in a warm oven to speed up the process of **evaporation**.

A desiccator can also be used to remove water. The substance being dried may be a product from a reaction, or a reactant that must be dry in order to take part in a reaction. Desiccators contain a substance that attracts moisture so a desiccator will keep the contents drier than if they are in the open atmosphere. Desiccators tend to have a seal so that air cannot get into them.

Distillation can be used as a method to remove water. (See Figure 4.8.) Chemical drying agents can also be used to remove water from a solution in an organic solvent. Organic liquids are considered wet if they contain water. The organic liquid is still a liquid once the water is removed, but it is now dry. Common drying agents are **anhydrous** calcium chloride, $CaCl_2$, anhydrous sodium sulfate, Na_2SO_4, anhydrous calcium sulfate, $CaSO_4$, and anhydrous magnesium sulfate, $MgSO_4$. These work in the same way as the silica gel you find in a box of new shoes.

Rotatory evaporation can be used to dry products by rapidly removing the solvent impurities. Rotatory evaporators essentially perform distillation, but at a lower pressure. There is a motor in the evaporator that turns the flask containing the substance to be dried. This means that the solvent covers the entire surface area of the flask, creating a greater surface area over which evaporation can occur, thus speeding up the process of drying. There is also a vacuum which speeds up the drying process by reducing the pressure. At a reduced pressure, the boiling point of the solvent is reduced and it therefore evaporates at a lower temperature. A water bath is used to control the temperature of evaporation, and cooling coils condense the solvent vapours that are given off. The solvent vapours turn into liquid and are collected, to either be recycled or disposed of properly. The solute/dissolved substance remains in the round-bottomed flask and can be collected at the end of the process.

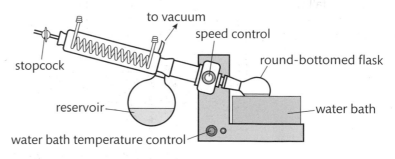

▶ **Figure 4.21:** Rotary evaporator

PAUSE POINT   Explain the methods you would use to dry an organic product such as aspirin.

> **Hint**   Consider how the liquid can be evaporated or removed.
> **Extend**   Give a reason why is it important to obtain dry products.

Industrial manufacturing techniques

As a chemist working in industry, you will need to be familiar with a range of different techniques to make a range of chemical substances. Chemists may work in a research and development laboratory or in a **chemical plant** where chemicals are produced on a large scale. You will not have used some of these chemicals in a school or college laboratory. This may be because they are too expensive or too dangerous to use outside of industry.

You will be familiar with some techniques, but they may be carried out in a different way in industry. Often this is due to the technique being performed on a larger scale in industry, as it is rare that small quantities of a product need to be manufactured, since this would not be cost effective.

Spray drying

In industry, large amounts of liquid or wet solid (slurry) need to be dried. One method is spray drying, which involves using a hot gas to produce a dry powder. It is often used for medicines and food stuffs that may be damaged if they are dried by heating. Hot air is used, unless the product is flammable or corrosive, in which case, nitrogen will be used.

A spray nozzle or atomiser is used to spray the liquid or slurry into the hot gas. The air is often blown in at the same time as the liquid. The hot drying gas can flow with the liquid or against the liquid. Flowing with the liquid gives a quicker drying time and can be more effective.

There are several different types of spray drying apparatus. Figure 4.22 shows one of them.

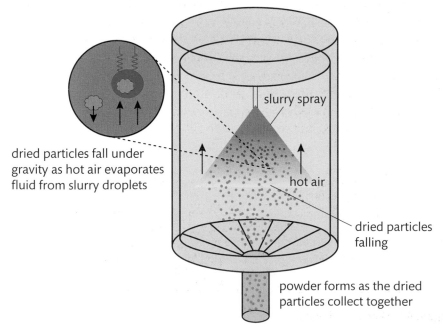

dried particles fall under gravity as hot air evaporates fluid from slurry droplets

slurry spray

hot air

dried particles falling

powder forms as the dried particles collect together

▶ **Figure 4.22:** Spray drying apparatus

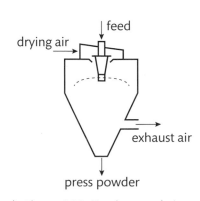

feed

drying air

exhaust air

press powder

▶ **Figure 4.23:** Simple spray drying method

Freeze drying

This is a method that is also often used in the pharmaceutical and food industries. Freeze drying a product means that it will have a longer shelf life, i.e. it will last longer and behave as if it is fresh for a longer period of time than if it had not been through the process. It also often means that the freeze-dried product takes up less volume because a lot of the original volume is made up of water. It is now easier and cheaper to store as it needs less storage space.

There are four stages to freeze drying.

▶ Pre-treatment – this happens before the freeze drying and could involve concentrating the product and/or adding preservatives.

▶ Freezing – this is usually done using a freeze-drying machine. In this step, it is important to cool the material below the lowest temperature at which the solid and liquid phases of the material can co-exist. This ensures that sublimation rather than melting will occur. Sublimation is where a solid becomes a gas without becoming a liquid first. Larger crystals are easier to freeze-dry as the large crystals give a more open structure for water vapour to escape from so the product should be frozen slowly. However, in the case of food, large ice crystals will break the cell walls of the product. For food, the freezing must be done rapidly to avoid ice crystals forming. Usually, the freezing temperatures are between -50 °C and -80 °C. The freezing phase is the most important in the whole freeze-drying process, because the product can be spoiled if it is not done properly.

▶ Primary drying – the pressure is lowered, and enough heat is supplied to the material for the ice to sublime. About 95% of the water in the material is sublimated. This phase may be slow (it can be several days in the industry), because, if too much heat is added, the material's structure could be altered. In this phase some proteins denature or break down with high temperatures. Pressure is controlled using a partial vacuum which speeds up the sublimation. Vapour is produced more quickly in order to fill the partial vacuum.

▶ Secondary drying – here unfrozen water molecules are removed. In this phase, the temperatures is raised higher than in the primary drying phase, to break any interactions that have formed between the water molecules and the frozen material. After the freeze-drying process is complete, the vacuum is usually broken with an inert gas, such as nitrogen, before the material is sealed. An inert gas is used so that there are no reactions between the materials and the gas.

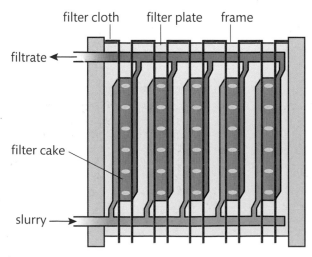

▶ **Figure 4.24:** Filter press

Use of a filter press

You can use a filter press for a range of reasons, including:

▶ to purify water
▶ to filter fresh fruit juice
▶ to filter blood plasma
▶ to remove unwanted chemicals from pharmaceutical product.

The filter press produces very dry products.

The press consists of a series of filter plates that are held together during the process. During filtration, slurry (a mixture of liquid and small solid particles) can flow in and filtrates (the purified liquid) flow out. The slurry is pumped into the middle of the plates where it spreads between the plates. Filter cake builds up on the filter cloth and this in turn acts as a filter.

You can use the press to filter large amounts of slurry. However, as filter cake builds up, eventually the cake is too thick and filtration stops. At this point the press is stopped and the cake has to be removed.

Manufacture of paracetamol

Paracetamol is one of the commonest pharmaceuticals used to relieve the symptoms of fever and pain.

It can be made in the lab as well as on a large scale in industry. You need to be able to compare the differences between production of chemical substances such as paracetamol in the lab and in industry.

There are three steps in the production of paracetamol. These steps have slight differences in industry from those carried out in a college or school lab. You need to know what these differences are and why there are differences.

Research the differences, then copy and complete Table 4.5 to show your understanding.

▶ **Table 4.5:** Production of paracetamol

	Production of paracetamol in a school/college lab	Differences in industry	Reasons for any differences
Step 1 starting materials	Phenol, sodium nitrate(V), water, concentrated sulfuric acid		
Step 1 techniques	Decanting, distillation, filtration, recrystallisation		
Step 2 starting materials	Sodium hydroxide, sodium tetrahydridoborate(III) 4-nitrophenol (made in step 1)	Hydrogen used in a hydrogenation process instead of sodium tetrahydridoborate(III)	
Step 2 techniques	Filtration Use of sodium hydrogencarbonate (I M NaOH solution) to remove excess acid		
Step 2 conditions/ catalysts	Keep reaction at about 13–17 °C Palladium on charcoal catalyst Acidify I M NaOH solution with hydrochloric acid	Platinum catalyst used	Hydrogenation gives a higher yield than using sodium tetrahydridoborate(III) Platinum is too expensive to use in school/college lab
Step 3 starting materials	4-aminophenol (made in step 2) Distilled water Ethanoic anhydride		
Step 3 techniques	Filtration under suction Crystallisation Recrystallisation		

Making aspirin

Aspirin is made by reacting ethanoic anhydride with 2-hydroxybenzoic acid. The first step is to prepare 2-hydroxybenzoic acid. The second step is to use ethanoic anhydride to convert 2-hydroxybenzoic acid into aspirin (2-Ethanoyloxybenzenecarboxylic acid).

Steps in the investigation	Pay particular attention to...	Think about this...
1. Set up reflux apparatus for heating about 30 cm³ of reaction mixture, oil of wintergreen with aqueous sodium hydroxide, using a water bath. Add a condenser.	Ensure the glassware fits snugly.	The condenser prevents any vapours escaping.
2. Put 2 g of oil of wintergreen into your flask and add 25 cm³ of 2 mol dm⁻³ sodium hydroxide along with anti-bumping granules.	The anti-bumping granules must be added before the reaction mixture is heated.	
3. Heat over a boiling water bath for 30 minutes.	**Safety tip**: never use a Bunsen burner as the reactants are volatile and flammable.	
4. Allow the mixture to cool.		
5. Pour the mixture into a small beaker surrounded by a mixture of ice and water.		The next step is to add hydrochloric acid. The mixture must be cool before you do this as otherwise it will be too vigorous.
6. Add concentrated hydrochloric acid to the mixture dropwise until it is just acidic, stirring all the time.	You can check pH with litmus paper or by removing a small amount of mixture and adding universal indicator.	
7. Filter the product using a Büchner funnel and suction apparatus.	Allow time for all the liquid to be sucked through the funnel.	You must have a dry product for the next step.
8. Wash the product (2-hydroxybenzoic acid) with a little ice cold water and transfer it to a watch glass. Allow to dry overnight.	The product must be dry before being used for the second step.	
9. Add 1 g of 2-hydroxybenzoic acid into a dry pear-shaped flask.		This is the product you obtained in steps 1–8.
10. Add 2 cm³ of ethanoic anhydride followed by 8 drops of concentrated phosphoric acid. Put a condenser on the flask.	**Safety tip**: take care with the reactants. You should be wearing safety goggles and a lab coat throughout.	

Steps in the investigation	Pay particular attention to...	Think about this...
11. In a fume cupboard, reflux the mixture in a hot water bath, with swirling, until all the solid has dissolved.	**Safety tip**: the fumes produced can cause irritation and damage to your lungs and eyes.	The solid will not dissolve in cold solvent.
12. Continue to warm for approximately 5 minutes.		This ensures all solid is dissolved and the reaction is complete.
13. Carefully add 5 cm³ of cold water to the solution. Use a teat pipette.	This reaction can be vigorous so take your time adding the water.	
14. Stand the flask in a bath of iced water until precipitation appears to be complete. You may need to stir vigorously with a glass rod to start the precipitation process.	The stirring allows the crystals to start to form.	Think about the process of nucleation.
15. Filter off the product using a Büchner funnel and suction apparatus.		
16. Wash the product (aspirin) with a little cold water, transfer to a watch glass and leave to dry overnight.	Wash while it is still in the funnel. Take care that the filter paper does not lift up when the water is added as you may lose some product.	

❚❚ PAUSE POINT   What are the risks in this preparation and how should they be controlled?

Extend   Why is the hydrochloric acid added slowly to the mixture sitting in an ice bath in step 6?

In industry, aspirin is made in a batch process. It is made in much higher quantities than you would make it in the lab as it is in very high demand. The flow chart in Figure 4.25 outlines the industrial process.

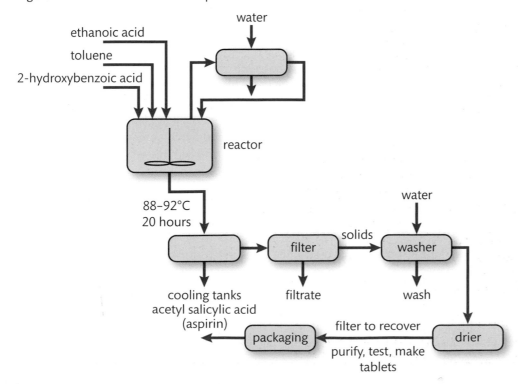

▶ **Figure 4.25:** Aspirin production in industry

Once the aspirin is made, it still needs to go through several more steps. It will need to be purified and tested for purity. Any impurities may cause side effects that may harm the patients that take them.

Once the aspirin crystals have been purified, they then have to be made into tablets that patients can swallow easily.

To produce hard aspirin tablets, corn starch and water are added to the active ingredient (acetyl salicylic acid). These bind the aspirin crystals together, and also add bulk to the tablet, as without them the size of the tablet would be too small to handle. Only a small part of the aspirin tablet is made of acetylsalicylic acid. Some lubricant such as vegetable oil is also added to stop the mixture sticking to the machinery.

Some aspirins are chewable, and contain different ingredients that may allow them to dissolve faster and have a nicer taste. These often contain small doses of the active ingredient and are mainly used for young children. Aspirin tablets can be in soluble or insoluble form, depending on the other ingredients present.

Not all chemical plants produce aspirin tablets in exactly the same way. Here is one possible method.

First the corn starch, the acetyl salicylic acid and the lubricant have to be weighed separately in sterile canisters to ensure the correct proportions of each are used and the dosage of the active ingredient will be correct.

The corn starch is added to cold purified water, then heated and stirred until a paste forms. The corn starch, the acetyl salicylic acid, and some of the lubricant are next poured into a sterile canister. The ingredients are then all mixed within the canister. This also removes any air from the product.

The mixture is portioned into small units called slugs. These are about 2 cm in size. Small batches of the slugs are then forced through a mesh screen, either by hand or by an automated process, depending on the size of the factory. The rest of the lubricant is added at this stage.

The mixture is compressed into tablets, either by a single-punch machine (for small batches) or a rotary tablet machine (for large-scale production).

The tablets are then tested for hardness and other quality controls. Only then can they be packaged or bottled.

Quality control is an important part of the production process. Regulations control all activities, equipment and products made in the pharmaceutical company. All machinery is sterilised before beginning the production process to ensure that the product is not contaminated or diluted in any way. In addition, operators assist in maintaining an accurate and even dosage amount throughout the production process by performing periodic checks, keeping meticulous batch records, and administering necessary tests. Tablet thickness and weight are also controlled.

Estimation of purity

Assessment of the appearance of crystals as an indicator of purity

The first thing a chemist should do when assessing purity is look at the product that has been made. You should know what you expect the product to look like. Here are some questions that should be asked.

▶ Should it be a solid?

▶ Should it be colourless?

▶ If not colourless, what colour should it be?

▶ Is it a crystalline?

▶ What shape of crystal should it be?

For example, copper acetate is a green monoclinic crystal.

If the crystal does not look like this, for example, the colour is different or there are crystals of more than one colour, or if the shape is different, then you can say that the product is probably impure.

In industry it is important to know how pure your product is as impurities will affect how the product will behave.

You can test the purity of your solid product in a number of ways.

Measurement of melting point

Melting points are known very accurately for most elements and compounds and, like boiling points, are listed in data books. A substance can also be identified as impure by comparing the experimental value of the melting point with that in a data book.

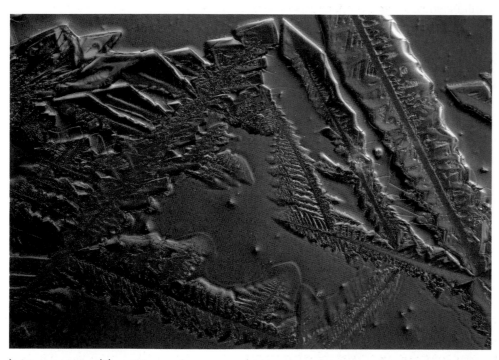

▶ Impure copper (II) acetate sample

> **Key term**
>
> **Melting point** – the temperature at which a solid becomes a liquid.

> **Link**
>
> *Unit 2: Practical Scientific Procedures and Techniques* has more information on melting points.

The melting point is the temperature at which a substance exists in both liquid and solid state, somewhere between the first signs of liquid to the total disappearance of the solid. Observations of melting point provide chemists with an indication of the purity of organic and inorganic substances.

Pure solids melt at a definite temperature (usually within 1 or 2 °C of the published value) but impure solids will become soft and melt over a range of temperature. In general, the addition of another element (impurity) lowers the melting point of the substance. Inorganic solids melt at very high temperatures and so their purity cannot be tested in school or college laboratories.

You can use a cooling curve to find the melting point of a solid. For example, to produce a cooling curve for naphthalene, the naphthalene is heated until it is completely melted. It is then allowed to cool and the temperature taken regularly as it becomes solid. The temperature is plotted against time on a graph and we can use this to determine the melting point of the naphthalene.

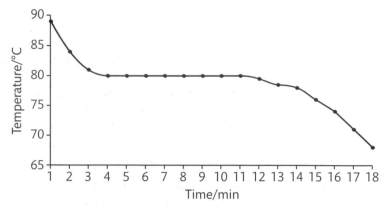

▶ **Figure 4.26:** Cooling curve of naphthalene

The melting point is where the line on the graph levels off. This is where the naphthalene is in both liquid and solid states (at 79.5 °C on the graph).

Melting point apparatus can be used to give a very accurate measurement of melting point. It is also useful when you only have a small sample as only a little is needed.

Step by step: Finding the melting point of synthesised benzoic acid

10 Steps

1 Apparatus – eye protection, benzoic acid samples, pestle and mortar, capillary tubes, Bunsen burner, electronic melting point apparatus, thermometer (0 °C to 200 °C), magnifying glass.

▼

2 Ensure that you put on eye protection first. Remember to take care, as apparatus gets hot.

▼

3 Grind the benzoic acid into a fine powder using the pestle and mortar.

▼

4 Prepare a suitable number of capillary tubes by breaking gently and sealing one end by heating in a hot Bunsen burner flame for a few seconds.

▼

5 Switch on the melting point equipment. Refer to the data tables for approximate temperature values for melting the compound. Set the equipment to a point below the expected melting point. Note that the rate of heating will influence the reading of the melting point. Read the thermometer.

▼

6 Dip the capillary tube into the benzoic acid powder until sufficient depth (a few millimetres) of sample is contained in the tube. Tap gently or scrape the glass to allow the powder to fall to the sealed bottom of the capillary tube.

▼

7 Place the sample tube into the hole next to the thermometer in the melting point equipment.

▼

8 Carry out trial at speed to observe the approximate melting point.

▼

9 Observe the melting of the benzoic acid sample through the magnifying glass (some melting point equipment has this built in). At the point when the sample begins to melt, record the temperature using the thermometer. When the melting is complete, record the final thermometer reading. This is the full temperature range of the melting point.

▼

10 Compare your results with data from the data tables. Repeat the procedure.

PAUSE POINT   Research different apparatus that you can use to measure melting point.

(Extend)   Consider the thermometer to be used. Explain your choices to your group.

Several pure substances have similar melting points so, although you may know the melting point of a pure substance, you may not know what that substance is. If you have an unknown pure substance you can determine what the substance is by carrying out a mixed melting point. Remember that there are other methods to test the purity of a substance, such as HPLC, GC and TLC. These are discussed earlier in the unit.

You do this by adding another known pure substance with the same reference melting point and testing the melting point. If the melting point does not change, then the substances are the same. If the melting point lowers, then you know that the substances are different. Sometimes a mix of substances does not lower the temperature, and this happens at certain compositions. To check, it is worth testing at a range of different ratios of the mixture, for example, 20:80, 50:50 and 80:20. If all three mixtures melt at the same temperature, then you can be confident the substances are the same and you have identified your unknown substance.

Case study

Quality assurance

Isabelle is a technical support worker at an organic chemical plant. She works on the quality assurance team. They take samples of all the chemical substances produced in the plant to ensure they are pure and dry.

Check your knowledge

1 What techniques might Isabelle use when testing the quality of the chemical substances produced?
2 What equipment might Isabelle need?
3 What health and safety regulations should Isabelle follow and how might this affect her working practice?

Assessment practice 4.3   C.P5 C.P6 C.M4 C.M5 C.D3

You work for a pharmaceutical company that produces painkiller tablets.

You must prepare a suitable painkiller tablet and test it for purity. You also need to research how the painkiller tablet would be manufactured and tested in industry.

You then need to produce a report containing:
- notes and results from preparing the painkiller tablet
- a description of the principles behind the preparative methods and tests used
- analysis of ways to improve yield and purity and the reliability of testing methods as a guide to purity and show their relevance in industrial manufacture
- an explanation of the principles behind the industrial manufacture and testing of the painkiller tablet comparing it to the methods you have used.

Plan
- What is the task? What am I being asked to do?
- How confident do I feel in my own abilities to complete this task? Are there any areas I think I may struggle with?
- Do I have all the information I need? Do I need to do more research?
- How can I ensure I carry out the practical work safely?

Do
- I know what it is I am doing and what I want to achieve.

Review
- I can explain what the task was and how I approached the task.
- I can explain how I would approach the hard elements differently next time (i.e. what I would do differently).

D Understand the importance of managing, storing and communicating scientific information in a workplace laboratory

Systems for managing laboratory information

The work carried out by scientists and technicians relies heavily on the structure of the team they work in, and the way each team member acts. In most workplaces there is a hierarchy.

This means the most senior person will have various levels of personnel reporting into them. How this is organised depends on:

▶ how large the team is
▶ the particular routines that are carried out in a workplace
▶ whether the team is spread out over a large area or different sites
▶ if the team is split into smaller groups carrying out a particular job at particular times of day or night.

No matter how people are organised, the way they communicate within their team or outside of it is crucial to the safe and smooth running of the organisation.

There is a need for traceability within any organisation, whatever the size of the team. This means that each member of the team has to take responsibility for their work. This may be that they have to sign off forms and records of work. For example, if a process is completed and the report has been written up, all staff responsible will sign the bottom of the report. In some cases, only a manager or team leader will sign off on a report. This again will depend on the size of the team and the style of management in the organisation. The organisation will have a policy for this.

Security is also important, so all staff will have a secure login password for all work recorded on computer systems. It is likely that this login will have to be changed regularly in order to keep it secure.

Records associated with laboratory work

It is important that records are kept in an organisation of any work that is carried out. Again the organisation will have its own policy on how these are kept. Nowadays, most records are kept on computer, although in some cases, for example, laboratory notes, they may still be hand-written records. Such records are often transferred to computer systems in order to keep them secure and safe.

Any samples of chemicals must be booked in when they are delivered to the lab. Many substances are controlled and so all regulations must be followed. A record is kept of all chemical substances stored in the lab and, when they are used, they are signed out of storage. There should always be a record of where the chemical substances are located. When the chemical substances arrive at the lab they must also be clearly labelled following COSSH guidelines. It must also be clearly recorded from where the chemical substance originated. If there are any issues with the sample, or if an identical sample is needed, then this tracking is essential.

Each sample must have a unique sample identification number. This is usually based on the date and time the sample was booked in and may have a randomly generated number at the end. This number should be recorded whenever the sample is being used. This allows for traceability of the sample.

Results generated in a workplace will be specific to that workplace. There may be results of research performed by colleagues or results generated for the use of outside agencies. Whatever the results are, they only need to be communicated to those who need to know them.

Internal day-to-day results will probably be reported via the laboratory notebooks, printouts from the laboratory equipment and team meetings. These results may be gathered together to produce a report on completion of the research.

Unless there are reasons for urgent results to be communicated directly to another person, results will normally go through an office procedure where they are written up and copied to the recipient, for example, a GP. In some cases results, such as scans, can be viewed via a computer screen along with test results.

Often the equipment used is computerised and the computer records the results. These can be added to files with other results, or printed off to be used in laboratory notebooks.

It is essential that scientific terminology is used and understood by all members of the team if effective communication is to take place. This is particularly important where research work or production is being carried out in different countries where language may cause confusion if standard terminology is not used. However, the language sometimes has to meet the needs of the client. In some cases, scientific terminology will have to be explained in order for the client to understand the report.

The report must also be in a format that meets the client's needs. This may be a computer-produced report. In some cases, a client may want a verbal presentation to give them the opportunity to ask questions.

Laboratory information management systems (LIMS)

In the course of work being carried out in a laboratory, large amounts of information will be produced. It has become increasingly important for data to be stored so that it can be retrieved at a later date.

 PAUSE POINT　　Can you think of reasons why you might want to store your BTEC data?

　　　　　Extend　　Consider what would happen if you lost the data for a practical you had carried out.

Storage of records has changed enormously since the computer arrived in the workplace. In most organisations, large boxes or filing cabinets of documents have been replaced by hard drives and the Cloud, where most of the data is stored. Organisations often buy secure cloud space to ensure that they cannot lose data.

There are many benefits to using computer and cloud storage.
- The amount of space needed for computer storage is much less than the space needed for paper storage.
- Computer storage is less of a fire risk than large amounts of paper.
- Data stored on computer or in the Cloud can be searched quickly.
- Records can be accessed from a number of sites. This means paper copies do not have to be made and delivered to other sites. This saves time and removes the risk of loss. It is also more secure as only employees with the correct login password can access them.
- Records can be updated quickly and there is less chance of technicians using out-of-date information.

There is a range of data that might need to be stored in a laboratory, including:

▶ data produced from research carried out in the laboratory
▶ data about staffing levels
▶ personal data about members of staff
▶ data about resources/equipment.

All types of data will need careful management if they are to be kept safe, secure and be available at a later date.

With more organisations using computers to store and process personal information, there is a danger that the information could be misused or get into the wrong hands. Organisations need to be able to answer these questions.

▶ Who could access this information?
▶ How accurate is the information?
▶ Could it be easily copied?
▶ Is it possible to store information about a person without the individual's knowledge or permission? (Organisations have to follow procedures to ensure data is protected.) Is a record kept of any changes made to information?

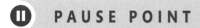

PAUSE POINT   What does the **Data Protection Act** cover and how might it be relevant in a science laboratory?

Extend   Research the Data Protection Act and produce a summary of how it affects the science workplace.

> **Key term**
>
> **Data Protection Act** – the Data Protection Act 1998 was passed by Parliament to control the way information is handled and to give legal rights to people who have data stored about them.

Table 4.6 gives a list of some of the types of data stored in a science workplace.

▶ **Table 4.6:** Data types in a science workplace

Type of data	Reason to keep data	Who should record it, have access and be able to make changes
COSHH records	To ensure awareness of health and safety issues with substances being used in the organisation.	Stores technicians and whoever is involved in ordering, storing and use of the substances.
Scientific data	In any scientific workplace it is vital to be able to safely store and then retrieve scientific data generated by that workplace and also data from other sources (scientific literature, for example).	Heads of department, deputies and those working in the laboratories.
Scientific apparatus	Data such as date of purchase, maintenance data and schedules for maintenance.	Heads of department, deputies and those involved in the schedules.
Waste disposal	To show what and how much waste is produced and the manner of disposal.	Stores technicians and those involved in disposal. Heads of department may need to authorise costs of disposal.
Health and safety checks	To show that health and safety is being monitored and to hold accident reports if necessary.	Heads of department, health and safety officers and possibly others who have special responsibilities.
Training records	To know the level of training or qualification of members of staff, and to keep and maintain a record of training required and completed by staff.	Training officer, heads of department, supervisors, human resource department and individual members of staff.

▶ **Table 4.6:** *Continued*

Quality assurance	To be able to show that quality procedures are being carried out (for audit purposes).	Head of department, quality officers and those with special responsibility.
Report records	Reports following tests for GPs or hospital records, or for use in developing new medicines, etc.	Office support personnel will usually be responsible for recording results, with access needed by clinical staff (in a clinical environment). Report records in this setting would not usually be subject to change by anyone.
Specification levels	This could be the level at which the organisation is allowed to work, for example, the danger levels of microorganisms in use.	The head of department and organisation management.
Sample throughput	This gives information about the number of samples going through processes in the laboratory in a given time and could be an indicator of the efficiency and effectiveness of the organisation.	The head of department and organisation management.
Management	This could cover the management hierarchy and their roles.	Organisation management and human resources department.
Security	Different types of laboratory might need different levels of security depending on the work being carried out.	Head of department, security staff, health and safety officer and all staff.

Ⅱ   PAUSE POINT   How might a computer system be used to store the data in Table 4.6?

(Extend)   What are the benefits and drawbacks of using the Cloud to store data?

Many laboratories have bought a laboratory information management system (LIMS), which is like an electronic filing cabinet. The system allows laboratories to input data in a useful form in order to use it and they can customise the system so that they can input information relevant to their organisation.

The LIMS can store text and graphical documents and can use the data to produce relevant information such as investigation results. It can also be used to monitor good laboratory practice by, for example, monitoring sample collection, testing, quality assurance and outgoing results.

The system can alert the laboratory of incoming samples so that when they are received into the laboratory they can be bar coded and devices can be used to generate labels for quick error-free processing. A hand-held device can be used to enter the samples onto the LIMS. The sample can then be put through the testing procedure with minimal work for the technical staff.

The LIMS can also be used to monitor stock levels so that ingredients or products do not fall below safe levels for the company to continue working. Depending on the organisation and how sophisticated the LIMS is, much of the laboratory documentation can be taken over by the system.

Communicating information in a scientific organisation

As already discussed, there are several ways in which information can be communicated in a scientific organisation. The way it is communicated depends on: company policy, who it is being communicated to, what is being communicated, and the purpose of the communication.

Many companies have their own intranet, which is similar to the internet, but only company employees can access it and it usually only contains information to do with the company. It may contain data as in Table 4.6 and may also contain other information. It will have a search tool that allows employees to find relevant data quickly.

Occasionally outside companies are invited to use the organisation's intranet. They may be restricted to particular areas. This may be because they are working on similar research projects and need to share resources and results.

A lot of information is communicated in documents. These can be hard (paper) copies or soft copies (files on the computer). They could be reports of results or documents relating to management issues, etc.

Most companies use emails to get important information to their employees. These can be sent to one specific individual or a group email can be sent out. Emails are a useful way of communicating as you can keep an email trail which means that it is always clear what has been discussed. Group emails are a quick way of ensuring everyone has the same information. Emails can also be used to communicate in the same way with clients and other organisations that the company works with. Emails are often shorter than more formal letters but they should still be written with a professional tone as they are a business document in this situation.

Many companies have their own website, which allows people from outside the organisation to find out about the organisation. It may give the organisational structure as well as contact details. The website will show what the organisation does. It may also have a method of ordering goods and services. Not all this information will be on the company website as much of it is confidential. Table 4.7 shows types of information used in organisations.

▶ **Table 4.7:** Types of information used in organisations

Types of information used in organisations	Description of information
Customer details	• Contact details • Order history
Product details	• What is produced • Quantity produced • Time to produce
Manufacturing data	• Methods and equipment used • Staffing • Raw materials • Yields • Dates of batches of product produced
Warehousing data	• Quantity of raw material stored • Quantity of product stored • Deliveries
Standard operating procedures	• All procedures followed to produce the product
Sample details	• Chemicals being stored in labs
Results of analysis of raw materials and products	• Purity/quality
Maintenance records	• When equipment was serviced and repaired • When equipment is due to be serviced
Safety data	• Health and safety procedures • Accident logs • Noise logs
Environmental records	• Waste disposal • Recycling policies • Production of pollutants

Channels of communication

Organisations will also have policies on how information is communicated and to whom.

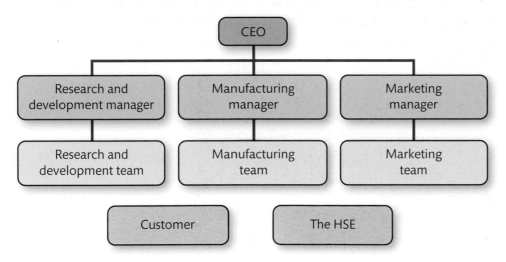

▶ **Figure 4.27:** Organisational chart for a chemical company

Figure 4.27 shows a simple organisational chart for a chemical company. There will be many communication channels within the company:

▶ within departments
▶ between departments
▶ with external customers
▶ with regulatory bodies
▶ with the wider scientific community.

Most organisations will have a policy on how information is communicated.

Communication within and between departments can be verbal, by email, through use of log books or via an intranet.

Organisations usually have a policy on how to share information with external customers. Email is the most common form of communication although phone calls will also be used. It is good practice to send an email documenting what was discussed after a phone call. This ensures that everyone agrees on what was said and that there is traceability.

Regulatory bodies such as the HSE can often ask to see documentation from the company. They may need access to areas of the organisation's intranet. This is so they can monitor the organisation's policies and work practices. This might be related to health and safety or control of dangerous substances.

Science organisations often communicate with the wider scientific community. This is so that research data can be shared and so the scientists working within the organisation are up to date on all the latest research.

The employees should also use some common sense on who to communicate to and how to communicate. For example, group emails should only go to the relevant recipients, e.g. those that are working on the project. Sending a group email to the whole company can be very frustrating for the person who receives it, if he or she is not working on the project, as checking if it is relevant will take up valuable work time.

Information is often very sensitive and needs to be kept secure. So employees need to be careful that only people who need to know are sent the information.

⏸ PAUSE POINT

Copy and complete Table 4.8 to show suitable methods of communication within an organisation.

Extend

Explain why the methods chosen are suitable.

▸ **Table 4.8:** Communication channels

Communication channel	Communication method	Reasons for communication choice
Within departments	E.g. intranet	
Between departments		
With external customers		
With regulatory bodies		
With the wider scientific community		

Case study

Recording and communicating data

Sam works for a pharmaceutical company. His department have successfully produced a new drug to help ease cold symptoms.

1 What processes might Sam have used to record all his results and data?

2 Who should Sam communicate this information to?

3 How do you think Sam should communicate this information?

4 What do you think might happen if Sam communicated this information to the wrong people?

Use of informatics for storage and retrieval of scientific information

Informatics databases

Informatics is the science of processing data for storage and retrieval. The data is collected and classified, which allows for large amounts of complex data to be shared easily.

Science data can be stored in large databases. There are some areas of science and some scientific workplaces where these large databases are particularly useful, including:

▸ DNA sequencing

▸ healthcare records

▸ data relating to population surveys (humans/animals/plants)

▸ fingerprints.

Informatics is so useful in these particular fields because a very large amount of data is produced, and this data can change regularly, meaning it needs to be managed. It is important that the data can be organised and so easily accessed. It can be made available to a variety of research organisations that may use the same data in different ways for different research, making the organisation of the data very important.

We can look at healthcare informatics as an example.

▶ Healthcare informatics is the study of how healthcare knowledge is created, stored, shared and applied. It is a study of how both professionals and patients organise themselves and the data to help manage healthcare organisations.

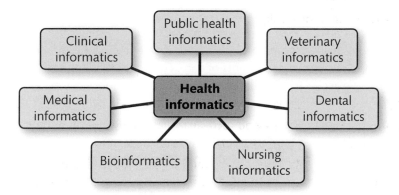

▶ **Figure 4.28:** Health informatics

The study of healthcare informatics is essential to the practice of medicine and the delivery of healthcare. All this data means that computers are an essential part of informatics. They will store information on clinical guidelines, electronic health records and communication systems. Healthcare informatics work with, and are used by, professionals and patients to ensure that the best healthcare possible is delivered.

Scientists working in healthcare informatics have to understand the principles behind the information and communication processes. They use this knowledge to develop new systems of storing and sharing information. They also evaluate the impact of any changes on the way healthcare organisations work as well as on the impact of the work. That is, has patient care improved due to the changes?

When designing informatics systems for any reason, you need to consider the following questions.

▶ What problem are we trying to solve? For example, how do we collate all clinical research?

▶ How will we know when we have succeeded? What are our success criteria?

▶ Have we used the simplest solution? For example, is technology the answer?

Informatics is particularly important in healthcare because so much clinical research is carried out in medicine. Research into diseases and drugs is carried out in thousands of laboratories across the world. Masses of data is generated every day. This can mean that there is too much information to deal with, which might mean that ground-breaking research cannot be used for years as it is not shared quickly enough.

It is important that all the information and data is collected, organised and shared effectively, as healthcare researchers cannot work in isolation.

The role of healthcare informatics is to develop a system where the organisational processes and structures support health professionals and allow for information to be pooled so that it benefits routine care in the most effective manner.

Healthcare informatics can be used in:

- the design of clinical decision support systems for medical professionals
- the design of patient information and decision aids
- online health services
- collation of data from clinical trials.

It can also be used in many other areas in a healthcare organisation.

Ⅱ PAUSE POINT   Research other uses of informatics.

      Extend   Explain examples of where information from large databases can be useful.

You can see from the healthcare example that there are several advantages to being able to store and retrieve large amounts of data. For example, when carrying out research you have access to a larger data set so you can be more certain in your conclusions.

Doctors have the experience of other medical professionals so if they come across a patient with an illness that they are not familiar with, they can understand how to treat the patient effectively. They will know what has or has not worked for other patients with similar symptoms.

You will have access to other relevant information and research to compare to your own. It may mean you do not have to carry out some research, as this will already have been carried out by other scientists.

Case study

Ebola

In 2014, there was a large outbreak of the Ebola virus in Africa and other parts of the world.

Scientists worked frantically to find a cure or ways to prevent more people being infected.

Check your knowledge

1 How could informatics have helped during the Ebola virus outbreak?

2 Who would have used informatics related to the Ebola outbreak?

3 Who would have benefited from the use of informatics at this time?

Bioinformatics

In general, informatics benefits those who use it as well as other parties such as patients.

However, there can be issues with gathering and sharing large amounts of information if it is personal information.

Bioinformatics is one area where there may be moral or ethical issues.

Bioinformatics is the application of computer technology to the management of biological information. Computers are used to gather, store, analyse and integrate biological and genetic information which can then be applied to gene-based drug discovery and development.

Research

Research the uses of bioinformatics. Focus on one beneficial use and present your findings to your group.

Analyse the advantages of bioinformatics with your group.

One example of a successful bioinformatics is the Human Genome Project. This determined the sequence of chemical base pairs which make up human DNA, and identified and mapped all of the genes of the human genome.

However, the following ethical issues come up from this.

▶ Privacy – who owns the information? Who is allowed to use the information? How easily is the information shared?

▶ Discrimination – can the information be used to prevent a person getting insurance because they have genetic markers for a disease? Can it prevent them from being allowed to adopt? Will people be treated differently because of their genetic profile?

▶ Genetic profiling – we can predict a person's future in terms of their health or their athletic ability.

▶ Development of drugs that target specific individuals – this can be a benefit as specific genes can be targeted to produce good health. Does it also mean that drugs or viruses may be targeted at sections of society with similar genetic profile in a detrimental fashion?

▶ Prediction of future genetic illness – again, this is about being able to get a job or insurance. It can also be about preventative measures as patients are forewarned about the possibility of getting ill, e.g. women who have healthy breasts removed because they have a high risk of breast cancer and want to minimise those risks.

Discussion

Work in pairs and research the issues that arise from bioinformatics, including:
- privacy
- discrimination
- genetic profiling
- targeted drugs
- prediction of illnesses.

Discuss whether the ethical issues are reasons not to have bioinformatics, or whether the advantages outweigh the disadvantages.

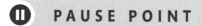 **PAUSE POINT**

In July 2002, the BBC ran a news article about a synthetic virus. Put 'synthetic virus' into a search engine and research this.

What ethical issue might this cause in relation to bioinformatics?

Hint   You might want to research bioweapons when considering this.

Resourcing informatics

You need to consider the following questions.

▶ What resources are available to you. This can be your computer capability as well as the skills of your Information Technology (IT) department. Your IT department may be able to build you an informatics system, or you may want to choose software that can be supported on your systems and can configure your systems appropriately.

▶ How flexible does your informatics solution need to be? Is your organisation changing? Are the needs of the organisation changing?

▶ Are there any regulations that control how you store and share information?

▶ What software will you use? Can the employees use the software? Does it improve their data storage and retrieval? Is it cost effective? Is it only used by the organisation or is it linked to the internet?

Assessment practice 4.4   D.P7 D.P8 D.M6 D.D4

You work for a large teaching and research hospital in the IT department.

You are investigating the informatics systems within the research laboratory at the hospital.

You have been asked to produce a report containing:

- a description of the type of information stored and used in the laboratory and explained how it is stored
- a description of how different departments useful information can be obtained from large data sets
- an analysis of the communication channels within the organisation and how data is stored
- an evaluation of the benefits and issues and challenges involved in making large volumes of data available to others inside and outside of the hospital. You can also consider the benefits and issues of receiving and making large volumes of data from outside the hospital.

Plan
- What is the task? What am I being asked to do?
- How confident do I feel in my own abilities to complete this task? Are there any areas I think I may struggle with?
- Do I have all the information I need? Do I need to do more research?

Do
- I know what it is I am doing and what I want to achieve.
- I can identify when I have gone wrong and adjust my thinking/approach to get myself back on course.

Review
- I can explain what the task was and how I approached the task.
- I can explain how I would approach the hard elements differently next time (i.e. what I would do differently).

Websites

www.rsc.org/learn-chemistry
A selection of resources to assist tutors and learners about chemistry topics.

www.hse.gov.uk
The website for the Health and Safety Executive.

www.cleapss.org.uk
The website for the organisation CLEAPSS, which gives advice on practical work in schools and colleges.

www.healthcareers.nhs.uk/explore-roles/health-informatics
This website gives more information about informatics in health care.

www.hse.gov.uk/coshh
This website gives further advice on COSHH.

www.sop-standard-operating-procedure.com
This website gives more detail about standard operation procedures.

THINK ▶FUTURE

Stuart Morley

Head of Research and Development at a large pharmaceutical company

I have been working in research and development (R&D) for 21 years now. I started as a laboratory technician and after working in several different companies I now run R&D at a large pharmaceutical company. I still spend a time in the lab, but not as much as I used to.

Every day can be different for me now. I now spend a lot of time coordinating the scientists who work in the lab. I have regular meetings with our marketing team who help guide us on the sort of products that clients are asking for. I then meet with my laboratory staff and we organise who is going to do what on each project.

It is important that all the research we do is coordinated so I make sure our IT department keep all our computer systems running smoothly. It is important that the scientists have access to current research. This can save them a lot of time and the company a lot of money as it ensures they know what drugs are already being used or tested and how effective they are.

Most mornings, the first thing I do is answer my emails. These can be from the scientists with queries about the projects or sometimes about personal issues like requests for holiday days. I also get emails from other departments and from the head of the company that may be about company policies or new regulations. It is my job to ensure that my department follows both company policies and any legal regulations related to our products.

Occasionally I do need to put on a lab coat and go into the lab. I have a lot of experience as a scientist and my team often need me to check results or go over data with them. Sometimes they get stuck on what to do next and often I can help them brainstorm new ideas or techniques.

My role now is mostly as a manager. I need good communication skills and excellent leadership skills. It is my responsibility to ensure my staff are happy and that they produce good results in the lab. I need to be able to motivate the staff at the same time as ensuring production is high.

Focusing your skills

Think about the role of a project manager.
- What types of people will you work with and how will you support them?
- What sort of tasks will you carry out each day?
- What scientific knowledge/skills will you need?

- What management skills will you need?
- What scientific knowledge/skills do you have now?
- What management skills do you have now?
- How can you acquire knowledge skills that you do not currently have?

Getting ready for assessment

Shau is working towards a BTEC National in Applied Science. He was given an assignment with the title 'Health and Safety in Scientific Organisations' for learning aim A. He had to write a report for the HSE on the health and safety policies in two science organisations. The report had to:

▶ include a description of the relevant health and safety legislation of each organisation
▶ include a description of the relevant hazards for each organisation
▶ include an evaluation of the measures taken to ensure high standards of health and safety within each organisation that comply with the legislation and comparing them.

Shau shares his experience below.

How I got started

First I collected all my notes on this topic and put them into a folder. I then carried out research on the two types of organisations. I put my research for each organisation into separate folders. I then made a table to compare health and safety in each organisation. I felt a table would help me see the information more easily and that I would be able to compare without having to keep flicking back and forth between my notes.

For each organisation I also produced a table linking health and safety measures in the organisation with health and safety legislation.

I managed to arrange to go into a chemical factory and talk to the health and safety officer there. He also gave me a tour of the site showing me how health and safety is managed. I had to wear a hard hat. I couldn't arrange a tour of the pharmaceutical research organisation, but I did email the health and safety officer with some questions and he sent me really detailed answers.

How I brought it all together

I wrote an introduction to the report describing each organisation and what the HSE inspector does.

For each organisation, I:
▶ described the potential hazards
▶ explained how the health and safety measures complied with legislation.

I then:
▶ produced a table comparing the health and safety measures in each organisation
▶ explained why there were differences in health and safety measures
▶ concluded by evaluating the measures in each organisation.

What I learned from the experience

I used examples from my visit to the chemical factory but I wish I had made clearer notes during my visit as I did not have all the information I needed. I also wish I had written down the questions I wanted to ask before I went on the visit as I was nervous and did not remember to ask everything I needed to. Next time, I would write a checklist of things I wanted to find out before a visit.

I gave very good descriptions of health and safety measures but did not explain how they were linked to legislation very well. I just stated which legislation they were linked to and not why.

I found evaluating the measures very difficult. I was not very sure what to include for the evaluation and I will make sure I ask for more help on this before doing another assignment.

Think about it

▶ Have you written a plan with timings so you can complete your assignment by the agreed submission date?
▶ Do you have notes on health and safety legislation for each type of organisation that will help you when explaining why the health and safety measures are used?
▶ Have you made sure you understand all the command words so that you produce work that shows the correct depth of understanding?
▶ Is your information written in your own words and referenced clearly where you have used quotations or information from a book, journal or website?
▶ For this unit there is a lot of practical work. Have you had the opportunity to practise all the techniques you may need in the assessments?
▶ How can you ensure that you are going to be confident in carrying out and writing up all the different techniques you are going to have to use?

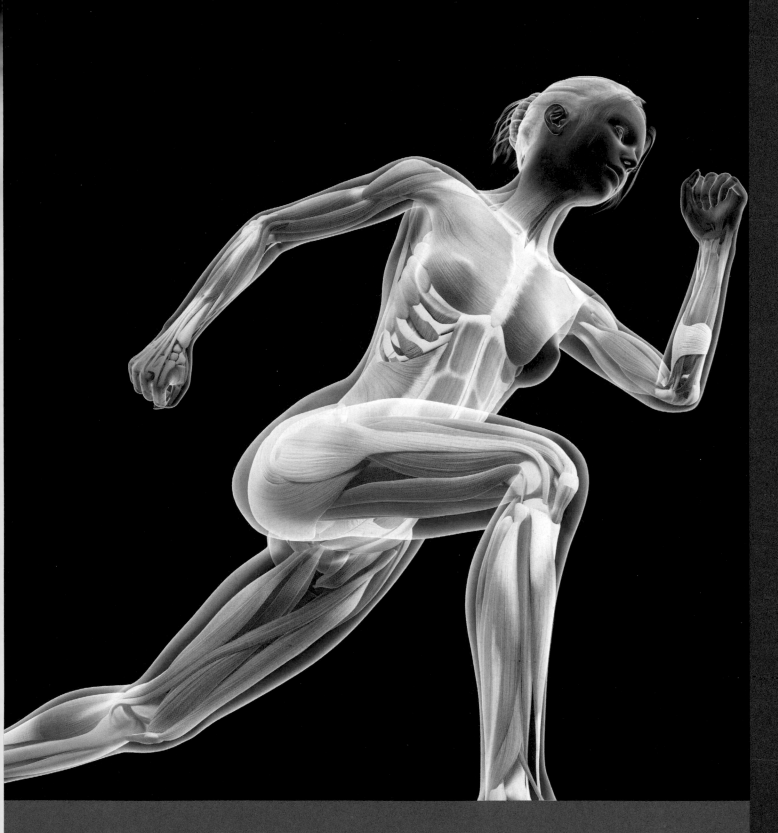

Physiology of Human Body Systems

8

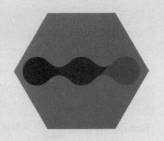

Getting started

Write down as many as you can of the names of the bones in your skeleton. Is your musculoskeletal system involved in helping any of the other body systems to work properly? Can you list the other body systems? Record what you know now. After studying this unit, see how well you can answer these questions.

A Understand the impact of disorders of the musculoskeletal system and their associated corrective treatments

Before you can understand how things might go wrong with the musculoskeletal system, you need to have a basic knowledge of its structure.

Structure of the musculoskeletal system

The skeleton

The skeleton forms a framework for your body. It supports your body and allows it to move.

The adult human skeleton consists of 206 bones organised into the **axial skeleton** (skull, backbone and rib cage) and the **appendicular skeleton** (limbs and limb girdles). You probably know some common names for bones, for example, the 'thigh bone' and the 'funny bone', but all parts of the skeleton have Latin or Greek names that you need to learn.

Figure 8.1 shows a human adult skeleton with all the names of bones that you need to know.

> **Key terms**
>
> **Axial skeleton** – this forms the longitudinal (lengthways) axis of the skeleton, which runs from your head to your feet. It consists of the cranium (top part of the skull) together with mandible and maxilla (upper and lower jaw bones); the vertebral column (backbone) with its different types of vertebrae (cervical, thorax, lumbar and, between them, the intervertebral discs); plus the rib cage and sternum (breast bone).
>
> **Appendicular skeleton** – this is the bones forming the appendages (limbs) and the limb girdles that join your limbs to the axial skeleton.

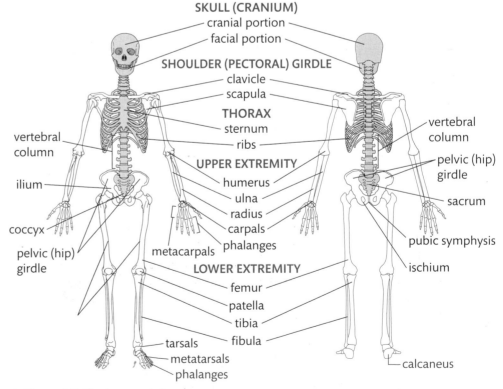

▶ **Figure 8.1:** The human skeleton

Discussion

The human skull consists of 22 bones. The joints in the cranium are fused to increase its strength. At birth a baby has membrane-filled spaces, called fontanels, between the cranial bones. These areas are soft but eventually become filled with bone.

Why do you think babies have cranial fontanels?

Bones

There are different types of bone. Each type is able to perform a certain function.

▶ **Long bones** form the limbs. These are cylinders of hard bone with soft spongy marrow inside. They are wider at each end than they are at the middle. This gives extra solidity at the joint where the bone articulates with (one moves in or on the other) another bone.

▶ **Short bones**, for example, in your wrist and ankle, have the same structure as long bones but are squat. This gives a greater variety of movement with no loss of strength.

▶ **Flat bones** are made up of a sandwich of hard bone with a spongy layer between. Some are protective – the cranium protects the brain or, in the case of the scapula, they give a large area for muscle attachment. The sternum is also a flat bone.

▶ **Irregular bones** have various forms – for example, the box-shaped vertebrae that form the backbone. Vertebrae are strong, contain marrow and protect the spinal cord. The facial bones are irregular and contain air-filled cavities, making them light. The hip bones are also irregular.

▶ **Sesamoid bones** are small bones in tendons at regions where there is a lot of pressure. Your knee caps (patellae) are sesamoid bones.

Figure 8.2 shows long, short, flat and irregular bones.

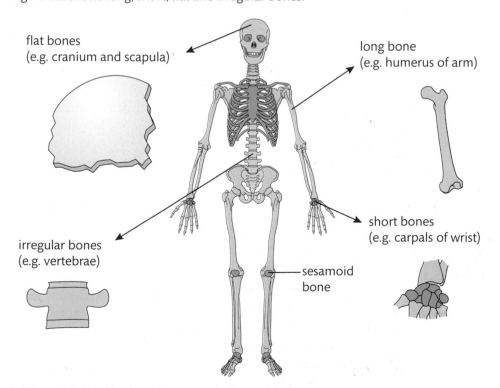

flat bones
(e.g. cranium and scapula)

long bone
(e.g. humerus of arm)

irregular bones
(e.g. vertebrae)

short bones
(e.g. carpals of wrist)

sesamoid bone

▶ **Figure 8.2:** Classification of bones on the basis of their shapes

Bone composition

Bone is a type of tissue known as connective tissue. Other types of connective tissue are:

▶ blood

▶ cartilage

▶ connective tissue proper – a primary tissue found in many parts of the body that forms the basic packaging of the body that holds organs in place.

All these tissues came from the same embryonic tissue and all have three parts:

▶ matrix – unstructured material that fills spaces between cells

▶ fibres in the matrix (such as the proteins collagen and elastin, and **reticular fibres**)

▶ cells that make the matrix and the fibres in it.

Osteoblasts make the matrix for bone. **Haematopoietic stem cells** in red bone marrow make blood, and **chondroblasts** make the matrix for cartilage. Once osteoblasts and chondroblasts have secreted the matrix, they become less active and *maintain* the matrix.

Most of an embryo's skeleton is made of cartilage which becomes ossified (turned to bone).

Adults have:

▶ hyaline cartilage covering the epiphyses (ends of the bones), between the ribs and sternum, in the ear lobes, in the trachea, bronchi, epiglottis and larynx, and at the ends of their noses

▶ fibrocartilage, that has many bundles of collagen and can withstand compression. It covers the intervertebral discs (discs between the vertebrae), and joins the two parts of the hips together at the pubic symphysis

▶ elastic cartilage, containing many fibres of elastin as well as collagen, in the epiglottis, the ear lobe and in the larynx.

Cartilage is smooth and tough and does not contain any blood vessels. Its cells receive nutrients via diffusion through the covering layer and, at joints, are lubricated by the synovial fluid that is made at joints.

A child's skeleton is made of bone and flexible cartilage, which gradually ossifies as the child grows. Over time the solid bones develop hollow centres. This reduces the weight of the bone but only slightly reduces the strength. The marrow inside the hollow centres is where blood cells are made. As a child grows, the bones of the back, arms and legs get longer. These bones have a growth plate at each end that is made of hyaline cartilage. Cells in the growth plate multiply and move down the bone, producing a calcified matrix. These cells then die, leaving spaces. Osteoblasts produce bone to fill the spaces and replace the cartilage matrix. Some osteoblast become trapped in the matrix and become mature and inactive cells called osteocytes.

Key terms

Reticular fibres – fibres made of collagen and coated with glycoprotein. They form a network around fat cells, nerve cells, muscle cells and in the walls of blood vessels.

Osteoblasts – cells that make bone.

Haematopoietic stem cells – stem cells that divide and give rise to blood cells.

Chondroblasts – cells in cartilage that are actively dividing by mitosis. They give rise to chondrocytes – mature cells in cartilage.

Link

See *Unit 1: Principles and Applications of Science I* for more on tissue cells.

Bone consists of an organic matrix made mostly of collagen fibres and some ground substance, both secreted by osteoblasts. The ground substance contains extracellular fluid, chondroitin sulfate, proteoglycans and hyaluronic acid. The collagen fibres are lined up along the lines of tension (pulling force) that bones sustain. This gives bones a lot of tensile strength.

Osteoblasts deposit bone, but phagocytic cells, formed in the bone marrow and called osteoclasts, break it down and absorb it. Normally the deposition and breaking-down processes are in balance and under the control of certain hormones. However, this bone remodelling responds to external conditions, such as how much stress bones are subjected to. This is how archaeologists can tell from skeletons whether someone was, for example, an archer, or can work out what bone injuries that person suffered when they were alive.

Ⅱ   PAUSE POINT   How does the structure and distribution of cartilage differ from that of bone? What are the roles of osteoblasts and osteoclasts?

> Hint   Think about what cartilage is made of and what bone is made of, i.e. their structures. Think about where in the body the cartilage is.

> Extend   Why do you think a woman's pelvis is wider than that of a man?

The periosteum

Surrounding the diaphysis (tubular shaft) of each bone is a white double-layered membrane called the periosteum. The outer layer consists of dense irregular connective tissue. The inner layer consists of osteoblasts and osteoclasts. Tufts of collagen attach this layer to the underlying bone. In the periosteum there are many blood vessels, lymphatic vessels and nerve fibres. All of these enter the bone tissue through special canals.

Compact bone and spongy bone

Just beneath the periosteum is **compact bone**. It appears hard and dense, but under a microscope you can see that it is full of Haversian canals which act as passageways for nerves, blood vessels and lymphatic vessels. Bone is living tissue and needs:

▸ nutrients and oxygen
▸ waste removal and sensitivity.

This bone is made of many structural units called osteons (see Figure 8.3). Each osteon is an elongated cylinder. Osteons run lengthways, acting like small weight-bearing pillars, inside the bone. Each osteon is a group of hollow tubes of bone matrix, one inside the next. Within each tube of matrix there are collagen fibres. Inside the bone matrix are osteocytes with cytoplasmic projections that, when these cells were active osteoblasts, connected to other bone-forming cells.

Spongy bone has more spaces between structures called trabeculae and is less dense than compact bone. The trabeculae do not have osteons but contain osteocytes and small canals (canaliculi). Nutrients diffuse from the marrow, through these tiny canals to the osteocytes. These osteocytes are still living, although they are not secreting bone matrix.

In flat bones, a layer of spongy bone is sandwiched between two outer layers of compact bone.

In long bones, the spongy bone is at the ends and in the shaft adjacent to the inner marrow cavity.

> **Key terms**
>
> **Compact bone** – one of the three layers of bone. Nearly 80% of a bone is this layer.
>
> **Spongy bone** – one of the layers of bone. Only 20% of the mass in bone is spongy bone but the surface area is ten times that of compact bone tissues.

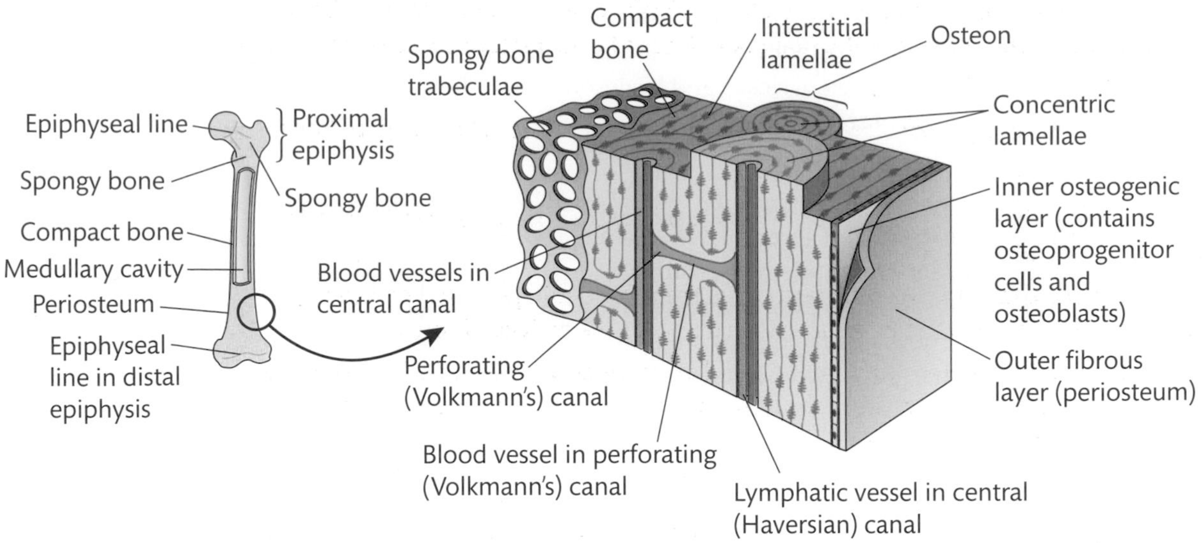

Epiphyseal line
Proximal epiphysis
Spongy bone
Spongy bone
Compact bone
Medullary cavity
Periosteum
Epiphyseal line in distal epiphysis
Spongy bone trabeculae
Compact bone
Interstitial lamellae
Osteon
Concentric lamellae
Inner osteogenic layer (contains osteoprogenitor cells and osteoblasts)
Outer fibrous layer (periosteum)
Blood vessels in central canal
Perforating (Volkmann's) canal
Blood vessel in perforating (Volkmann's) canal
Lymphatic vessel in central (Haversian) canal

Figure 8.3: Section through part of a long bone showing the structure of the structural units of bone, the osteons

Reflect

Children and teenagers may suffer bone fractures and need an X-ray. Why do you think the growth plates at the ends of their long bones do not show up on X-rays?

Research

Research and find out if there are any other medical imaging technologies that show these growth plates.

Bone marrow

This consists of haematopoietic (blood forming) tissue or red marrow. It is in the central cavity of long bones and in small cavities within the spongy bone of flat bones. Stem cells in the red marrow divide and give rise to all the types of blood cells (see *Unit 1: Principles and Applications of Science I*).

Yellow marrow is mainly for fat storage. In adults, blood cells are made mainly in the head of the femur and the humerus, flat bones and irregular bones. If a person becomes very anaemic, yellow marrow can change to red marrow.

Mineral use

When the protein matrix of cartilage becomes ossified (turns to bone), crystals of the mineral calcium phosphate, along with magnesium, sodium, potassium and carbonate ions, are deposited in the organic collagen matrix. The mineral content of bone gives it a lot of *compressive* strength – it is able to withstand pushing forces. The collagen fibres in the organic matrix give *tensile* strength – it is able to resist pulling forces.

The calcium in bones can also be a store and used to top up the levels of calcium ions in blood (where it is needed for clotting) and muscles (where it is needed for contraction).

Bone is continually being remodelled – broken down by osteoclasts and deposited by osteoblasts. Your skeleton is completely remodelled about every ten years. Some bone parts, such as the knee end of the thigh bone, are remodelled much more frequently, such as every six months. Your overall bone mass stays the same, because bone is reabsorbed and remade at the same rate. After a bone fracture, more bone deposition occurs.

Bone remodelling is controlled by hormones and by forces acting on your bone.

Hormones

Hormones that control bone remodelling are secreted from the parathyroid and thyroid glands.

▶ Parathyroid hormone is secreted from the parathyroid glands when levels of calcium ions in the blood fall and it stimulates the activity of osteoclasts.

▶ Calcitonin is secreted from the thyroid gland when your blood calcium ion level increases. It causes excess calcium salts to be removed from your blood and deposited in your bone.

Forces

Your bones respond to compression and tension forces, for example, muscles pulling on them.

▶ The trabeculae in spongy bones form buttresses along compression lines.

▶ Long bones are thick in the middle where bending stress is greatest.

▶ The compact bone becomes thicker and stronger and bone density increases when people do weight-bearing exercise, such as walking and running.

Ⅱ   **PAUSE POINT**   Distinguish between red marrow and yellow marrow.

What part of a bone's structure gives it compressive strength and what part gives it tensile strength? What are the functions of the calcium ions stored in bone?

    Hint   If you are asked to distinguish between two things, say in what ways they are *different* from each other.

    Extend   What are the possible problems with respect to the skeleton when astronauts working on the International Space Station are exposed to low gravity for long periods of time?

Worked Example

If you place a small chicken bone in a beaker of 1M hydrochloric acid and leave it for a day, you can then use forceps to remove the bone, wash it in lots of water and then feel how bendy but strong it is. The mineral part has been removed by the acid, leaving just the mainly collagen matrix. If you weigh the bone before and after putting it in the acid, then you can calculate the percentage of bone that is organic matrix.

Mass of chicken bone before being placed in acid = 20.34 g

Mass of chicken bone after being placed in acid = 16.53 g

Mass of mineral in bone = (20.34 − 16.53) g = 3.81 g

Percentage of mineral in bone = 3.81/20.34 × 100 = 18.73%

Percentage of protein in bone = (100 − 18.73) = 81.27%

If you hold a small piece of chicken bone in a Bunsen flame and burn it, you can smell the burnt protein. What is left after burning is the mineral content.

How could you calculate the percentage mineral content of a bone, without using acid?

Joints

Joints (articulations) are sites where two or more bones meet. They hold the bones of the skeleton together while allowing movement.

Classification of joints

Joints can be classified according to their:

▶ structure – joints may be fibrous, cartilaginous or synovial (see Table 8.1)

▶ motion – they may be immoveable, slightly moveable or freely moveable. Limbs have mainly freely moveable joints whereas the axial skeleton contains mainly immoveable or slightly moveable joints.

Table 8.1 shows the characteristics and some examples of each type of joint.

▶ **Table 8.1:** Characteristics and examples of different joints

Type of joint	Characteristics	Examples
Fibrous	• Bones are joined by fibrous tissue • There is no joint cavity • Immoveable or slightly moveable	• Sutures – immoveable joints, e.g. those found between the bones of the skull • Syndesmoses – bones are connected by a ligament; slightly moveable, e.g. between distal (far) ends of tibia and fibula • Gomphoses – only one example – teeth embedded in their sockets; the fibrous connection is the periodontal membrane
Cartilaginous	• Articulating bones are joined by cartilage • There is no joint cavity • Immoveable or slightly moveable	• Synchondroses – bones joined by a plate of hyaline cartilage which may ossify with maturity, e.g. joint between first rib and sternum; joint between epiphyseal plate and shaft of long bone; immoveable • Symphyses – articulating surfaces of bones are covered with hyaline cartilage and sandwiched between this is fibrous cartilage which is resilient and compressible and acts as a shock absorber; these joints give strength and flexibility (slightly moveable), e.g. intervertebral discs; pubic symphysis
Synovial	• Articulating surfaces of bones are covered by articular cartilage and separated by a fluid-filled joint cavity • Freely moveable • Present in appendicular skeleton • Allow movement • Joints surrounded by a double-layered capsule – the outer tough flexible fibrous coat and the inner synovial membrane that secretes hyaluronic acid into the synovial fluid, making it viscous • Synovial fluid is a slippery lubricating layer, derived by filtration from the blood plasma, that warms during activity and becomes less viscous, reducing friction at the moving joint • Reinforced by ligaments either outside of or forming part of the joint capsule • Some, e.g. knee and jaw, contain cushioning fatty pads called articular discs that make the joint more stable. Muscle tendons that cross joints also aid their stability (see Figure 8.4)	Six subtypes according to the shape of the articulating surfaces. • Gliding – e.g. between carpals, between tarsals, and between scapula and clavicle • Hinge – elbow and knee • Pivot – between atlas and axis • Condyloid or Ellipsoidal – in wrist, between radius and carpals • Saddle – between wrist and thumb • Ball and socket – hip and shoulder joints • Socket

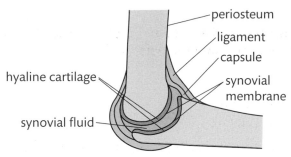

▶ **Figure 8.4:** The structure of a synovial joint

Ligaments and tendons

Table 8.2 shows the roles and composition of ligaments and tendons.

▶ **Table 8.2:** Role and composition of ligaments and tendons

Structure	Role	Composition
Ligaments	Bind bones to bones at synovial joints	Dense regular connective tissue containing bundles of collagen giving large tensile strength, and elastin fibres giving flexibility
Tendons	Join skeletal muscles to bones	Dense regular connective tissue with collagen bundles meaning they are also strong but contain much less elastin than ligaments do – an example is the Achilles tendon (see Figure 8.5)

Major muscle groups

Skeletal muscle tissue is packaged into muscles that attach to and cover your skeleton (see Figure 8.5). They can contract and relax, enabling bones to move. They also help to stabilise your joints. Your brain consciously controls their actions, so these muscles are also called voluntary muscles. This tissue appears striped when viewed under a microscope, and is described as striated.

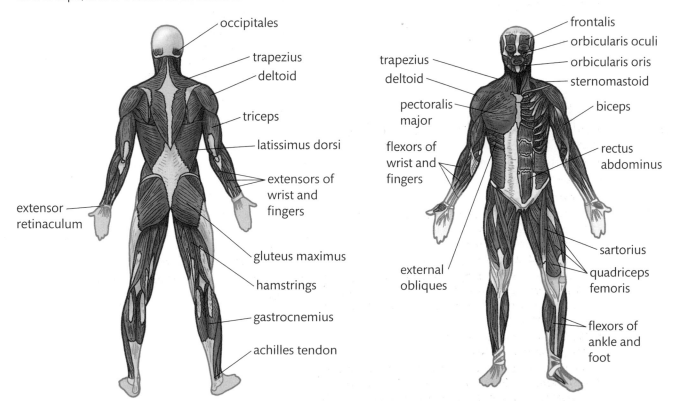

▶ **Figure 8.5:** Anterior and posterior views showing the main muscle groups of the human musculoskeletal system

Smooth muscle and cardiac muscle are not part of the musculoskeletal system as neither type of muscle is attached to bones. Neither type is under voluntary control.

▸ Smooth muscle is found in the walls of digestive, urinary, reproductive and respiratory tracts. Its contractions are slow and sustained and involved in peristaltic (contracting and relaxing) movements of substances through some of these tracts. In the respiratory tract, it can change the diameter of the airways.

▸ Cardiac muscle makes up the heart walls. It resembles skeletal muscle as it is striated.

Muscle fibres

Each skeletal muscle is an organ and consists of many muscle fibres (muscle cells), connective tissue, blood vessels and nerve fibres (see Figure 8.6). All the connective tissue sheaths are attached to each other and to the tendon that joins each muscle to a bone.

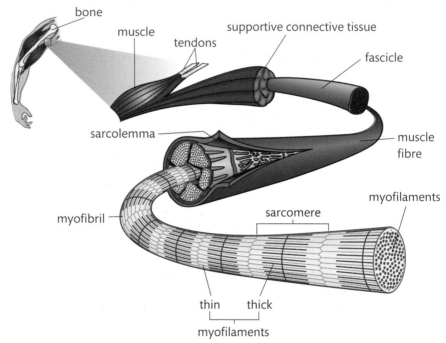

▸ **Figure 8.6:** Gross structure of a muscle. Muscles are made from bundles of fibres, each of which contains many myofibrils made from thick and thin myofilaments.

Muscle cells are called sarcomeres (from the Greek *sarx*, flesh, and *meros*, part). They contain organelles made of the proteins actin and myosin. Molecules of actin join together to make thin filaments.

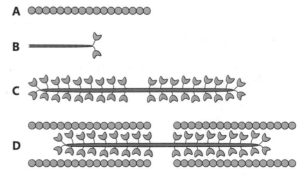

▸ **Figure 8.7:** The arrangement of actin and myosin filaments in a muscle fibre. A: Actin monomers join together to make thin filaments. B: two myosin molecules with their tails intertwined. C: Myosin molecules join together to make a thick filament with heads and flexible neck regions projecting out from the backbone region. D: The overlapping arrangement of thick myosin and thin actin filaments in a muscle fibre.

Myosin molecules consist of head and tail regions. Their tails intertwine to make thick filaments, with the heads sticking out. Figure 8.7 shows how actin and myosin filaments are arranged in a muscle fibre.

There is a Z line at each end of the sarcomere to which the actin filaments are attached. Muscle fibres are made of thousands of sarcomeres in lines. In 1 mm of muscle fibre there are about 400 sarcomeres. Figure 8.8 shows how the sarcomeres shorten during muscle contraction.

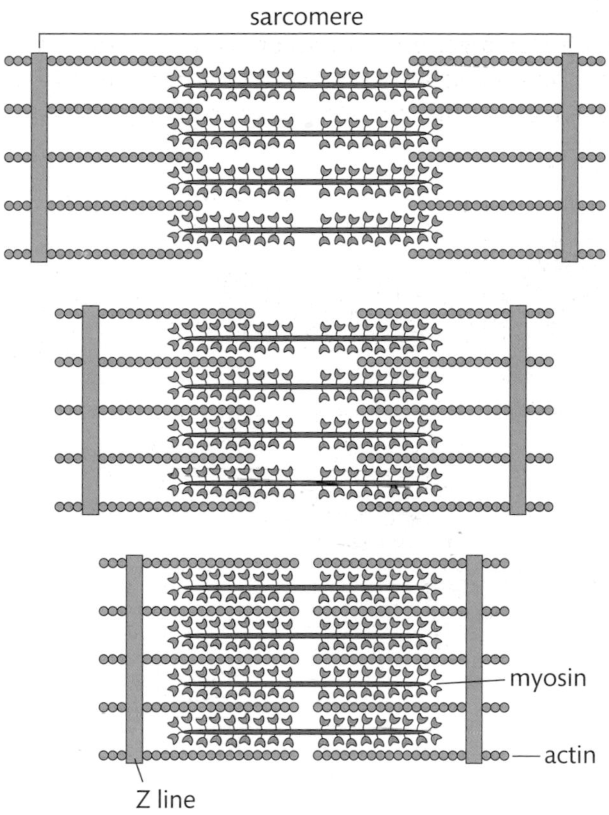

▶ **Figure 8.8:** How the sarcomeres shorten during muscle contraction. The thin actin filaments slide over the myosin filaments. The Z lines anchor the actin filaments.

Worked Example

Pathologists may need to examine muscle cells to see if there are any abnormalities that indicate diseases. They need to be able to measure the sizes of structures as seen under a microscope. Many biological structures are small and are measured in micrometres (μm). There are 1000 μm in 1 mm.

Muscle fibres are made of thousands of sarcomeres in series. In 1 mm of muscle fibre, there are about 400 sarcomeres. What is the length of one sarcomere?

400 sarcomeres measure 1 mm

There are 1000 μm in 1 mm

So 400 sarcomeres measure 1000 μm

1 sarcomere measures 1000/400 μm = 2.5 μm.

Adult male human thigh muscles are about 1 metre long. How many sarcomeres would be present in one linear row of sarcomeres in this muscle?

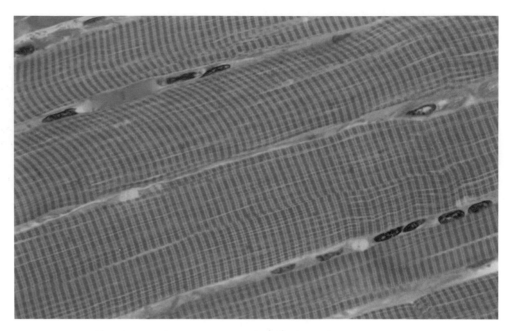

▶ Light micrograph of skeletal muscle fibres. The light (I) bands consist of areas where there are only actin filaments. The dark (A) bands are where actin and myosin filaments overlap.

Ⅱ PAUSE POINT

1. Use the light micrograph of skeletal muscle fibres to draw annotated diagrams to describe the microscopic structure of skeletal muscle and explain how the actin and myosin are arranged within the sarcomeres.

2. Arrange the following in order of decreasing size:

    actin filament   biceps muscle   myosin filament   sarcomere.

Hint Think about whether each structure is a cell, an organelle or an organ.

Extend How do you think you could measure the length of a sarcomere in the light micrograph?

Functions of the musculoskeletal system

The musculoskeletal system consists of many organs all working together so that the system functions effectively.

Table 8.3 gives an overview of the functions of various components of the musculoskeletal system.

▶ **Table 8.3:** Functions of the musculoskeletal system

Structure	Functions
Skeleton	<ul><li>Support – humans are land-dwelling animals and the skeleton gives support to resist forces of compression and tension, shearing and gravity. The skeleton prevents your internal organs from being squashed by the force of gravity.</li><li>Protection – the rib cage protects your internal organs such as heart, lungs and liver; the cranium protects your brain and the vertebral column protects your spinal cord.</li><li>Producing blood cells – haematopoietic stem cells inside red marrow of some bones divide and differentiate into the different types of blood cells – see *Unit 1: B2 Cell specialisation*.</li><li>Storing minerals – calcium ions in bone can be used to raise the levels of calcium ions in blood and muscle.</li><li>Maintaining mineral homeostasis – the release or absorption of calcium ions helps maintain the mineral content of blood, muscle and other body fluids in balance.</li><li>Attachment for skeletal muscle – the skeleton provides areas for attachment of the tendons of skeletal muscles.</li><li>Movement – the skeleton provides areas for attachment of the tendons of skeletal muscles. Bones articulate at joints, allowing movement.</li></ul>

▶ **Table 8.3** *continued*

Ligaments	• Ligaments strengthen joints and allow flexibility of their movement.
Skeletal muscles	• Movement – skeletal muscles work in antagonistic pairs (one contracts while the other relaxes). Skeletal muscles are responsible for your being able to move (locomotion), breathing and ability to manipulate objects. • Maintenance of posture – some muscle fibres are always contracted and produce muscle tone. • Stabilise joints. • Generate heat – when muscle cells contract, during movement or shivering, their rate of respiration increases, producing ATP for contraction and releasing some energy as heat which helps to maintain the body temperature.
Tendons	• Join muscle to bone. Because tendons are fairly inelastic, when a muscle contracts, the force exerted by the contracting muscle pulls on the tendon and this contraction moves bones. This is how you can flex and extend limbs.

Slow and fast twitch muscle fibres

Humans have a mixture of two general types of skeletal muscle fibres, slow twitch (type I) and fast twitch (type II). Table 8.4 shows their characteristics.

▶ **Table 8.4:** Comparison of slow and fast twitch skeletal muscle fibres

Characteristic	Slow twitch fibres	Fast twitch fibres
Number of mitochondria	Many	Few/none
Type of respiration	Aerobic	Anaerobic/glycolysis
Colour	Dark due to many electron transport proteins that contain iron, and also to a lot of myoglobin (oxygen store)	Pale due to lack of electron transport proteins that contain iron, and lack of myoglobin (oxygen store)
Length of contractions	Long duration	Short duration
Fatigue	Slow to fatigue	Fatigue quickly
Used for	Endurance activities	Activities needing short burst of power

Discussion

ATP does many things in the body. One of them is that it joins to receptors in bone tissue and stimulates osteoblasts while suppressing osteoclasts. What will be the effect on bone density of exercising, which increases the rate of respiration and therefore the rate of ATP production?

Skeletal muscle contraction

Muscle contraction can be reviewed on a step-by-step basis (see Figure 8.9).

1 Myosin heads can quite readily bind strongly to actin filaments. In the resting state, the site where the myosin head can bind to the actin is obscured by another protein molecule called *troponin*.

2 In the resting state, ATP is bound to another binding site on the myosin head. ATP is hydrolysed, by the myosin head acting as an enzyme, to ADP and P_i. The myosin head is now primed, ready to attach to an actin binding site.

3 When a nerve impulse arrives at a neuromuscular junction it sets up an action potential in the sarcolemma (muscle membrane).

4 This causes the release of calcium ions from the sarcoplasmic reticulum (modified endoplasmic reticulum) in muscle fibres.

5 Calcium ions bind to troponin, causing it to move and expose the myosin binding site.

6 Myosin heads bind to actin binding sites.

7 ADP and P$_i$ are released. This releases energy and allows the flexible neck region to rotate. This pulls the actin filament towards the centre of the sarcomere.

8 A new ATP enters the binding site on the myosin head.

9 The myosin head now detaches from the actin filament.

10 The ATP is hydrolysed. The chemical energy released when it is hydrolysed is converted to potential energy and stored by straightening the neck region of the myosin.

11 The myosin head can now attach further along the actin filament and the cycle is repeated. The effect is that the actin threads are slid towards the centre of the sarcomere.

12 The sarcomeres shorten because there is more overlap between actin and myosin filament but the actin and myosin filaments themselves do not alter their lengths.

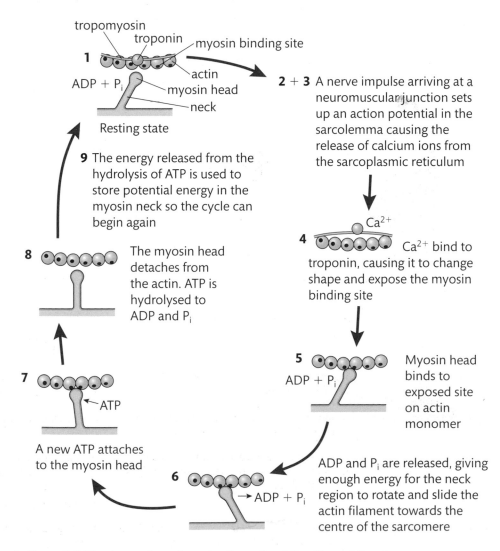

▶ **Figure 8.9:** The stages of muscle contraction – the sliding filament theory

PAUSE POINT

Make a 3D model to show the sliding filament theory of muscle contraction.

Hint You may find the following materials useful: pipe cleaners, modelling clay, drinking straws.

Extend Which type of muscle fibres, slow or fast twitch, do you think are involved in each of the following activities?

a tennis serve   **b** marathon run   **c** golf swing   **d** penalty kick in football
e swimathon   **f** Tour de France cycle race   **g** weightlifting
h 100 m sprint   **i** 800 m run   **j** 1500 m run.

Types of movement

There is limited or no movement at *fibrous* or *cartilaginous* joints.

At synovial joints, the range of movements for each type is limited by the:

▶ shape of the articulating bone surfaces – the way in which they fit together at the joint

▶ tension of the ligaments – for example, the major ligaments at the knee joint are tense when the knee is straightened and slack when the knee is bent

▶ muscle arrangement and tension – for example, when the thigh is raised with the leg extended, the tension exerted by the hamstring muscles restricts the movement at the hip joint.

The simplest type of movement at joints is *gliding* movement, where one surface moves back and forth or side to side over another surface, as happens between the articulating surfaces of vertebrae.

Angular movements

These increase or decrease the angle between the articulating bones. They include:

▶ flexion/extension
▶ abduction/adduction.

Table 8.5 shows the characteristics and examples of angular movements.

▶ **Table 8.5:** Characteristics and examples of angular movements

Type of angular movement	Characteristics	Examples
Flexion	Decrease in angle between surfaces of articulating bones	• Bending the knee or elbow • Bending the head forward (the joint is between occipital bone at the base of the skull and the atlas)
Extension	Increase in angle between the surfaces of articulating bones	• Straightening leg or arm after flexion • Bringing the head back up after flexion
Hyperextension	Continuation of extension beyond the normal anatomical position	• Bending the head backward
Abduction	Movement of a bone away from the midline of the body or away from adjacent structures, such as the middle finger	• Moving the arm upward and away from the body, so that it is held straight out sideways at right angles to the chest • Spreading the fingers out
Adduction	Movement of a part towards the midline of the body or towards, e.g. the middle finger	• Bringing the arm back down to your side after abduction • Bringing the fingers back together

Rotation

This is the movement of a bone around its longitudinal (lengthways) axis. When you shake your head from side to side you are rotating your atlas (first cervical vertebra) around the peg of your axis (second cervical vertebra).

You can also rotate your arm so the humerus turns in towards your body or outwards away from your body.

Circumduction

This is the type of movement where the **distal** end of a bone moves in a circle while the **proximal** end stays stable. The bone describes a cone shape in the air. An example is moving your outstretched arm in a circle (360°).

Special movements

These occur only at specific joints and include:

- inversion (internal) – such as movement of the soles of the feet so they both face each other
- eversion (external) – such as movement of the feet so the soles face away from each other
- dorsiflexion – bending the foot towards the upper part of the body
- plantar flexion – bending the foot downwards towards the sole
- protraction – moving the jaw/chin forwards
- retraction – moving the jaw backwards
- supination – moving the forearm (while it is by your side) so that the palm faces forwards
- pronation – moving the forearm (while it is by your side) so the palm faces backwards
- elevation – moving a part of the body, e.g. mandible (lower jaw), upward
- depression – moving a body part, e.g. mandible, downward.

PAUSE POINT

In pairs or small groups, carry out each type of movement described in Table 8.5, and rotation and circumduction. Also carry out each type of special movement listed above.

In larger groups, produce a poster showing diagrams to illustrate each of these types of movements.

Hint Do 'matchstick men' drawings as your partner makes a particular movement. Label each diagram properly with the type of movement it is illustrating.

Extend Choose two types of movement and suggest an activity where would you use each type of movement.

Disorders of the musculoskeletal system

There are different ways that the musculoskeletal system can malfunction (not work properly).

- Deficiencies of vitamins or minerals. Vitamin D is a hormone made in the skin that regulates deposition of calcium salts in bone. Lack of vitamin D and/or calcium can lead to rickets in children and osteomalacia (bone softening) in adults.

Key terms

Distal – situated away from the centre of the body or from the point of attachment.

Proximal – situated nearer to the centre of the body or to the point of attachment.

- Over- or undersecretions of the hormones that regulate bone homeostasis. If the bone building and bone reabsorption are out of balance, loss of bone density can lead to osteoporosis. This can be made worse by age, lack of oestrogen in females after the menopause, being thin (because adipose [fat storage] tissue is a source of oestrogen), smoking and lack of weight-bearing exercise when young.

- Degenerative – for example, arthritis – see Table 8.6.

- Infections, for example, *Staphylococcus aureus* – bacteria that can infect the bone marrow, site of a fracture, or cause a tooth abscess. Infections of bones are called osteomyelitis.

- Tumours – bone cancer.

- Trauma – for example, sprains, strains, fractures, dislocations, ruptured (herniated) discs, repetitive strain/stress injury (RSI), muscle trauma – see Table 8.7.

- Congenital (developmental) e.g. spina bifida, hip dysplasia, abnormal spine curvature – scoliosis, kyphosis, lordosis.

- Hypermobility (see below).

Arthritis

Arthritis describes any condition where joints are inflamed. It can apply to different diseases, such as rheumatoid arthritis, osteoarthritis and gouty arthritis. These are usually chronic. This means they have slow onset, get steadily worse and can be treated but not cured.

Table 8.6 shows the main causes, characteristics and treatments of the types of arthritis.

- **Table 8.6:** Causes, characteristics and treatments of arthritis

Type of arthritis	Cause	Characteristics	Treatment
Rheumatoid arthritis	Autoimmunity – the body's immune system attacks its own cells and tissue – in this case, cartilage and joint linings.	Inflammation of synovial membrane that may lead to thickening of the membrane, pain and swelling at joints. Articular cartilage may be destroyed and fibrous tissue joins the bone ends, making the joint deformed and immoveable. More common in women.	Anti-inflammatory drugs; steroids; rest; heat treatment; weight-loss to reduce stress on joints; physiotherapy and exercise to keep joints mobile; joint replacement.
Osteoarthritis	Degenerative – due to ageing, wear and tear and irritation of joints. Genetic factors are often involved.	Articular cartilage deteriorates and bony spurs grow at the bone ends. These spurs decrease the space in the joint cavity and restrict joint movement. Synovial membrane is *not* usually damaged.	Anti-inflammatory drugs; corticosteroid injections; pain killers; exercise to keep joints mobile; weight-loss to reduce stress on joints; surgery – removal of damaged tissue joint replacement; fusion of vertebrae.
Gouty arthritis (Gout)	Excess uric acid in blood forms crystals of sodium urate deposited at joints of extremities (e.g. big toe) where body temperature is slightly lower. Genetic link – certain alleles may lead to excess production of uric acid. Some types of food may aggravate the condition.	Sodium urate crystals irritate the joint cartilage, causing inflammation and severe pain. If untreated can lead to destruction of joint tissues and loss of mobility of affected joints; may also lead to blindness. More common in men.	Drugs to block production of uric acid. Avoid certain foods, e.g. purine-rich foods such as offal, that are known to aggravate the condition.

John is a health promotion specialist. One of his ideas is to encourage all schools to provide skipping ropes for children to use during break activities. The rationale behind this initiative is that children need to strengthen their bones by weight-bearing exercise so that when they attain maximum bone density as they near late adolescence, they will have reduced their risk of getting osteoporosis later in life.

He has also helped to introduce a health promotion initiative in another primary school nearby, where the headmistress has recently introduced a regime where all children have to do some running. This is about 30 minutes per day, with each child running at his/her own pace. After two terms the school has noticed that none of its children are obese and that their attention span and behaviour in lessons has improved. Running uses up calories and also increases the blood circulation and the amount of oxygen reaching the blood and the brain.

Check your knowledge

1   Discuss the pros and cons of skipping as a weight-bearing exercise in schools.

2   Discuss the advantages, both physical and mental, for the children who run every day in school.

3   Do you think dancing would be a good form of exercise to introduce into schools? Explain your answer.

Trauma

Trauma injuries of the musculoskeletal system refer to physical injuries caused by actions that disrupt the structure of tissues. They include damage to muscles, tendons, ligaments and bones. They are usually acute (sudden onset) and heal when the tissues repair. Table 8.7 shows the types, causes, characteristic and treatments for various trauma injuries.

▶ **Table 8.7:** Types, causes, characteristics and treatments for trauma injuries

Type of trauma	Cause	Characteristics	Treatment
Sprain	Traumatic injury to tendons, ligaments or muscles at a joint	Pain, swelling and discoloration of skin over the joint; loss of mobility	**RICE:** **Rest** the injured area. **Ice** – apply ice pack wrapped in a towel, 10–20 minutes at a time every 2–3 hours to reduce pain and swelling. After 2 days, if swelling has gone, apply heat to the painful area. **Compression** – wrap the joint using an elastic bandage to support the joint and help reduce swelling. **Elevation** – elevate the injured area on pillows and try to keep the area above the level of your heart (e.g. by lying down with the leg elevated) to reduce swelling.
Strain	Damage to muscles due to excess physical force or overexertion	Pain and loss of mobility	RICE – see above.

Table 8.7 *continued*

Type of trauma	Cause	Characteristics	Treatment
Fractures	Any break in a bone. Types:	Pain, swelling and immobility	Closed reduction – the bone is restored to normal position by manipulation without surgery. A plaster or sling keeps the bones in place and immobilised while it is healing.
• Partial	• Moderate shearing force	• break across bone is incomplete	Open reduction – surgery is used to expose the fracture and the broken bone ends are joined.
• Complete	• Severe shearing force	• bone broken in two or more places	Bone may take several months to heal.
• Green stick	• Force due to fall	• one side bone broken, the other bends; occurs in children as their skeleton less ossified and more flexible	1 Blood from broken blood vessels within the bone forms a clot (fracture haematoma). This happens 6–8 hours after the break.
• Closed	• Moderate force	• broken bone does not break through skin	2 A callus (new bone tissue) develops in and around the fractured area. This forms a bridge between the separated bone ends. The external callus forms from osteoblasts in the torn periosteum. The internal callus develops from osteoblasts in between the two marrow cavities.
• Open	• Moderate to severe force	• broken ends of bone protrude through skin	Forty-eight hours after the break, osteoblasts and osteoclasts divide by mitosis.
• Spiral	• Twisting force	• bone twisted apart	During the first week after the fracture, osteoblasts form new bone trabeculae in the marrow cavity near the line of the fracture. This is the internal callus.
• Transverse	• Direct or indirect force	• break at right angles to long axis of bone	During the next few days a collar of new bone cells forms around each new bone fragment. This is the external callus.
• Stress	• Repeated stress from running on hard surfaces	• partial fracture where bone cannot withstand repeated stress; involving fibula	Bone where the fracture has mended is remodelled and most surplus tissue is reabsorbed.
• Compression	• Compression forces	• fractures squeezed together	
• Pathologic	• Bone weakened due to osteomyelitis, osteomalacia, osteoporosis or tumour	• break due to weakening of bone; often compression fractures	
Dislocations	Displacement of bone from its normal place in a joint	Pain and immobility	Immobilise the joint and take patient to hospital where a doctor can manipulate the bone back to its normal position under anaesthetic.
RSI	Tissue damage due to repeated movements, e.g. playing a musical instrument, using a keyboard	Inflammation of tendons; nerve and joint pain	Anti-inflammatory drugs; steroids, rest the affected area.
Muscle problems	• Cramp	• Sudden painful spasm/ involuntary contraction; may be due to exertion, coldness, excess heat or arthritis	• Manipulation; application of heat; avoid overexertion.
	• Fatigue	• Reduction of pH due to excess lactic acid; reduces enzyme activity and ability of muscle to contract	• Avoid overexertion of muscle.
	• Bruising	• Damaged blood vessels and bleeding in the muscle	• Rest affected muscle.
	• Tearing	• Damage to muscle tissue	• Rest affected muscle.
	• Dystrophy	• Genetic; lack of specific proteins; progressive loss of strength	• Support and help for patient – e.g. dog to carry out certain tasks.
	• Atrophy	• Loss of muscle tissue due to lack of use	• Keep muscles active.

Arthroscopy is surgery used to diagnose and treat problems with joints such as knees, ankles, shoulders, elbows, wrists and hips.

Joint replacement

Erica is 67 years old and has suffered loss of mobility of her left hip joint. Despite regular exercise and losing some weight, she is finding walking increasingly painful and difficult. Her GP has organised for her to be X-rayed at her local hospital and they will then decide the best course of treatment. Hip replacement, which involves replacing the head of the femur, is fairly common these days, and patients are often up and about within 36 hours after surgery. Provided there are no complications, such as infection, she will be able to go home after a few days and resume normal activities, except that she will not be able to drive her car for 4–6 weeks.

Check your knowledge

1   Suggest why, despite being able to move soon after surgery, Erica should not drive for 4–6 weeks.

2   How do you think the risk of infection after such surgery can be reduced?

3   Why do you think it is good for patients to be up and about as soon as possible after surgery?

Hip dysplasia

This refers to abnormal development of the hip joint. The acetabular cavity may be too shallow for the head of the femur to fit properly, resulting in a dislocated joint. It may affect one or both hip joints and sufferers, if untreated, develop osteoarthritis. There are many possible causes, including genetic factors (some cases run in families) breech births (where the baby is born feet first rather than head first), and certain swaddling techniques used in some cultures. In the UK, all new-born babies undergo a hip check examination for dysplasia. If their hips 'click', they will be further tested with ultrasound scans. Early diagnosis gives better prognosis and babies can be placed in special harnesses to enable their hip joints to align properly.

Hypermobility

Hypermobility is the ability to move joints beyond the normal range of movement. It is a type of inherited connective tissue disorder but many with it do not suffer adverse effects. In fact, it is an advantage to gymnasts, athletes and dancers. It is more common in children, whose joints are supple, and in females. Those with it are often described as 'double-jointed' or 'loose jointed'. Hitchhiker's thumb is one example.

However, for a small percentage of the population, hypermobility may be associated with joint and ligament injuries, pain and fatigue. It is a feature of an inherited condition, Marfan syndrome, and can be life threatening. The joints are loose because they have looser and more elastic connective tissues of ligaments and tendons. The joint can move too much and dislocate or partially dislocate (subluxation). Such injuries lead to acute pain and can cause chronic pain. Muscles have to work hard to stabilise joints, and this can lead to muscle fatigue.

(see *Unit 11: Genetics and Genetic Engineering*)

Case study

The role of service dogs

Robert has Duchenne muscular dystrophy. This is a genetic inherited sex-linked condition (see *Unit 11: Genetics and Genetic Engineering*) that is more common among males. Robert's symptoms first appeared when he was three years old. He fell over more than is usual for a child of that age and often had difficulty in getting up again. His symptoms have got progressively worse as his muscles have wasted. He may eventually need a wheelchair. Robert has recently been given an assistance dog, a black Labrador called Poppy. Poppy has been trained as a service dog. She assists Robert by accompanying him when he is moving around at home and while outside in his wheelchair. She can pull the wheelchair and pick up objects for Robert. When he gets out of bed and has to get into his wheelchair, she positions herself so that he can lean on her and pull himself into the chair. Poppy is also a loyal friend and companion for Robert. Service dogs can also help other people with severe mobility problems.

Check your knowledge

1. What sort of mobility problems do you think service dogs can assist with?

2. How does Poppy help Robert with his mental health as well as his physical health?

Theory into practice

Justine Aguayo, care worker

I work in a care home for the elderly. Some patients are quite immobile and we are all trained in lifting, transferring and repositioning patients. We need to consider:

- the force or physical effort needed to lift and move a patient or to control the lifting equipment
- how often we repeat the action
- how to avoid awkward positions that put stress on our bodies, such as leaning over a bed while lifting.

We have small handling aids such as low-friction fabric sheets, ergonomic belts, and trapeze bars above the bed. Sometimes we use electro-mechanical lifting equipment.

Assessment practice 8.1

A.P1   A.P2   A.M1   A.D1

A senior paramedic working in a teaching hospital is responsible for helping to train paramedics in first aid. She needs to produce a leaflet or poster explaining and evaluating the treatments for trauma injury to the musculoskeletal system.

A physiotherapist working at a large GP practice needs to produce a poster to inform patients about disorders of muscles and joints and their corrective treatments.

(a) Produce a poster showing the basic structure of the human skeleton and the main skeletal muscles and indicate the functions of the musculoskeletal system.

(b) Produce a leaflet to describe and evaluate the treatments for trauma injury to the musculoskeletal system.

(c) Produce a leaflet describing the effect of chronic and degenerative disorders of muscle and joints and describe and evaluate the corrective treatments.

Plan
- I know what the task is.
- I know how confident I feel in my own abilities to complete this task.
- I know if there are any areas I think I may struggle with.

Do
- I know what it is I am doing and what I want to achieve.
- I can identify when I have gone wrong and can adjust my approach to get back on course.

Review
- I can explain what the task was and how I approached it.
- I can explain how I would approach the difficult parts differently next time.

Understand the impact of disorder on the physiology of the lymphatic system and the associated corrective treatment

Structure of the lymphatic system

The lymphatic system, shown in Figure 8.10, consists of the following.

▶ Vessels that transport lymph fluid.

▶ Other structures and organs that contain specialised lymphatic tissue.

▶ Lymphatic organs (see Figure 8.11) – lymph nodes, spleen and thymus, all contain lymphatic tissue enclosed by a capsule.

▶ Lymphatic nodules. These are not enclosed by a capsule. They are oval in shape and contain lymphatic tissue. In the central regions are large **lymphocytes** and around the periphery are smaller lymphocytes. These are found in special areas of the gastrointestinal (GI) tract, such as tonsils, Peyer's patches in the ileum wall, and in the appendix wall.

▶ Diffuse lymphatic tissue in the **mucous membranes**, walls of the GI tract, airways, urinary and reproductive tracts and, in small amounts, in all organs.

▶ Bone marrow – as it produces lymphocytes.

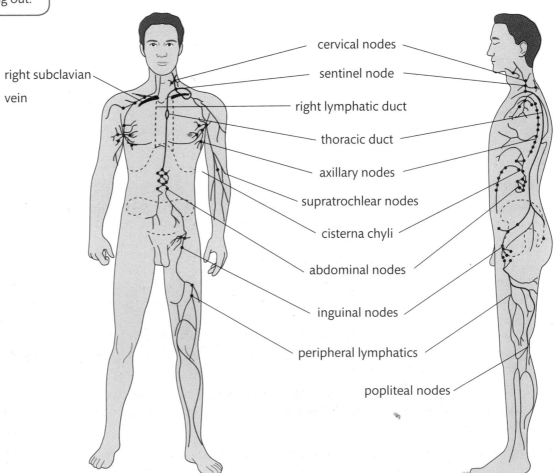

▶ **Figure 8.10:** The lymphatic system

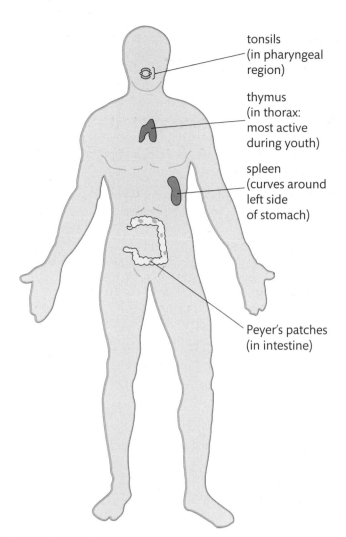

▶ **Figure 8.11:** The positions of the lymphatic organs

tonsils
(in pharyngeal
region)

thymus
(in thorax:
most active
during youth)

spleen
(curves around
left side
of stomach)

Peyer's patches
(in intestine)

Lymphatic organs

Lymph nodes

At various points along their path the vessels join with a knot of tissue called a lymph node (also called a lymph gland). These nodes are found around major arteries and you can feel them:

▶ in your neck, armpits (axillary nodes) and groin (inguinal nodes)

▶ at the back of your knee (popliteal nodes)

▶ above your elbow on the inner side of the upper arm (supratrochlear nodes) where arteries run close to the body surface.

There is also a ring of lymph nodes (tonsils and adenoids) circling the oesophagus and airways in the throat region. These are thought to be used for filtering infecting organisms out of food and inhaled air.

In the lymph nodes, bacteria, cancer cells and other foreign particles are filtered out and ingested by **macrophages**. As the fluid leaves the nodes, it picks up lymphocytes and some antibodies.

Dendritic cells in lymph nodes trap antigenic material circulating in the lymph and blood and present it to the resident lymphocytes. This causes production of the appropriate T and B cells, which can then mount an immune response against infecting organisms that have these **antigens** on their surfaces.

Key terms

Macrophage – type of white blood cell that ingests foreign material; found in liver, spleen and connective tissues.

Dendritic cells – antigen presenting cells; they process antigen material and present the antigens to T cells.

Antigens – molecules, often proteins, on the surface of all cells, for example, on the surface of pathogens, and viruses.

Spleen

The spleen is the largest lymphoid organ and is about the size of your fist. It is an oval-shaped organ in your abdomen. It is found under your rib cage on the left side, between your stomach and diaphragm. Around the spleen is a capsule of dense connective tissue and some smooth muscle fibres. Covering the capsule is a membrane, similar to the **peritoneum**. Inside the spleen is lymphatic tissue and red pulp – spaces filled with blood. There are blood vessels, the splenic artery and splenic vein, taking blood to and from the spleen. Here lymphocytes can divide by mitosis to mount an immune response.

Key term

Peritoneum – membrane that lines the internal body cavity and organs within it.

The spleen also has important blood cleansing functions. It:

▸ extracts old and defective blood cells and platelets from the blood and breaks them down
▸ removes foreign matter, bacteria, viruses and toxins from the blood
▸ stores some of the products (such as iron) of old broken-down red blood cells for later reuse, or releases them to the blood to be taken to the liver or bone marrow for reuse
▸ stores blood platelets.

In the fetus and in adults with bone marrow disease, new erythrocytes (red blood cells) are made in the spleen.

Case study

Spleen injuries

Nathan works in the A and E department of a large hospital. When patients are brought in with severe knocks or crushing to their lower left chest region or upper left abdominal region, he helps to check their spleen in case it has ruptured. If it has, it has to be surgically removed (this is called a splenectomy) to prevent the patient from bleeding to death. A ruptured spleen bleeds internally and this leads to physiological shock (a severe fall in blood pressure) and death. The spleen is the most frequently damaged organ during abdominal trauma, including sporting and traffic accidents.

Check your knowledge

1 Why do you think that splenectomy patients have an increased post-operative risk of suffering infections, such as sepsis?
2 Explain why internal bleeding leads to physiological shock.

Thymus gland

This bi-lobed organ (the two lobes are arranged in a shape rather like a bow tie) lies between your lungs, above the heart. In infants it is larger, compared to the rest of the body, than in adults. At puberty, your thymus gland is at its largest size, weighing about 40 g. Each lobe is covered by a connective tissue capsule and subdivided into smaller lobules. Each lobule has a central medulla and peripheral cortex. The cortex is tightly packed with lymphocyte cells. The medulla contains epithelial cells, some lymphocytes and thymic corpuscles that make some chemical messengers.

T cell lymphocytes mature in the thymus gland where they are 'educated'. T cells that contain receptors for self-antigens (antigens on the surface of your own cells or tissues), or that do not contain any receptors, are destroyed. This lowers the risk of autoimmune disorders. The thymus gland also secretes various hormones that encourage the reproduction and maturation of T cells.

Link

See *Unit 12: Disease and Infection* for more details about the immune response.

Functions of the lymphatic system

Lymph vessels

Lymph vessels are found in all parts of the body except the central nervous system, bone, teeth and cartilage.

There is no pump associated with lymph vessels (there is no structure like a heart, which is a pump, generating a force to drive the movement of lymph fluid).

Formation and transport of lymphocytes and lymph

Lymphocytes divide and increase in numbers at lymph nodes and in the thymus gland.

The lymph capillaries are the smallest lymph vessels. They run alongside the body's arteries and veins. Their walls are very thin and permeable (they contain small holes). Some lymph vessel walls contain smooth muscle which contracts rhythmically to propel the fluid. The skeletal muscles surrounding the vessels also help propel the fluid, and semi-lunar valves prevent backflow (fluid travelling in the wrong direction). Skeletal muscle contractions compress lymph vessels and propel lymph towards the subclavian veins that are in the lower neck region.

Lymph capillaries join together to form larger vessels that are similar to veins but with thinner walls and more semi-lunar valves (see Figure 8.12).

Lymph vessels join up to form two main **ducts**, the thoracic duct and the right lymphatic duct (see Figure 8.13). The lymph fluid from these two ducts drains into the blood vessels in the neck region.

Key term
Duct – tube, canal or vessel that carries a body fluid, secretion or excretion.

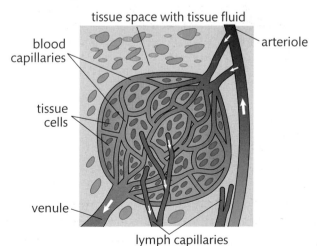

▶ **Figure 8.12:** Structural relationship between a capillary bed of the blood vascular system and lymphatic capillaries. Between the tissue cells and in tissue space is tissue fluid – plasma from the blood capillaries. There are also proteins and some excretory products from cells in the tissue fluid.

Removal of interstitial fluid from tissues

Interstitial fluid is the fluid found in between cells within tissues. It is also called tissue fluid. It has come from plasma that has been forced out of blood capillaries at the arterial end. It bathes cells and exchange of materials takes place. Oxygen and nutrients diffuse from tissue fluid into cells; carbon dioxide, other wastes and some proteins pass from cells into tissue fluid.

At tissues, *excess* tissue fluid that has leaked out of blood capillaries passes into lymph capillaries. As the small holes in the lymph capillary walls are larger than those in the walls of blood capillaries, the large protein molecules in tissue fluid can also pass into lymph capillaries. These protein molecules are then carried away from tissues. If these proteins were not carried away, they would exert osmotic effects and prevent the removal of tissue fluid, leading to swelling (oedema).

Maintenance of hydrostatic pressure

When lymph fluid drains into the blood vessels in the neck region, the blood volume is increased and this helps maintain the hydrostatic blood pressure (blood pressure generated by the fluid in the blood vessels).

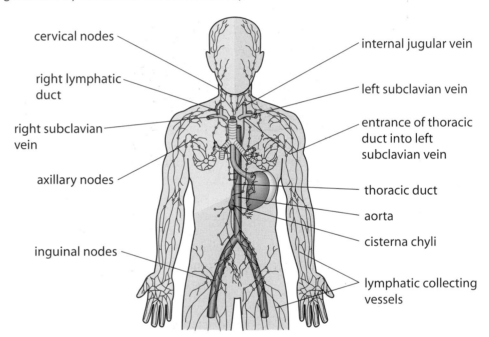

cervical nodes

right lymphatic duct

right subclavian vein

axillary nodes

inguinal nodes

internal jugular vein

left subclavian vein

entrance of thoracic duct into left subclavian vein

thoracic duct

aorta

cisterna chyli

lymphatic collecting vessels

▶ **Figure 8.13:** Distribution of lymphatic collecting vessels and lymph nodes. The area of the body shaded blue is drained by the right lymphatic duct. The rest of the body is drained by the thoracic duct.

Absorption of fats from the digestive system

The products of **digestion** are absorbed across the ileum wall. The lining of the ileum is highly folded and contains finger-like projections called villi (see Figure 8.14).

Within each villus is a lacteal. Products of fat digestion pass into each lacteal and this **chyle** then passes along lymph vessels. Eventually it will go into the blood system when the lymph ducts drain into blood vessels in the neck region.

Link

Go to Learning Aim C to learn more about the absorption of digested food.

Key terms

Digestion – break-down of large organic molecules to simpler soluble molecules that can be absorbed by a living organism/cell.

Chyle – milky body fluid, consisting of lymph and emulsified fats and fatty acids, formed in the small intestine during the digestion of fatty foods.

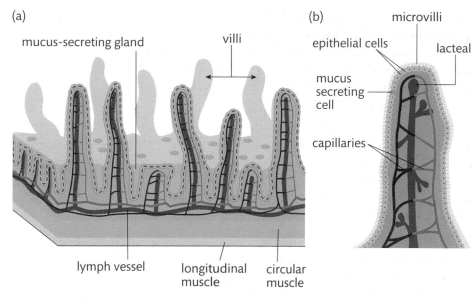

(a)

mucus-secreting gland   villi

(b)   microvilli

epithelial cells   lacteal

mucus secreting cell

capillaries

lymph vessel   longitudinal muscle   circular muscle

▶ **Figure 8.14:** (a) Part of the ileum wall (b) A villus

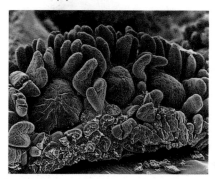

▶ False colour scanning electron micrograph (SEM) showing Peyer's patches (red) in the wall of the ileum. They defend against infection as the contents of the gut lumen are external to body tissue. Peyer's patches are named after the Swiss anatomist who described them in 1677.

Health matters and treatments related to the lymphatic system

Disruption or dysfunction of the lymphatic system can lead to diseases such as autoimmunity, severe combined immune deficiency, allergies, lymphadenitis, lymphedema and Hodgkin's lymphoma.

Lymphadenitis

Lymphadenitis is when the lymph glands (nodes) become swollen as a result of infection. The number of microbes that are collected from tissue fluid and circulate in the lymph vessels and then pass through the lymph nodes may be too large for the macrophages in the nodes to ingest. This causes the nodes to become infected, enlarged and tender.

Usually the swelling subsides with treatment (antibiotics to combat bacterial infection) but in some cases lymphadenectomy (removal of the infected lymph nodes, such as tonsils or adenoids) is carried out.

Glandular fever is an acute infection caused by the Epstein-Barr virus. Symptoms include:

▶ fever

▶ sore throat

▶ swollen lymph glands

▶ abnormal lymphocytes

▶ enlarged liver and spleen.

In young children it is fairly mild, but in adolescents and adults it may be severe and could lead to a ruptured spleen that would have to be surgically removed immediately.

Lymphedema

This is where lymph vessels are obstructed and tissue fluid cannot be sufficiently drained from tissues. It leads to oedema.

Causes

The causes of lymphedema could be the following.

▶ Milroy's disease or hereditary lymphedema is caused by chronic lymphatic obstruction.
▶ Women can suffer bouts of lymphedema during menstruation or pregnancy.
▶ Obesity or prolonged standing.
▶ Tumours that obstruct lymph vessels.
▶ Elephantiasis is caused by filarial worm infection where the parasitic worms block the lymph vessels.
▶ Secondary lymphedema occurs following surgery for removal of lymph vessels during mastectomy (breast removal).

Treatment

Lymphedema has no cure. However, lymph drainage from the extremities can be improved if the patient:

▶ sleeps with the foot of the bed elevated to 10–20 cm
▶ wears elastic stockings
▶ takes regular moderate exercise
▶ avoids spicy or salty foods
▶ lightly massages the limbs in the direction of the lymph flow
▶ takes diuretics (drugs that increase urination and therefore loss of body fluid).

In severe cases, lymph vessels may be surgically removed.

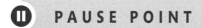 **PAUSE POINT** What are the three main functions of the lymphatic system? Explain how the lymph system is adapted to carry out its functions.

 Hint Describe how the structure of the lymph system enables it to carry out its functions.

Extend For each of the treatments listed above for disorders of the lymph systems, suggest how and why it works.

Hodgkin's lymphoma

This is a malignant (cancerous) disorder. Lymphocytes either divide abnormally or fail to die. They build up in lymph nodes which then enlarge due to the tumours. This usually occurs first in the lymph nodes of the neck region and there is no pain. Other lymph nodes may be affected (see Figure 8.15). The spleen gets bigger and the macrophages become abnormal, containing many lobed nuclei and prominent nucleoli. Other symptoms include:

▶ weight loss
▶ night sweats and fever
▶ anaemia
▶ an abnormal increase in the number of circulating white blood cells, by a factor of 100.

Diagnosis involves identifying a cell called Reed-Sternberg in lymphoma as seen under a microscope. Treatment involves chemotherapy and sometimes radiotherapy as well. The cure rate is quite high.

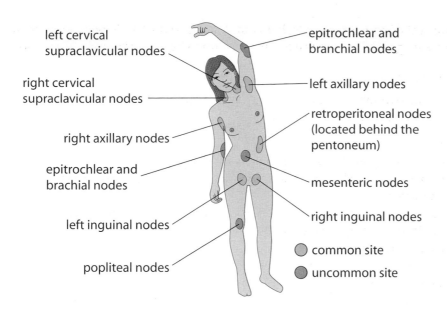

left cervical supraclavicular nodes

right cervical supraclavicular nodes

right axillary nodes

epitrochlear and brachial nodes

left inguinal nodes

popliteal nodes

epitrochlear and branchial nodes

left axillary nodes

retroperitoneal nodes (located behind the pentoneum)

mesenteric nodes

right inguinal nodes

○ common site

○ uncommon site

▶ **Figure 8.15:** Lymph node sites for Hodgkin's disease (Hodgkin's lymphoma)

This disease usually affects people between the ages of 15 and 35 years of age but it can affect older people. Close relatives of a patient with Hodgkin's lymphoma have a 1–3 times increased risk of also developing the disease, which suggests an underlying genetic mechanism.

Theory into practice

Theory into practice

Roland Kessler is a health promotion practitioner within the NHS. He wants to raise awareness among the general public about glandular fever.

Glandular fever is caused by a virus called the Epstein-Barr virus (EBV). It causes sore throat (sometimes difficulty with swallowing), fever, swollen glands in the neck, and fatigue. In about half the cases the spleen swells. In many cases, with rest, the patient recovers after a few weeks but in some cases chronic fatigue ensues. In children the symptoms are less severe. Most cases occur in teenagers and young adults. A blood test helps to confirm the diagnosis. In some cases glandular fever can lead to chronic fatigue, anaemia (reduced red blood cell count), neutropenia (reduced white cell count) and reduced platelet levels, headaches and joint pain. In rare instances the spleen ruptures and has to be removed. In less than 1% of cases, this virus can affect the nervous system, causing Bell's palsy, viral meningitis, encephalitis and Guillain-Barre syndrome. Patients who are immunocompromised – for example, those who are HIV positive or who are being treated with chemotherapy – may develop secondary infections such as pneumonia following glandular fever. In some developing countries, patients who have suffered EBV infection and malaria can develop Burkitt's lymphoma (a non-Hodgkin's lymphoma).

1 Why do you think it is important that the general public should be better informed and more aware about glandular fever?

2 Why do you think patients who are immunocompromised may develop secondary infections?

3 What is the difference between anaemia and neutropenia?

Research

Use the Internet to find out more about the lymphatic system.

Assessment practice 8.2

You have been asked to write an article for a family health magazine to explain what the lymph system is and to indicate some disorders of the lymphatic system and their treatments.

Prepare an illustrated account for this magazine that has a general readership. Many of its readers will not have much biological knowledge. This should be a maximum of two sides of A4 including diagrams.

Submit a list of references showing the sources of information you have used.

You need to explain clearly what the lymph system is and what it does, what disease can affect the lymph system and how these diseases are treated, with evaluation of at least one treatment.

Plan
- I know what the task is.
- I know how confident I feel in my own abilities to complete this task.
- I know if there are any areas I think I may struggle with.

Do
- I know what it is I am doing and what I want to achieve.
- I can identify when I have gone wrong and can adjust my approach to get back on course.

Review
- I can explain what the task was and how I approached it.
- I can explain how I would approach the difficult parts differently next time.

C

Explore the physiology of the digestive system and the use of corrective treatment for nutritional deficiency

Dieticians and nutritionists need to know about food and health. This helps them to design diets for individuals and groups such as:

▶ vegetarians and vegans

▶ pregnant women

▶ children

▶ athletes and other sportspeople

▶ the elderly

▶ cancer patients

▶ patients who have had parts of their intestine removed.

In this section, you will learn about the digestive system and some treatments for nutritional deficiencies.

Structure of the digestive system

Your digestive system (shown in Figure 8.16) consists of many organs where food is broken down to smaller soluble molecules and absorbed into the blood stream or lymph system. There are also accessory organs – the pancreas, the liver and the gall bladder. Food does not pass through these structures but they may secrete chemicals that aid digestion or deal with some of the products of digestion before they are assimilated into (used by) body tissues.

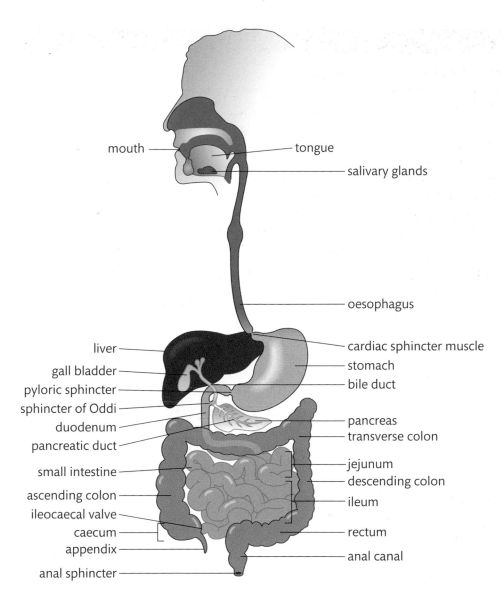

mouth

tongue

salivary glands

oesophagus

cardiac sphincter muscle

liver

stomach

gall bladder

bile duct

pyloric sphincter

sphincter of Oddi

duodenum

pancreas

pancreatic duct

transverse colon

small intestine

jejunum

ascending colon

descending colon

ileocaecal valve

ileum

caecum

appendix

rectum

anal canal

anal sphincter

▶ **Figure 8.16:** The organs and associated organs of the digestive system

Table 8.8 shows the main structural features of the parts of the digestive system and associated organs.

▶ **Table 8.8:** Structure and features of the digestive system

Structure	Main features
Buccal cavity (mouth)	Cavity bounded by lips. Tongue (taste buds) and soft palate enable taste. Tonsils filter bacteria from food. Jaws (mandibles and maxillae), facial muscles and teeth enable chewing (mechanical digestion). Mucous membrane lining mouth cavity secretes mucus which, together with saliva from salivary glands, keeps mouth moist. Saliva enables tastes to be identified by tongue and palate as food needs to be in solution; it also contains hydrolytic enzymes. Mouth, together with larynx, vocal cords and facial muscles, enables speech.
Pharynx	Muscle-lined cavity at back of throat. Tongue rolls chewed food into a bolus (ball) and pushes it against the roof of the mouth (hard and soft palates) to the pharynx where swallowing occurs as an automatic cranial reflex. Muscles of pharynx contract, tongue raises up against roof of mouth, epiglottis closes over glottis, thus closing off airway.
Oesophagus	Muscular tube. About 25 cm long, 2.5 cm diameter. Wall made of four layers: mucous membrane to secrete mucus enabling smooth passage of food; submucosa holding mucous membrane in place; a relatively thick layer of muscle consisting of circular and longitudinal smooth muscle fibres, and an outer protective covering. By **peristalsis** – alternate contraction and relaxation of the two muscle layers – the oesophagus *pushes* the food to the stomach.

Structure	Main features
Stomach	Muscular bag in upper part of abdomen, just below diaphragm. Wall consists of a thick layer of muscle, consisting of longitudinal, circular and oblique smooth muscle fibres, lined with epithelial cells. Food enters from oesophagus, at cardiac end of stomach. Epithelial cells produce gastric juice containing acid and enzymes. Muscular wall of stomach generates peristaltic movement to churn the food and mix it with enzymes to form **chyme**. Food remains in the stomach for about 1–3 hours.
Small intestine: • duodenum • jejunum • ileum	At the pyloric end of the stomach the acidity of the chyme causes the pyloric **sphincter muscle** to relax. Chyme, in small quantities, can pass into the duodenum. The duodenum is about 25 cm long, has a diameter 2.5 cm and is fixed to the dorsal abdominal wall. Consists of layers of smooth muscle cells lined with epithelium. Receives pancreatic juice with hydrolytic enzymes; receives bile from liver. These secretions enter the duodenum at the sphincter of Oddi. The jejunum is about 2.5 m long, has a diameter 3.8 cm, and extends from duodenum to ileum. The ileum is about 3.6 m long. Walls are thinner than those of jejunum and highly folded; epithelium contains villi (finger-like projections). These features greatly increase the surface area for absorption of the products of digestion. Jejunum and ileum are supported on a membrane called the **mesentery**.
• Pancreas	Soft pink gland supported by mesentery, within the loop of the duodenum. It produces a wide range of hydrolytic enzymes to aid digestion of all food types. Clusters of **acini** (secretory cells), surround ducts. Inside the acinar cells are large amounts of rough endoplasmic reticulum (*see Unit 1: B1 Cell structure and function*) and vesicles containing the newly made enzyme molecules. Epithelial cells lining the pancreatic ducts secrete hydrogencarbonate ions that make the pancreatic juice alkaline (pH 8). The pancreas also has an endocrine function. Scattered among the acini are islets of Langerhans. Beta cells in these islets secrete the hormone insulin, in response to increased blood glucose level. Insulin helps liver, muscle and many other cells take up more glucose, thereby lowering blood glucose level. In liver and muscle cells, the glucose is converted to glycogen. Glucagon is secreted by alpha cells in the islets, in response to lowered blood glucose level. It causes stored glycogen in the liver to be broken down to glucose and released into the blood.
Gall bladder	Thin-walled green muscular sac, about 10 cm long, snuggled into a depression (fossa) on the ventral surface of the liver. It stores bile made in the liver and releases the bile, via the bile duct, into the duodenum at the sphincter of Oddi, when food enters the duodenum from the stomach.
Liver	Large gland in the abdomen, ventral to (in front of) the stomach. Consists of hexagonal-shaped liver lobules (see Figure 8.17), inside which are hepatocytes (hepato = liver, cytes = cells). Oxygenated blood enters the liver from the *hepatic artery* and nutrient-rich blood from the ileum enters the liver via the *hepatic portal vein*. Deoxygenated blood leaves the liver in the *hepatic vein*. Hepatocytes make bile that enters canaliculi and passes to the gall bladder. It is stored in the gall bladder until needed. Bile contains: • salts that emulsify fats to increase their surface area for digestion • hydrogencarbonate ions to neutralise acidic chyme • products (bilirubin and biliverdin) of broken-down red blood cells and cholesterol. It does *not* contain enzymes. The liver also stores glycogen, helps regulate blood glucose level, makes plasma proteins, stores fat-soluble vitamins (for example, vitamin A) and metabolises (chemically alters by reactions inside cells) alcohol, drugs and other toxins. It breaks down excess amino acids to make urea for removal at the kidneys.
Large intestine • caecum • appendix • colon • rectum • anal canal	Caecum is the sac like first part of the large intestine. Branching from it is the appendix that contains much lymphoid tissue and bacteria that may help recolonise the **gut microbiota**. The colon consists of four parts: ascending colon, transverse colon, descending colon and sigmoid colon. Colon mucosa consists of columnar epithelial cells, no villi or folds and very few/no digestive enzyme-secreting cells; many goblet cells secrete mucus that protects wall from acids and gasses produced by bacteria that live there. The mucus lubricates passage of faeces to rectum and anal canal. The gut bacteria are essential for our wellbeing and make vitamins B and K as well as certain appetite-regulating hormones. Water is absorbed from the undigested food in the colon. Faeces, containing undigested fibre, bacteria, gut epithelial cells and excretory products such as bilirubin and biliverdin, pass into the rectum. More water is then absorbed and the faeces pass into the anal canal to be expelled.
Anus	Stretching of the rectum wall initiates the defaecation reflex. As faeces are forced into the anal canal, impulses reach the brain and we can make voluntary decisions as to whether or not to open the external anal sphincter.

Peristalsis – involuntary contraction and relaxation of smooth muscles of the intestine (and other canals in the body) creating wave-like movements that push forward the contents of the canal.

Chyme – semi-fluid mass of partly digested food formed in the stomach.

Sphincter muscle – circular muscle that surrounds an opening and acts as a valve.

Mesentery – double-layered extension of the peritoneum able to support organs within the abdominal cavity.

Acinus (plural acini) – cluster of cells resembling a berry, for example, raspberry.

Gut microbiota – all the microbes that live in the human gut.

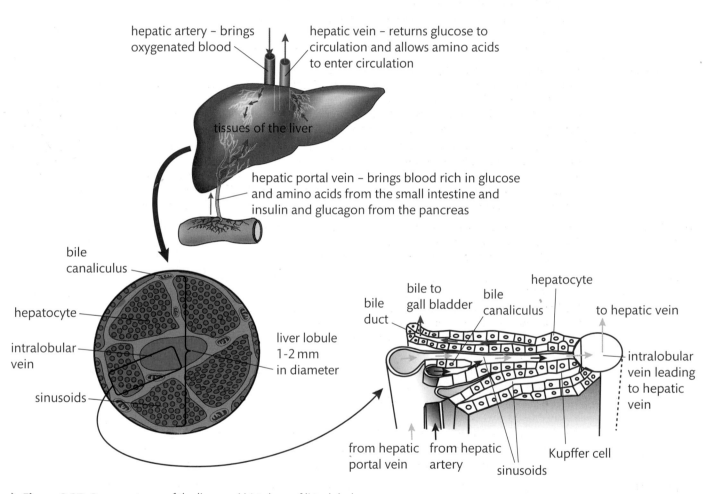

▶ **Figure 8.17**: Gross anatomy of the liver and histology of liver lobules

---

ⅠⅠ PAUSE POINT   Examine and identify the parts of the digestive system in an anatomical model. Make a large annotated drawing of the digestive system and accessory organs to show the positions and the functions of the structures listed in Table 8.8.

Hint   Use the anatomical model, diagrams from biology and anatomy texts or the Internet. By each label, add annotations to show, in concise form, its main functions.

Extend   How do you think the gall bladder and pancreas 'know' when to release their secretions to aid digestion, even though they are never in direct contact with food?

Appendicitis

Shiloh Moreno was aged 40 when he suffered from appendicitis. He felt pain in the umbilical (belly button) region of his abdomen, felt sick and vomited. He thought that he was suffering from food poisoning, after having eaten at a restaurant. He did not think it could be appendicitis as, like many people, he believed that only children suffer from appendicitis, thinking that this organ shrivels and disappears by adulthood.

Shiloh lost his appetite and after several hours the continuous, severe pain he felt was localised in the right lower part of his abdomen. He visited his GP, who called an ambulance to take him to the hospital.

Further investigations suggested that Shiloh had appendicitis and he was taken to the operating theatre for an appendectomy. The surgeon said that the operation should be carried out as soon as possible as an infected appendix can rupture, leading to peritonitis and gangrene, with potentially fatal consequences.

Check your knowledge

1   Which fairly widely held misconception/belief does this case study refute?

2   Explain how a ruptured appendix can lead to infection of the peritoneum and gangrene.

Function of the digestive system

While food is in the gut lumen (space) it is still outside of the body proper. Large molecules undergo **hydrolysis** and are digested to smaller molecules that can be absorbed across the gut wall into the blood stream for use in the body. Assimilation is where digested food molecules move into the cells of the body to be used.

Mechanical and chemical digestion

When you bite and chew food, this breaks large pieces down into smaller ones. This is mechanical digestion. The churning action of the stomach is also an example of mechanical digestion. In the stomach the food is also warmed, and so fats melt.

Enzymes that hydrolyse macromolecules are also present in saliva, gastric juice, enteric juice and pancreatic juice. The hydrolysis of macromolecules to smaller molecules is chemical digestion.

Actions of enzymes

Table 8.9 shows the action of digestive juices and their enzymes.

> **Key term**
>
> **Hydrolysis** – chemical reaction that splits, by adding water, large molecules into smaller molecules.

> **Link**
>
> See *Unit 3: Science Investigation Skills Learning Aim D* for more about enzymes and the factors that affect their rates of action.

▶ **Table 8.9:** Action of digestive juices

Region of digestive tract	Gland	Digestive juice	Enzymes	Substrate	Products of hydrolysis	Notes
Mouth	Salivary glands	Saliva	Salivary amylase	Starch	Maltose	Optimum pH around 6.5
Stomach	Gastric glands in epithelium	Gastric juice Parietal cells secrete HCl (hydrochloric acid)	Peptic cells secrete pepsin, a protease enzyme	Protein	Peptides	HCl kills microbes and provides the optimum pH of between 1 and 2 for pepsin.
			Gastric lipase	Fats	Fatty acids and glycerol	
Duodenum and jejunum	Pancreas	Pancreatic juice	Trypsin	Proteins and peptides	Amino acids	Trypsin is secreted in an inactive form, trypsinogen.
			Amylase	Starch	Maltose	
			Lipase	Fats	Fatty acids and glycerol	
	Liver	Bile	none			Bile emulsifies fats and neutralises chyme.

▶ **Table 8.9** *continued*

Ileum	Glands between villi secrete mucus	Enteric (intestinal) juice	Pancreatic enzymes	Peptides	Amino acids	Maltase, sucrase and lactase enzymes are present within the plasma membranes of the epithelial cells of villi. Pancreatic enzymes still present in the ileum may also be adsorbed onto epithelial cell surface membranes.
				Fats	Fatty acids and glycerol	
			Maltase	Maltose	Glucose	
			Sucrase	Sucrose	Glucose and fructose	
			Lactase	Lactose	Glucose and galactose	The main function of the ileum is the absorption of the products of digestion.
Colon	Enzymes made by microorganisms of the **gut microbiota**, using genes in the gut microbiome, may digest cellulose in fibre					Water is absorbed. Bacteria of the gut microbiota secrete hormones that help regulate appetite.

Case study

Stomach ulcers

Doctors used to think that gastric ulcers were caused by stress or eating too much spicy food. Between 1979 and 1982, two Australian doctors, Dr Barry Marshall and Dr Robin Warren, showed that stomach ulcers were caused by an infection with a stomach-dwelling bacterium, *Helicobacter pylori*. They had seen these Gram negative (see *Unit 17: Microbiology and Microbiological Techniques*), curved, rod-shaped, microaerophilic (able to live at low oxygen concentrations) bacteria in all the ulcers they had examined, but the medical profession was slow to accept a new idea. At the time, doctors believed that bacteria could not survive in the acid environment of the stomach. One of the Australian doctors deliberately infected himself with the bacteria, suffered an ulcer, then cured it with antibiotics. Later, in 2005, they received the Nobel Prize.

These *H. pylori* bacteria have flagella and can burrow through the layer of mucus lining the stomach into the tissues and cells below. They produce oxidase enzymes that they use to obtain energy by oxidising hydrogen produced by intestinal bacteria. They also make urease

enzymes that break down urea in the stomach, releasing alkaline ammonia. *H. pylori* also produce protease enzymes that allow them to enter stomach cells and cause those cells to die. Where the bacteria have burrowed through the mucus layer, stomach acid can access and damage or kill the underlying cells.

Due to this important piece of research, stomach ulcers can now be treated with antibiotics and other effective drugs. Scientists are trying to develop a vaccine, and a substance called sulforaphane that is in broccoli and cauliflower is being investigated as a possible treatment.

Check your knowledge

1 How do you think the doctor infected himself?

2 Why do you think there was such a long period of time between the discovery of *H. pylori* as the causative agent of stomach ulcers and the awarding of the Nobel Prize for that discovery?

3 Explain how *H. pylori* bacteria are adapted to living in the stomach, which has little oxygen and a lot of acid.

Nutrient absorption

Some small molecules such as glucose may be absorbed from the stomach. However, the main site of nutrient absorption is the ileum (see Figure 8.18). You have already learned that the ileum is long, its wall is folded and the epithelium of the ileum wall contains villi. All of these features increase the surface area for absorption of the products of digestion. The cells covering the surface of the villi have projections on the plasma membrane, called microvilli, which increase the surface area even more (see Figure 8.19).

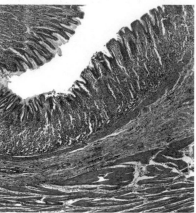

▶ Light micrograph of a section through the stomach wall. The upper layer is the glandular mucosa which has gastric pits (white areas between purple-stained cells) where gastric glands secreted digestive juices and enzymes into the lumen (large white area). The orange/pink layer is the submucosa and contains blood vessels, lymph vessels and nerves. Beneath this are three layers of smooth muscle.

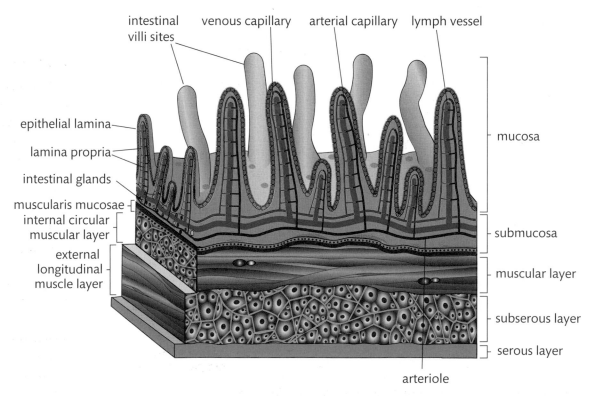

▶ **Figure 8.18:** Section through wall of ileum. Inside each villus is a lacteal surrounded by capillaries. The epithelial cells lining each villus contain microvilli on their cell surface membranes. The submucosa contains blood and lymph vessels and beneath that are layers of smooth muscle.

Worked Example

Scientists study anatomy, histology and cytology to understand the structures that make up bodies. They also study biochemistry to help work out the functions of these structures.

They need to examine specimens under a microscope and have to calculate the true size of the structures. They know how much the various lenses in the microscope magnify objects.

The villi shown in Figure 8.18 have a true length of 1 mm. Calculate the magnification of this picture.

Hint: Magnification = $\dfrac{\text{image size}}{\text{actual size}}$

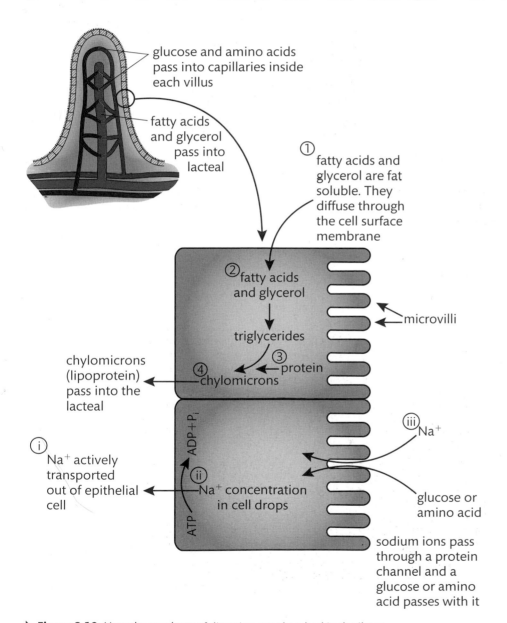

glucose and amino acids pass into capillaries inside each villus

fatty acids and glycerol pass into lacteal

① fatty acids and glycerol are fat soluble. They diffuse through the cell surface membrane

② fatty acids and glycerol

microvilli

triglycerides

③ protein

④ chylomicrons

chylomicrons (lipoprotein) pass into the lacteal

(iii) Na⁺

glucose or amino acid

ADP+Pᵢ

(i) Na⁺ actively transported out of epithelial cell

(ii) Na⁺ concentration in cell drops

ATP

sodium ions pass through a protein channel and a glucose or amino acid passes with it

▶ **Figure 8.19:** How the products of digestion are absorbed in the ileum

Some of the absorption in the ileum is by **diffusion**, some by **facilitated diffusion** and some by **active transport**.

Key terms

Diffusion – movement of molecules down their concentration gradient. This may or may not be through a partially permeable membrane. It uses only the kinetic energy of the molecules, and does not use energy from ATP.

Facilitated diffusion – diffusion that is enhanced by the presence of carriers or channels made of protein in the cell surface membrane.

Active transport – movement of molecules into or out of cells against their concentration gradient. It uses carrier proteins in the cell surface membrane and energy from ATP.

Link

See *Unit 3: Science Investigation Skills Learning Aim E* for more about diffusion.

Key term

Cotransporter – a type of transport protein that transports two or more substances at the same time across a cell membrane.

Glucose and amino acids are absorbed by active transport using a **cotransporter** mechanism. Sodium ions are actively transported out of the base of the epithelium cells lining the villi. This uses energy from ATP and reduces the concentration of sodium ions in these cells. Sodium ions from the contents of the ileum lumen pass, through special protein channels in the epithelial cell surface membranes, down their concentration gradient, and glucose molecules or amino acid molecules pass into the cells with them.

Fatty acids and glycerol are fat-soluble and diffuse through the phospholipid bilayer of the epithelial cell surface membranes of the villi. Inside the epithelial cells they are converted back to triglycerides on the smooth endoplasmic reticulum. These triglycerides are then transported to the Golgi apparatus and a protein coat is added, forming a type of lipoprotein (fat and protein complex) called a chylomicron. The chylomicrons diffuse out through the plasma membrane of epithelial cells on the inner side of the villi and enter the lymph fluid within the central lacteal. Fat-soluble vitamins (A, D and E) diffuse across the plasma membranes of epithelial cells.

Inorganic ions pass through the epithelial cell membranes by facilitated diffusion. Water passes down its water potential gradient by osmosis.

PAUSE POINT

Summarise the roles of protease, lipase and pancreatic enzymes.

Explain how digested food is absorbed and how water is absorbed from the gut.

Hint

Use Table 8.9 and reorganise that information by focusing on the types of enzymes. Proteases hydrolyse proteins; lipases hydrolyse fats. You also need to know how the products of digestion enter the villi of the ileum.

Extend

Molecules moving by diffusion, facilitated diffusion and osmosis do not use energy from ATP. However, the movement needs energy. What type of energy is used for these processes?

Chemical tests for the presence of nutrients in food

Tests for macronutrients

You can carry out lab tests for the presence of macronutrients in various foods. Table 8.10 shows the tests for different macronutrients.

▶ **Table 8.10:** Tests for macronutrients

Macronutrient	Test	Positive result
Starch	Add iodine in potassium iodide (KI) solution	Colour change from brown to blue/black
Protein	Add biuret reagent (dilute sodium hydroxide and dilute copper sulfate)	Colour change from blue to mauve/purple
Lipids	Shake food with ethanol. Allow to settle and pour ethanol into a test tube containing distilled water	A white milky emulsion seen near the top of the water
Reducing sugars	Add Benedict's reagent and heat to 80 °C for 10 minutes	Colour change from blue to green/yellow/brick red
Non-reducing sugars	If test for reducing sugar is negative, hydrolyse any sucrose in the food by heating with dilute hydrochloric acid. Cool and add sodium hydrogen carbonate to neutralise. Now add Benedict's reagent and heat	First Benedict's test: no colour change. After heating with acid and carrying out second Benedict's test there is a colour change from blue to red

Test for vitamin C (ascorbic acid)

The tests described above are all *qualitative*. They tell you if a certain type of food macromolecule is present, but not how much of the food type is present.

The test described here for vitamin C is *quantitative*. Not only does it show that the vitamin is present, but you can also calculate how much. Vitamin C is ascorbic acid. It decolorises blue DCPIP (dichlorophenolindophenol) by reducing it. Ascorbic acid in the process becomes oxidised.

Step by step: Test for vitamin C   `5 Steps`

1   Pipette 2 cm³ 1% ascorbic acid solution into a test tube.

▼

2   Note the level of 1% DCPIP solution in a burette or graduated pipette.

▼

3   Using the burette or graduated pipette add 1% DCPIP solution, drop by drop, to the 2 cm³ ascorbic acid solution. Shake the tube after each drop. The blue colour will disappear. Continue until the blue colour of the last drop added does NOT disappear.

▼

4   Record the exact volume of DCPIP used (final burette reading – initial burette reading).

▼

5   Repeat this procedure twice more and find the average volume of DCPIP used.

Worked Example

1% ascorbic acid solution contains 1 g (1000 mg) solid ascorbic acid dissolved in 100 cm³ distilled water. Therefore each cm³ of the solution contains 10 mg ascorbic acid.

1 cm³ 1% DCPIP solution contains 10 mg DCPIP.

During standardisation, 2 cm³ 1% DCPIP solution is decolorised by 2 cm³ 1% ascorbic acid solution.

So 20 mg DCPIP is decolorised by 20 mg ascorbic acid.

Therefore 1 mg DCPIP reacts with 1 mg ascorbic acid.

Half a yellow pepper is juiced in a blender and distilled water added to make up the volume to 10 cm³.

2 cm³ juice made from a yellow pepper decolorises 1 cm³ 1% DCPIP solution.

So 1 cm³ pepper juice would decolorise 0.5 cm³ 1% DCPIP solution.

So 1 cm³ pepper juice contains 5 mg ascorbic acid.

The concentration of ascorbic acid in this pepper juice is 5 mg cm⁻³.

The half yellow pepper when juiced gave 10 cm³ juice, so it contained 10 × 5 = 50 mg ascorbic acid.

So if you ate a whole large yellow pepper, you would obtain 100 mg ascorbic acid. Or if you ate 0.6 of the yellow pepper, you would obtain your daily requirement of vitamin C.

Health matters and treatments related to the digestive system and diet

Balanced diet

Humans need to eat a **balanced diet** to maintain health. Macronutrients include carbohydrates, fats and proteins. These may all provide energy. Micronutrients include vitamins and minerals. They are essential for proper body function but do not provide energy. Humans also need dietary fibre and water.

Tables 8.11 and 8.12 show dietary sources of macronutrients and micronutrients, and their importance in the body.

▶ **Table 8.11:** Dietary sources of macronutrients, and their importance

Nutrient	Examples of sources	Use in body	Result of deficiency or excess
Carbohydrate • Starches • Sugars	*Starches:* Potatoes, rice, maize, quinoa, sorghum, bread, cereals, muscle meat (glycogen) *Sugars:* Fruit, honey, milk, table sugar, processed foods with added sugar; fizzy drinks, fruit squashes and juices	Makes up the staple/main part of diet; energy source	As carbohydrate foods are usually cheap and plentiful, you are unlikely to suffer from a deficiency. Too much carbohydrate may lead to weight gain. Too much sugar can lead to tooth decay or type 2 diabetes.
Lipids (fats)	Meat, oily fish, oils, nuts, butter, cream, cheese, margarine	Energy source; stored in body as energy store, and for protection of internal organs and under skin for insulation; source of fat soluble vitamins; some fatty acids are essential to make cell membranes and nerve tissue; cholesterol needed to make sex hormones and to strengthen membranes	Deficiency of fat-soluble vitamins (A, D and E). Too much saturated fat may lead to weight gain/obesity, fatty plaques in artery walls (atherosclerosis) and increased risk of heart attack and stroke.
Proteins	Meat, fish, cheese, eggs, milk, soya beans, nuts, beans, quinoa, tofu, yoghurt	Growth and body structures such as bone, muscle, skin, internal organs; enzymes, haemoglobin, antibodies, neurotransmitters	Lack of protein can lead to kwashiorkor – stunted growth, muscle wasting and tissue oedema.
Fibre	Fruit and vegetables, porridge	Soluble fibre can lower blood cholesterol level; fibre adds bulk, prevents constipation and encourages growth of bacteria in the gut	Lack of fibre can lead to: • constipation • potentially bowel cancer, due to increased time faeces spends in the large intestine • lack of desirable bacteria in the gut microbiota.
Water	Water, drinks such as coffee and tea, milk	To make body fluids such as gastric juice; to remove excretory waste products such as urea in urine; keeps body hydrated; keeps eye surface moist; blood plasma; provides medium for metabolic reactions within cells; helps regulate body temperature (sweat); humans are about 80% water	Lack of water leads to dehydration – disruption of electrolyte (ions) balance. Loss of water from blood leads to osmotic imbalances and water leaving body and blood cells; enzyme-catalysed reactions cannot take place in dehydrated cells; sweat cannot be produced so leads to hyperthermia.

▶ **Table 8.12:** Dietary sources of micronutrients, and their importance

Nutrient	Examples of sources	Use in body	Result of deficiency or excess
Vitamin A (retinol) and beta carotene	Liver Carrots, sweet potatoes, squash, pumpkin, spinach, green vegetables, apricots, mango, egg yolks, peppers	Colourful vegetables supply beta carotene that the body changes to retinol. Vitamin A is needed for rod cells in the retina of the eye, healthy epithelial cells, resisting infections, growth and acting as an antioxidant reducing risk of cancer	Lack of beta carotene and vitamin A leads to poor night vision, xerophthalmia (dry hard cornea) and eventually blindness; severe deficiency is fatal. Excess vitamin A can lead to nerve disorders and during pregnancy can lead to abnormal development in the fetus.
Vitamin D	Formed in skin when exposed to UV light; a form of cholesterol in the skin is changed to vitamin D Milk, salmon, tuna, mackerel and herrings, egg yolks, liver	Is a hormone and regulates calcium phosphate deposition in bone; also helps protect against heart disease, cancer, multiple sclerosis, depression and schizophrenia	Too much in the diet by supplements can lead to calcium deposits in kidney, brain, heart and muscle, and learning difficulties in children. Negative feedback prevents formation of too much in skin. Lack leads to rickets in children, osteomalacia in adults and may contribute to osteoporosis.
Vitamin E	Nuts, prawns, wholemeal bread, sweet potatoes, oils	Antioxidant so may help reduce risk of cancer and heart disease	Deficiency very rare – poor nerve transmission, muscle weakness and degeneration of retina.
Vitamin K	Green leafy vegetables; made by gut microbiota bacteria	Needed to help blood clot during injury	Lack leads to easy bruising and internal bleeding.
Vitamin C (ascorbic acid)	Fruits and green vegetables, potatoes, kiwi fruits, blackcurrants and green peppers are very good sources	Helps body make collagen protein – important for muscles, bone, blood vessel walls and cartilage; aids absorption of dietary iron; is an antioxidant – by becoming oxidised itself it protects molecules such as DNA from damage due to oxidation by free radicals	Excess is passed out in urine. Lack leads to scurvy – poor bone and teeth development; delayed wound healing; weakened blood vessels and increased haemorrhaging, tender sore gums, loss of teeth and hair; painful joints due to internal bleeding. Death if untreated.
Vitamin B group B_1 thiamine	Bran, rice husks, meat, peas	Activates enzymes in respiration	Lack leads to mental confusion, Beriberi.
B_2 riboflavin	Green vegetables, meat	Activates enzymes used in respiration	Lack leads to decreased growth, cracked dry skin.
B_3 Niacin	Meat, fish, brown rice	Activates enzymes used in respiration	Lack leads to pellagra – depression and confusion; dementia and death.
B_6	Meat, fish, green vegetables, bananas	Activates enzymes used in protein metabolism	Lack leads to large irregularly shaped red blood cells.
Folic acid	Dark green vegetables	Activates enzymes for DNA replication and protein synthesis	Lack leads to pernicious anaemia.
B_{12}	Meat and fish	Activates enzymes involved in making nerve, blood and other cells	Lack leads to confusion and dementia-like symptoms. Pernicious anaemia, if caused by autoimmunity, leads to vitamin B_{12} deficiency as the cells making intrinsic factor are destroyed and vitamin B_{12} cannot be absorbed from the gut even if it is present in the diet.
Iron	Meat, soya beans, fish, whole wheat bread, prunes, plums	To make haemoglobin and myoglobin	Lack leads to anaemia.
Iodine	Seafood, iodised salt, egg	To make the hormone thyroxine	Lack leads to goitre; during pregnancy can lead to mental and physical development abnormalities in fetus – cretinism.
Calcium	Milk, cheese, yoghurt, ice cream, cream, green vegetables	Bones, muscle contraction, blood clotting, nerve function	Lack leads to problems with bone density and muscle contraction.
Magnesium (Mg), Sodium (Na), Phosphorus (P) Potassium (K)	Milk, meat, seeds, vegetables, Salt Tuna, potatoes Bananas, avocado, fish	Maintaining electrolyte balance for body fluids; nerve function (Na^+); bone formation (Mg^{2+} and P); heart function (K^+)	Lack leads to nerve and heart dysfunction.

The importance of B$_{12}$

Jemima is 40 years old and has been suffering from severe fatigue and inability to concentrate. She also has painful finger and wrist joints. After seeing her GP she was referred to the rheumatology department of a hospital for tests, which found she has pernicious anaemia due to autoimmunity. Her own immune system had attacked and damaged the cells in her stomach that make intrinsic factor needed for her to absorb vitamin B$_{12}$ from her food. Therefore the pernicious anaemia has led to a vitamin B$_{12}$ deficiency and she needs a B$_{12}$ injection three times a week. She is also in the early stages of rheumatoid arthritis, another autoimmune disease and she is taking non steroidal anti-inflammatory drugs for that.

PAUSE POINT

Keep a food diary for a week. Write down everything you eat at each meal and also for snacks. Include drinks as well.

Analyse your diet to see which food groups, vitamins and minerals you have eaten. Are you eating a balanced diet?

Hint   Make a table with the names of food groups and nutrients in the column headings and then for each food you eat you can tick its components.

Extend   Although scurvy is rare today, when it does occur, dentists are often the first to notice it. Why do you think this is?

The importance of zinc

Nutritionists have recently found that small amounts of zinc are essential in the diet. One use of zinc is to form part of the structure of many enzymes and the hormone insulin. Lack of dietary zinc can lead to reduced cell division and protein synthesis. In many countries where people do not eat much meat they are deficient in zinc.

Africa Harvest is a small local biotech company in Kenya, founded by Dr Florence Wambugu with some funding from Monsanto. One of their projects is to genetically modify plantains (a type of banana and a staple food in Kenya) to contain more zinc.

Check your knowledge

1 Explain why lack of zinc in the diet leads to reduced cell division and protein synthesis.

2 Discuss the ethics of making more zinc available to people in this way and discuss the ethics of those who oppose the introduction of all GM crops.

3 Use the Internet to find out more about Dr Florence Wambugu and Africa Harvest.

Discussion

About 500 000 children in the developing world each year go blind or die due to vitamin A deficiency. Golden Rice™ is genetically modified rice that contains beta carotene. It is a cheap staple food in developing countries, where many people do not have access to the other foods that contain beta carotene. Many people in the UK have opposed the introduction of GM crops such as this rice and these people have been described as 'anti humanitarian'.

Discuss the pros and cons of Golden Rice™. Use the Internet to help you with your research.

▶ Fruit and vegetables are good sources of vitamins and of antioxidants that help to protect us from cancer and heart disease

Case study

Studying human nutrition

Nadima Hossain has obtained a degree in nutrition and is hoping to become a registered nutritionist. She will have to take a qualification accredited by the Association for Nutrition (AfN). There are various specialist areas that she could study such as human nutrition, public health nutrition, food and nutrition and sports nutrition. However, she has chosen human nutrition and plans to work in the NHS for at least five years, before perhaps becoming a private consultant in a health clinic.

Nadima has the necessary personal qualities to work as a nutritionist. She is interested in science and food, is able to motivate others, understands other people and why they may choose certain lifestyles, understands how crucial good nutrition is, is not judgemental and can explain complex things in a simple way. She also has a good understanding of science and the scientific methodology, has good organisational and communication skills and good business skills which she will need for the freelance or private work that she hopes to do later.

Check your knowledge

1 Why does Nadima need to know about food and health for her role?

2 What nutritional deficiencies might Nadima encounter in her work?

Digestive system disorders

There are many digestive system disorders and eating disorders. Three will be considered here: coeliac disease, colitis and irritable bowel syndrome.

Coeliac disease

This is an autoimmune disorder. It is not an allergy.

▶ In people with a certain genetic make-up, their immune system identifies gluten, a protein in wheat, rye and barley, as a threat to the body and mounts an immune response.

▶ This damages the surface of the small intestine and reduces the body's ability to absorb products of digestion.

▶ Long-term complications include:
  - osteoporosis
  - anaemia caused by iron deficiency
  - vitamin B_{12} and folic acid deficiencies.

▶ Symptoms are bloated abdomen, vomiting, diarrhoea, muscle wasting and, because some food is not being digested and therefore not being absorbed, extreme lethargy.

▶ Most patients respond well to a change of diet – omitting all foods with gluten (beer, bread, pasta, cereals, ready meals and some sauces), and replacing instead with corn (bread and biscuits made with corn flour) and rice.

There is no screening programme, but anyone with symptoms or with a relative known to have coeliac disease can be tested on request.

Colitis

Colitis, ulcerative colitis and Crohn's disease all cause inflammation and are all examples of inflammatory bowel disease (IBD).

Colitis may be caused by infection, invasion of the colon wall with lymphocytes, or reduced blood supply to the colon.

Symptoms include chronic watery diarrhoea that may contain blood, abdominal pain and bloating.

▶ **Ulcerative colitis**: Small ulcers may develop on the colon and rectum lining. This may be an autoimmune condition in people of a certain genetic makeup.

▶ **Crohn's disease**: This is a chronic inflammatory disease of the colon and ileum, associated with ulcers and fistulae. Symptoms include diarrhoea that contains blood, abdominal pain, weight loss and extreme fatigue. Sufferers experience bouts of remission and flare-ups. Genetics, the immune system, infection and environmental factors such as smoking are all implicated.

IBD is treated with anti-inflammatories such as corticosteroids, and with immunosuppressants. In some cases, the inflamed section of the intestine is surgically removed.

Irritable bowel syndrome (IBS)

This is a common long-term condition of the digestive system. It may be linked to increased sensitivity of the gut and problems digesting food. Stress may also play a part.

IBS can be managed by:

- avoiding foods known to trigger it
- increasing the amount of dietary fibre
- taking regular exercise
- reducing stress levels.

Research

Visit the following websites to find out more about research into these digestive system disorders.

Go to *http://www.nhs.uk/conditions/Coeliac-disease/Pages/Introduction.aspx* to find out more about coeliac disease.

Go to *http://www.medicinenet.com/colitis/article.htm* to find out more about colitis.

Go to *http://www.nhs.uk/Conditions/Irritable-bowel-syndrome/Pages/Introduction.aspx* to find out more about IBS.

Ⅱ PAUSE POINT   Make a table to compare coeliac disease, colitis and irritable bowel syndrome.

    Hint   When you are asked to compare, you need to cover the similarities as well as the differences.

    Extend   Why do you think coeliac disease, IBD and IBS all cause patients to feel lethargic?

Case study

Clostridium difficile

Shannon Wiley caught an infection of *Clostridium difficile* while in hospital being treated with intravenous antibiotics for several weeks to treat endocarditis. Everyone has these bacteria inside their digestive tract, but the other bacteria usually keep it in check so that it does no harm. However, if you are on a long course of antibiotics, some of your gut bacteria can be killed and the *Clostridium difficile*, known as *C. diff*, can multiply and release toxins that cause swelling and irritation of the colon. This inflammation is known as colitis and the symptoms are diarrhoea, fever and abdominal cramps.

This infection can be difficult to treat and Shannon had tried various antibiotics (vancomycin and fidaxomicin), but with no positive result. She suffered from recurring bouts of *C. diff*, which was very debilitating and could eventually prove fatal. She had read about the gut microbiota – the 1000 or so different species of bacteria and other types of microbes that live in humans' (and other mammals') guts. These bacteria help digest some food and they make certain vitamins (for example,

vitamin K) that the body can use. They also make hormones that help to regulate appetites, as well as keeping some infectious bacteria at bay. She also learned that *C. diff* flourishes when this balance of gut microbes is upset (for example, after long exposure to antibiotics) and that putting this bacterial balance back to normal can cure *C. diff*. She talked this over with her GP, who suggested that she have a faecal transplant. This involves a doctor or nurse placing a sample of faeces taken from a healthy donor and that had been screened, into her colon, using a catheter. It worked (as it does in over 90% of cases) and she is now well again. Her GP has advised her to maintain a healthy diet to encourage the growth of the good bacteria in her colon. This diet involves plenty of fruit and vegetables, especially leeks and onions.

Check your knowledge

1 Why do you think the donor of the faeces has to be healthy?

2 Which diseases do you think are being looked for when the donated faeces is screened?

Assessment practice 8.3

A nutritionist works for the NHS in a hospital where she advises patients about their special dietary needs. She needs a leaflet explaining to elderly patients who may be at risk of being poorly nourished, the importance of good nutrition, how to recognise the symptoms of nutritional deficiency and how to prepare simple and inexpensive but nourishing meals.

Produce a leaflet that she could use. It should contain some information about:

- the role and location of organs of the digestive system
- the role of digestive enzymes in the stomach and small intestine
- the role of the small intestine in nutrient absorption
- what a balanced diet is
- the symptoms of nutrient deficiencies
- how such deficiencies can be treated.

It should also contain some evaluative information on how nutritional deficiencies, over-nutrition and under-nutrition impact on human health, and the effectiveness of treatments.

Plan

- I know what the task is.
- I know how confident I feel in my own abilities to complete this task.
- I know any areas I think I may struggle with.

Do

- I know what it is I am doing and what I want to achieve.
- I can identify when I have gone wrong and can adjust my approach to get back on course.

Review

- I can explain what the task was and how I approached it.
- I can explain how I would approach the difficult parts differently next time.

Further reading and resources

Marieb, E. N. (2014). *Essentials of Human Anatomy and Physiology*. San Francisco: Pearson/Benjamin Cummings (ISBN 9781292057200).

Palastagana, N. and Soames, R. W. (2012). *Anatomy and Human Movement*. Edinburgh: Elsevier/Churchill Livingstone (ISBN 9780702053085).

Tortora, G. J. and Derrickson, B. H. (2008). *Principles of Anatomy and Physiology*. Hoboken, NJ: John Wiley (ISBN 9780471718710).

Waugh, A. and Grant, A. (2014). *Ross and Wilson Anatomy and Physiology in Health and Illness*. Edinburgh: Elsevier/Churchill Livingstone (ISBN 9780702053252).

Websites

www.innerbody.com
A website exploring anatomy.

www.visiblebody.com/ap/pc
A website for Windows Desktop on anatomy and physiology.

https://www.collin.edu/ce/courses/basicanatomy.html
A website covering basic anatomy and physiology.

https://www.ashoka.org/fellow/florence-wambugu
Information about Dr Florence Wambugu and Africa Harvest.

http://www.theguardian.com/environment/2013/feb/02/genetic-modification-breakthrough-golden-rice
Information about golden rice.

THINK ▶FUTURE

Ellie Mitchell
Clinical Technician
for NHS national
bowel cancer
screening
programme

Ellie has worked in this department for three years. Bowel cancer (a generic term covering colon, rectal and colorectal cancers) is the third most common cancer in the UK and the second leading cause of cancer deaths. In its early stages, people may have no symptoms, so a screening programme has been in place to detect those early cases as early diagnosis often leads to better prognosis (outcome) as treatment can be given early before the cancer spreads to other organs. This programme can also detect non-cancerous polyps in the bowel which, although harmless at that stage, can progress to cancer later on.

Focusing your skills

Preparing yourself

Screening consists of three stages:

1 Identify the people in the population most at risk, for example, those over 60 years old.

2 Offer them a test – the faecal occult (hidden) blood test.

3 If blood is in their faeces, offer them a colonoscopy examination which can determine if there is cancer in their bowel. Blood in the faeces can be a symptom of something else such as piles.

Presently in England and Northern Ireland all men and women aged 60–69 (soon to be extended to ages 60–74) are sent bowel screening kits. Once they have completed the test the kits are sent to a lab where technicians such as me analyse the results. In Scotland, people aged 50–74 are sent the kits. If there is blood in the faeces, the patient is sent another test kit and if that also shows blood in the faeces, then they are offered a colonoscopy examination. Anyone outside of the above age groups can request a test. As with all tests, they are not infallible and sometimes a cancer may be missed. There is also some risk associated with colonoscopy but as this type of cancer is fairly common, we think that the benefits of screening outweigh the risks.

Getting ready for assessment

Layla Anwar is studying for a BTEC National in Applied Science. She was given an assignment as part of her practical portfolio. She was asked to explain the structure and functions of organs of the digestive system and investigate the nutrition content of some foods. Layla shares some aspects of her experience below.

How I got started

I gathered all my notes on the digestive system and also found some relevant textbooks in the library and some useful websites via the Internet.

How I brought it all together

I decided to make a large annotated diagram of the digestive system, explaining what each part does. I then examined prepared slides showing the histology of various parts of the system, for example the stomach wall and the ileum lining. I placed the detailed histology drawings in pockets placed next to the relevant structures on my large digestive system diagram. On the diagram, I also showed which enzymes are made in the various regions and what their functions are and I showed how digested food is absorbed. I used different colours where appropriate.

I then selected a range of foods and tested each one for starch, reducing sugar, non-reducing sugar, fats and vitamin C. I set out my results in a large table. I observed health and safety rules and carried out a risk assessment. I then investigated from book sources such as the *Manual of Nutrition* and from websites on the Internet to find out about other vitamins and minerals present in the foods I tested.

What I learned from the experience

I was too ambitious with the food testing and carrying out so many tests took a long time. I should have carried out each test on one type of food and used secondary data to complete the information for the rest of the foods.

I did not follow the conventions for drawing specimens properly. I should have used a sharp HB pencil and drawn clear unbroken non-overlapping lines. I should have drawn low power plans showing areas but no cells and then high power drawings showing some cells. I used ruled label lines but did not always state the magnification.

Think about it

▶ Have you thought about all the information you will need to include and which could be based around practical activities you have carried out during your course?

▶ Can you base your drawing on the dissection you have seen and use textbook diagrams for the labels?

▶ Should you use the Internet as well as textbooks for the annotations on aspects of the organs' structures and their functions to include some very up-to-date information?

▶ Can you write things in such a way that you include all the correct technical terms but keep them uncomplicated and easy to understand?

▶ Is your information written in your own words and referenced clearly where you have used quotations or information from a book, journal or website?

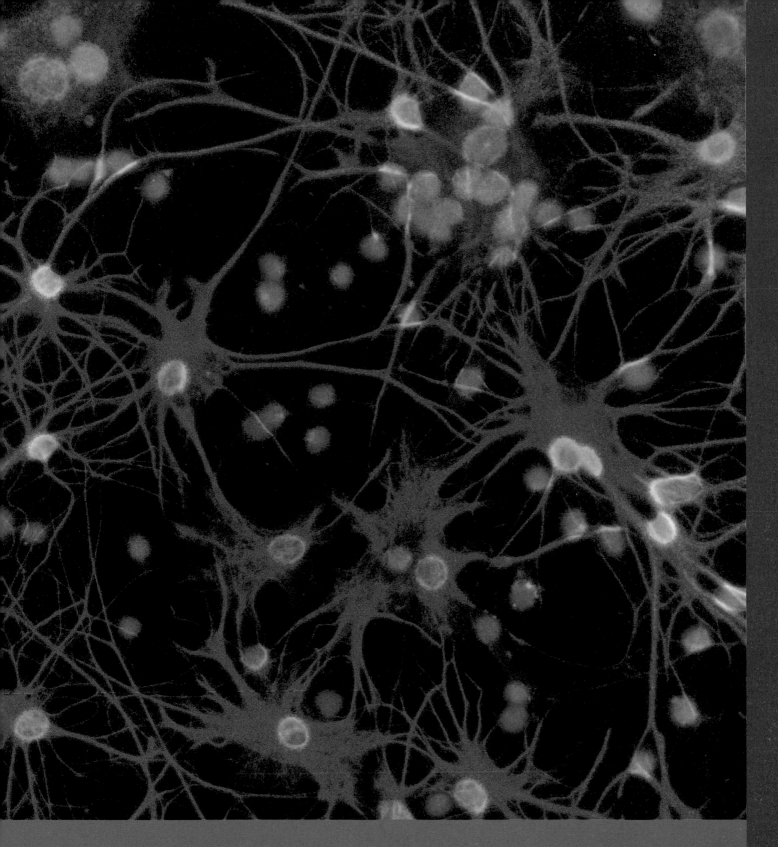

Human Regulation and Reproduction 9

Getting to know your unit

Regulation

The human body is a complex organisation of systems that each need to be controlled in different ways. This unit will help you understand how the human body keeps its internal conditions in a steady state.

Reproduction

There have been many advances in human fertility in recent years. In this unit you will be able to consider these and the hormonal control of the reproductive system. You will also look at fertility treatments.

How you will be assessed

This unit will be assessed by a series of internally assessed tasks set by your tutor. Throughout this unit you will find assessment activities that will help you work towards your assessment. Completing these activities will not mean that you have achieved a particular grade, but you will have carried out useful research or preparation that will be relevant when it comes to your final assignment.

In order for you to achieve the tasks in your assignment, it is important to check that you have met all of the Pass grading criteria. You can do this as you work your way through the assignment.

If you are hoping to gain a Merit or Distinction, you should also make sure that you present the information in your assignment in the style that is required by the relevant assessment criterion. For example, Merit criteria require you to analyse and explain, and Distinction criteria require you to assess, analyse and evaluate.

The assignment set by your tutor will consist of a number of tasks designed to meet the criteria in the table. This is likely to consist of a written assignment but may also include activities such as:

- creating a fact sheet about how a body system is controlled
- analysing tables and graphs of data relating to physiological measurements
- analysing case studies or observations from practical activities.

Assessment criteria

This table shows what you must do in order to achieve a **Pass**, **Merit** or **Distinction** grade, and where you can find activities to help you.

Pass	Merit	Distinction
Learning aim **A** : Understand the interrelationship and nervous control of the cardiovascular and respiratory systems.		
A.P1 Describe the organisation and function of the nervous system in relation to cardiovascular and respiratory requirements. **Assessment practice 9.1**	**A.M1** Explain how nervous impulses are initiated, transmitted and coordinated in the control of the cardiovascular and respiratory systems. **Assessment practice 9.1**	**A.D1** Assess the role of the nervous system in coordinating the cardiovascular and respiratory systems. **Assessment practice 9.1**
Learning aim **B** : Understand the homeostatic mechanisms used by the human body.		
B.P2 Describe how homeostatic mechanisms maintain normal function. **Assessment practice 9.2**	**B.M2** Explain the role of hormones in homeostatic mechanisms. **Assessment practice 9.2**	**B.D2** Analyse the impact of homeostatic dysfunction on the human body. **Assessment practice 9.2**
Learning aim **C** : Understand the role of hormones in the regulation and control of the reproductive system.		
C.P3 Describe the structure and function of reproductive anatomy. **Assessment practice 9.3**	**C.M3** Explain how the regulation of male and female reproductive systems can affect human reproductive health. **Assessment practice 9.3**	**C.D3** Evaluate how conception may be prevented and promoted. **Assessment practice 9.3**
C.P4 Describe how hormones are involved in gamete development and conception. **Assessment practice 9.3**		

The systems inside your body interact to respond to changes on the outside and the inside. On a large sheet of paper, draw a spider diagram to show all of the body systems and what they do. When you have completed this unit, add the interrelationships between the systems and the mechanisms by which the systems communicate with each other.

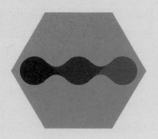

A Understand the interrelationship and nervous control of the cardiovascular and respiratory systems

The human body is able to control the activities of its different tissues and organs through detecting stimuli and generating appropriate responses. This is done through hormones, nerve impulses or a combination of these.

The need to respond to changes

The ability to respond to internal and external changes, and so avoid harmful situations, increases the chances of survival. In the human body, some nerve cells have become highly sensitive to particular stimuli. These are called **receptor** cells. Responses are brought about by body structures called **effectors**, usually muscles or glands.

Nervous system organisation

The nervous system consists of the brain, spinal cord and a network of neurons. It sends, receives, and processes information from all parts of the body. The central nervous system has two main organs: the brain and the spinal cord. The peripheral nervous system has sensory cells that send information to the central nervous system from external stimuli or internal organs, and motor nervous system cells that carry information to organs, muscles and glands from the central nervous system.

The nervous system can be divided into the **somatic nervous system** and **autonomic nervous system**. The somatic nervous system is sometimes referred to as the voluntary nervous system because many of its actions are under conscious control. The somatic nervous system includes sensory neurones which transmit impulses to the central nervous system from receptors all over the body and motor neurons which transmit impulses to the muscles.

The autonomic nervous system is often referred to as the involuntary nervous system because it enables the functioning of internal organs without conscious control. The autonomic nervous system controls involuntary responses, but it is possible to gain some voluntary control over these responses. Emptying the bladder and opening the anal sphincter are examples of activities that are controlled by the autonomic nervous system but can be brought under voluntary control through a process of learning called conditioning.

The autonomic system has two distinct parts:

▶ the parasympathetic nervous system, which maintains the body's functions on a day-to-day basis

▶ the sympathetic nervous system, which prepares the body to react in emergency situations.

Key terms

Receptor – a specialised cell or group of cells that respond to changes in the surrounding environment.

Effector – a muscle, organ or gland that is capable of responding to a nerve impulse.

Somatic nervous system – the part of the nervous system that brings about the voluntary movements of muscles as well as involuntary movements such as reflex actions.

Autonomic nervous system – the part of the nervous system that controls bodily functions which are not consciously controlled, such as the heartbeat and breathing.

These two systems act antagonistically. Some actions are shown in Table 9.1.

▶ **Table 9.1:** Actions of the sympathetic and parasympathetic nervous systems on body structures

	Sympathetic	**Parasympathetic**
Eyes	Dilates pupil	Constricts pupil
Salivary glands	Inhibits flow of saliva	Stimulates flow of saliva
Lacrimal glands	–	Stimulates flow of tears
Lungs	Dilates bronchi	Constricts bronchi
Heart	Accelerates heartbeat	Slows heartbeat
Liver	Stimulates conversion of glycogen to glucose	Stimulates release of bile
Stomach	Inhibits peristalsis and secretion	Stimulates peristalsis and secretion
Adrenal glands	Stimulates secretion of adrenaline and noradrenaline	–
Intestines	Inhibits peristalsis and anal sphincter contraction	Stimulates peristalsis and contraction of the anal sphincter
Bladder	Inhibits bladder contraction	Stimulates bladder contraction

Nerve cells

What are nerve cells like?

The nervous system is made up of two types of cells. **Neurons** are cells that transmit electrical impulses to and from the brain and nervous system. There are two types of neuron – myelinated and unmyelinated. Myelinated neurons conduct electrical impulses much faster than unmyelinated neurons. Myelinated neurons are found in the peripheral nervous system. They carry impulses from sensory receptors to the central nervous system, or from the central nervous system to the effectors. **Glial cells** provide support for the neuron by carrying out processes such as the digestion of dead neurons and manufacture of the components of neurons.

Neurons are the basic functional unit of the nervous system. They are highly specialised cells and can transmit impulses around the body at up to 200 mph. There are different types of neuron, motor and sensory, but their basic structure is the same. Figure 9.1 shows a motor neuron and a sensory neuron, and Table 9.2 shows the structures and functions of neurons.

> **Key terms**
>
> **Neuron** – a cell that transmits electrical impulses and is located in the nervous system.
>
> **Glial cells** – cells that provide support for neurons by carrying out processes such as manufacturing neuron cell components and digesting dead neurons.

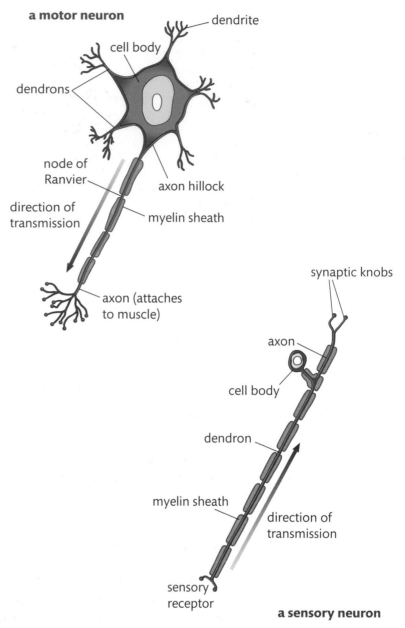

a motor neuron

dendrite

cell body

dendrons

node of
Ranvier

axon hillock

direction of
transmission

myelin sheath

axon (attaches
to muscle)

synaptic knobs

axon

cell body

dendron

myelin sheath

direction of
transmission

sensory
receptor

a sensory neuron

▶ **Figure 9.1:** The structure of a motor neuron and a sensory neuron

▶ **Table 9.2:** The structures of neurons and their functions

Structure	Function
Cell body	• Contains the cell nucleus and other organelles, such as the mitochondria and ribosomes.
Dendrites	• Very thin extensions of the cytoplasmic membrane that conduct impulses to the cell body and link with surrounding neurons.
Axon	• Long process that extends from the cell body to transmit impulses away from the cell body to form connections with a muscle or a gland. Axons and dendrites are collectively referred to as nerve fibres.
Myelin	• An insulating material that prevents loss of electrical impulse and rapid transmission in some types of neuron. (Unmyelinated neurons do not have this.)

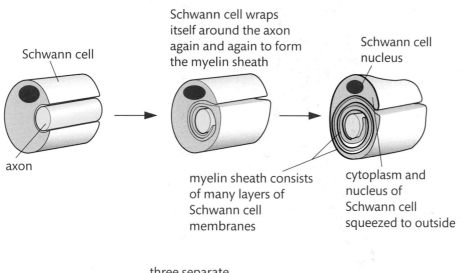

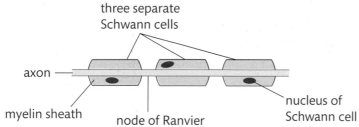

▶ **Figure 9.2:** The myelin sheath, an insulating layer, is created when Schwann cells grow around the axon

Figure 9.2 shows a Schwann cell, which is a type of glial cell. It produces the insulating myelin layer that can be seen on the axons of some neurons. (Several unmyelinated neurons may be surrounded by just the Schwann cell.)

⏸ PAUSE POINT	Can you describe the structure of motor neuron and a sensory neuron?
Hint	Draw a diagram of each type of neuron and label the structures.
Extend	Squids can escape quickly from danger because they have nerve fibres with a very large diameter. How can a larger nerve fibre enable faster movement than a small one?

How are impulses generated?

The body is able to produce electrical impulses by the movement of positively charged metal ions (Sodium, Na^+, and Potassium, K^+) in and out of nerve cells in a controlled manner. By moving certain ions into a cell, it is possible to change the potential difference (voltage) and cause an impulse to be transmitted.

Research

Most of our knowledge of nervous impulse transmission comes from the work of two scientists, Alan Hodgkin and Andrew Huxley, who conducted experiments on axons from the squid. Squids possess exceptionally large axons, termed giant axons, measuring a millimetre in diameter which were big enough to work on.

Find out more about their experiments.

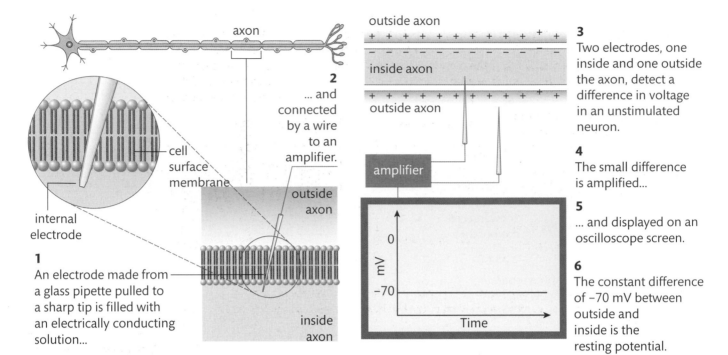

1
An electrode made from a glass pipette pulled to a sharp tip is filled with an electrically conducting solution...

2
... and connected by a wire to an amplifier.

3
Two electrodes, one inside and one outside the axon, detect a difference in voltage in an unstimulated neuron.

4
The small difference is amplified...

5
... and displayed on an oscilloscope screen.

6
The constant difference of −70 mV between outside and inside is the resting potential.

▶ **Figure 9.3:** This apparatus, with an internal and external electrode, is used to investigate how neurons work. Here you can see the resting potential of a neuron being measured. The resting potential is the potential difference across the membrane in millivolts.

Resting potential

When the neuron is resting (that is, between impulses), proteins in the axon cell membrane, called carrier proteins, pick up sodium ions and transport them out of the cell. This is known as the sodium pump. At the same time, potassium ions are actively transported into the axon cell cytoplasm. This is referred to as the potassium pump.

As approximately three sodium ions are carried out of the cell for every potassium ion that is brought in, the net result is that the outside of the axon membrane is positively charged compared to the inside. When in this resting state, the axon is said to be polarised. Figure 9.4 shows how the resting potential is maintained by the sodium pump.

We call the difference between the inside and outside potentials the resting potential and it is approximately −70 mV. This means that the electrical potential inside the axon is 70 mV lower than the outside when the axon is resting.

Action potential

A nerve impulse is initiated when a neuron is stimulated. In everyday situations, the stimulus can be chemical, mechanical, thermal or electrical. When scientists experiment on nerve impulses they use electrical impulses.

An impulse will travel along the axon when the neuron is stimulated. In experiments, the stimulus is an electrical current because scientists can control its strength, duration and frequency. This prevents the axon from being damaged.

When an electrical current is applied to the axon, there is a brief change in the potential from −70 mV to +35 mV. This means that the inside of the axon becomes positively charged relative to the outside. This change in potential is called the **action potential** and lasts about three milliseconds.

Key term

Action potential – a sudden and rapid increase in the positive charge of a neuron caused when sodium and potassium ions move across the cell membrane.

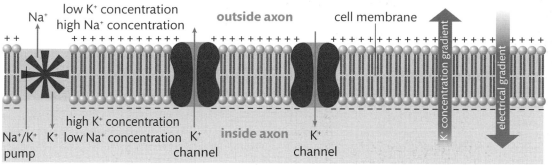

1 Na⁺/K⁺ pump creates concentration gradients across the membrane.

2 K⁺ diffuse out of the cell down the K⁺ concentration gradient, making the outside of the membrane positive and the inside negative.

3 The electrical gradient will pull K⁺ back into the cell.

4 At –70 mV potential difference, the two gradients counteract each other and there is no net movement of K⁺.

▶ **Figure 9.4:** The resting potential of the axon is maintained by the sodium pump, the relative permeability of the membrane and the movement of potassium ions along concentration and electrochemical gradients

During the action potential, the axon is **depolarised**. If the electrodes are connected to a cathode ray oscilloscope, the action potential shows as a peak in the trace. Figure 9.5 shows the changes in sodium ions and potassium ions during the excitation of an axon in an action potential.

> **Key term**
>
> **Depolarisation** – when the axon is stimulated, channels in the axon membrane open. This allows sodium ions to diffuse into the axon. This creates a positive charge in the axon and causes the action potential.

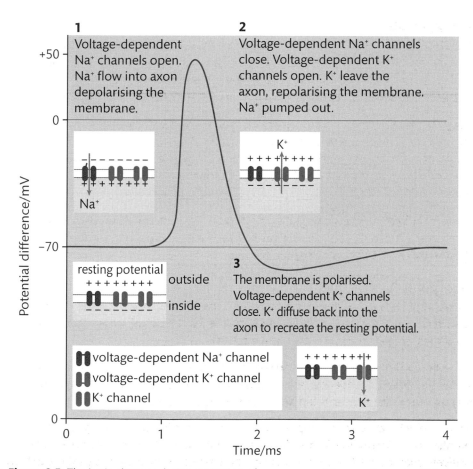

1 Voltage-dependent Na⁺ channels open. Na⁺ flow into axon depolarising the membrane.

2 Voltage-dependent Na⁺ channels close. Voltage-dependent K⁺ channels open. K⁺ leave the axon, repolarising the membrane. Na⁺ pumped out.

3 The membrane is polarised. Voltage-dependent K⁺ channels close. K⁺ diffuse back into the axon to recreate the resting potential.

▶ **Figure 9.5:** The ionic changes during excitation of an axon result in an action potential

Depolarisation

When the axon is stimulated, channels in the axon membrane open. This allows sodium ions to **diffuse** into the axon. This creates a positive charge in the axon and causes the action potential. Channels then open in the membrane to allow potassium ions to diffuse out of the axon.

Repolarisation

Sodium channels close. This prevents any further movement of sodium ions into the axon. This re-establishes the resting potential and the axon membrane is said to be repolarised.

The diffusion of potassium ions is so rapid that, for a brief period, the potential difference drops below that of the resting potential. This is termed an overshoot or hyperpolarisation, which helps to ensure that the action potential travels in one direction along the neuron. This recovering region of the axon membrane would require greater depolarisation than the 'downstream' region to initiate an action potential.

The potassium channels close and the **sodium-potassium pump** begins. The normal concentration of sodium and potassium ions is restored and the resting potential is re-established.

How does an impulse travel along a neuron?

Once an action potential is set up in response to a stimulus, it will travel the entire length of that nerve fibre. The length of a nerve fibre can range from a distance of a few millimetres to a metre or more.

The movement of the nerve impulse along the fibre is the result of local currents set up by the movements of sodium and potassium ions at the action potential. These ion movements occur both in front of and behind the action potential.

The effect is that the membrane in front of the action potential is depolarised sufficiently to cause the sodium ion channels to open. The sodium ion channels behind the action potential cannot open due to the **refractory period** of the membrane behind the spike. In this way the impulse can only travel in one direction along the axon of the neuron.

Figure 9.6 shows how changes in ions set up small local currents enabling the impulse to travel in one direction along the axon.

The all-or-nothing principle

Action potentials obey the all-or-nothing principle. This means that the size of the action potential is always the same despite the strength of the stimulus.

Information about the strength of the stimulus is carried along the neuron as changes in the frequency of the impulses. A stronger stimulus will result in a greater frequency of impulses being transmitted along the neuron.

1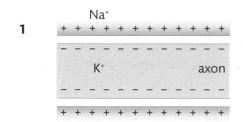

At resting potential there is positive charge on the outside of the membrane and negative charge on the inside, with high sodium ion concentration outside and high potasssium ion concentration inside.

2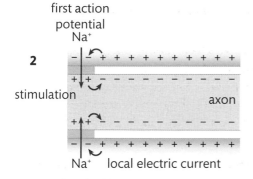

When stimulated, voltage-dependent sodium ion channels open, and sodium ions flow into the axon, depolarising the membrane. Localised electric currents are generated in the membrane.

3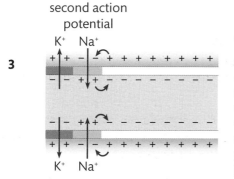

The potential difference in the membrane adjacent to the first action potential changes. A second action potential is initiated. At the site of the first action potential the voltage-dependent sodium ion channels close and voltage-dependent potassium ion channels open. Potassium ions leave the axon, repolarising the membrane. The membrane becomes hyperpolarised.

4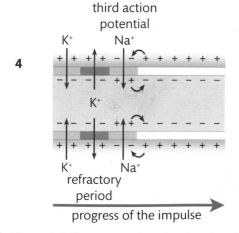

A third action potential is initiated by the second. In this way, local electric currents cause the nerve impulse to move along the axon. At the site of the first action potential, potassium ions diffuse back into the axon, restoring the resting potential.

▸ **Figure 9.6:** The transmission of an impulse along a neuron

Ⅱ PAUSE POINT   What are the main mechanisms that maintain the resting potential of a neuron?

    Hint   Draw diagrams to show how a resting potential is maintained and how an action potential is initiated.

    Extend   What will happen to the frequency of the action potential when a stimulus is increased above the threshold level?

Saltatory conduction

In neurons that are insulated by myelin, the ions can only pass in and out of the axon freely at the nodes of Ranvier, which are about 1 mm apart. This means that action potentials can only occur at the nodes and so they appear to jump from one to the next. This is shown in Figure 9.7.

As the movement of ions associated with the action potential occur much less frequently, the process takes less time. The effect is the increased speed of the impulse.

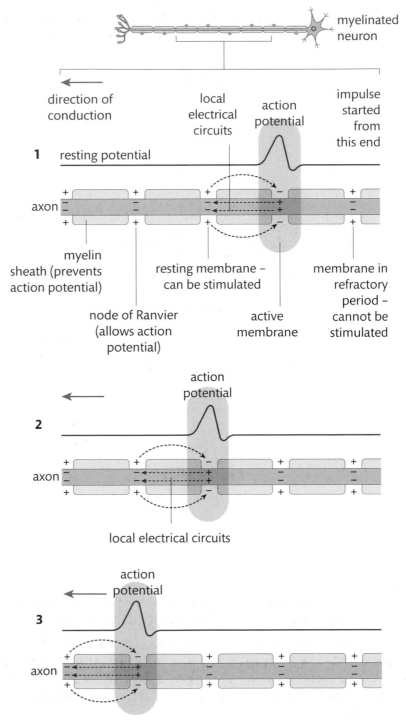

▶ **Figure 9.7: Saltatory conduction.** By 'jumping' from node to node along a myelinated nerve fibre, the nerve impulses in vertebrate neurons can travel very rapidly along very narrow nerve fibres. This allows for the development of complex, but compact, nervous systems.

The role of the synapse

The junction where two neurons meet is called a synapse. Figure 9.8 shows the structure of a synapse revealed by using an electron microscope. At the synapse, the neurons do not touch and are separated by a narrow gap called the synaptic cleft.

The neuron carrying the impulse to the synapse is termed the presynaptic neuron and the one carrying the impulse away is termed the postsynaptic neuron.

Information passes across the synaptic cleft from the presynaptic neuron to the postsynaptic neuron in the form of chemicals called neurotransmitters.

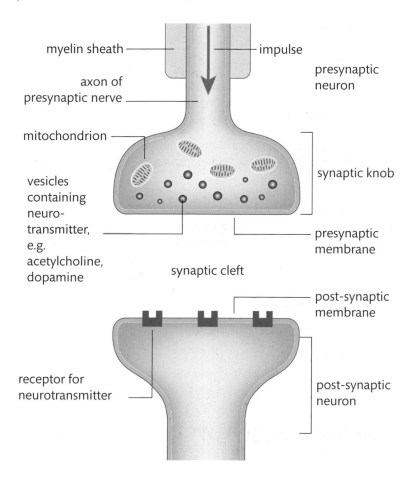

▶ **Figure 9.8:** The structure of the synapse

Neurotransmitters

Different neurons release different neurotransmitters, which diffuse across the synaptic cleft to trigger an action potential in the postsynaptic neuron.

Neurons that produce neurotransmitters which decrease the potential of the **postsynaptic membrane** and make it more likely to produce an impulse are termed excitatory presynaptic cells. Inhibitory presynaptic cells release neurotransmitters which increase the postsynaptic membrane potential and make it less likely to produce an impulse.

The minimum level of neurotransmitter required to produce a postsynaptic action potential is called the **threshold level**.

> **Key terms**
>
> **Postsynaptic membrane** – the membrane of the cell body or dendrite of the neuron carrying the impulse away from the synapse. It contains a number of channels to allow ions to flow through, and protein molecules which act as receptors for the neurotransmitter.
>
> **Threshold level** – the point at which increasing stimuli trigger the generation of an electrical impulse.

Drugs that affect the nervous system do so by speeding up and slowing down the transmission of nerve impulses across the synapse. They are classified as excitatory or inhibitory drugs.

Research examples of excitatory and inhibitory drugs. Find out how they act on the synapse. Find examples of these drugs that have been misused by sports-people to enhance their performance. How do they improve performance and what are the side effects on health?

Synaptic transmission

Acetylcholine and dopamine are examples of neurotransmitters released by excitatory presynaptic cells.

When the action potential arrives at the **axon terminal**, it causes calcium channels in the **presynaptic membrane** to open. As the concentration of calcium ions is greater in the synaptic cleft than the axon terminal, they diffuse into the axon terminal.

The increased presence of calcium ions in the axon terminal causes the synaptic vesicles to move towards the presynaptic membrane. The vesicles fuse with the membrane and release the neurotransmitter, acetylcholine, into the synaptic cleft.

Acetylcholine diffuses across the synaptic cleft and attaches to the receptor site on the postsynaptic membrane. The binding of the neurotransmitter to the receptors causes sodium channels to open in the postsynaptic membrane. As synaptic vesicles are only present in the axon terminal of the presynaptic neuron, impulses can only travel in one direction.

Sodium ions diffuse into the postsynaptic cell, causing depolarisation and an action potential to be set up. Enzymes split acetylcholine into acetate and choline so that it is removed from the receptor sites. The sodium channels close so that further action potentials stop.

The presynaptic cell takes up the choline by active transport using energy from ATP, where it is combined with acetyl coenzyme A to reform acetylcholine inside the axon terminal.

Key terms

Axon terminal – the axon of a neuron ends in a swelling called the axon terminal. It contains mitochondria which provide energy for active transport, and synaptic vesicles which release the neurotransmitter into the synaptic cleft.

Presynaptic membrane – the axon terminal membrane of the neuron carrying the impulse to the synapse.

 PAUSE POINT　　What are the processes that take place at a synapse?

Hint　　Draw an annotated diagram of a synapse to explain the function of each structure.

Extend　　Why does the synapse have a high concentration of mitochondria?

Responding to a stimulus

Being able to respond to changes in our environment is essential to our safety and survival. It is the function of the nervous system to enable us to detect changes and coordinate actions in response to these changes.

The nervous system enables us to respond to changes by:

▶ detecting changes (stimuli) inside the body and in the external environment
▶ interpreting the change and deciding how to respond to it
▶ coordinating actions or behaviours that bring about a response to the change, such as moving away from something dangerous.

Figure 9.9 shows the sequence of events that occur in a **voluntary response**.

Key term

Voluntary response – a conscious action taken in response to a stimulus (change in the environment).

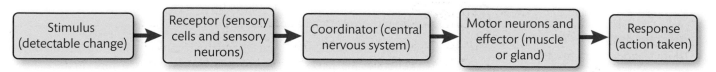

| Stimulus (detectable change) | → | Receptor (sensory cells and sensory neurons) | → | Coordinator (central nervous system) | → | Motor neurons and effector (muscle or gland) | → | Response (action taken) |

▶ **Figure 9.9:** The stages of a voluntary response

A reflex action is a rapid and unconscious response brought about by the nervous system. Many reflex actions are protective actions and occur in response to harmful stimuli but many of the actions that your body performs without you thinking about them, such as coughing and swallowing, are also reflex actions.

The neurons involved in a reflex make up a reflex arc. Figure 9.10 shows the reflex arc involved in removing your hand from a hot object.

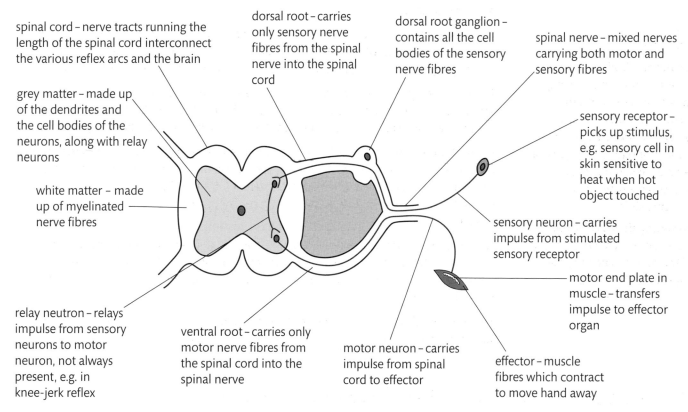

spinal cord – nerve tracts running the length of the spinal cord interconnect the various reflex arcs and the brain

dorsal root – carries only sensory nerve fibres from the spinal nerve into the spinal cord

dorsal root ganglion – contains all the cell bodies of the sensory nerve fibres

spinal nerve – mixed nerves carrying both motor and sensory fibres

grey matter – made up of the dendrites and the cell bodies of the neurons, along with relay neurons

sensory receptor – picks up stimulus, e.g. sensory cell in skin sensitive to heat when hot object touched

white matter – made up of myelinated nerve fibres

sensory neuron – carries impulse from stimulated sensory receptor

motor end plate in muscle – transfers impulse to effector organ

relay neutron – relays impulse from sensory neurons to motor neuron, not always present, e.g. in knee-jerk reflex

ventral root – carries only motor nerve fibres from the spinal cord into the spinal nerve

motor neuron – carries impulse from spinal cord to effector

effector – muscle fibres which contract to move hand away

▶ **Figure 9.10:** The reflex arc showing the structures and sequence of events involved in a reflex action

As you can see in the diagram, the brain is not involved in the reflex arc. This is why the response is unconscious. Instead, the sensory neuron forms a synapse with an **interneuron** (or relay neuron) which forms a synapse with the motor neuron. Figure 9.11 shows the structure of an interneuron.

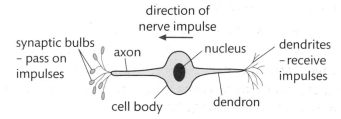

direction of nerve impulse

synaptic bulbs – pass on impulses

axon

nucleus

dendrites – receive impulses

cell body

dendron

▶ **Figure 9.11:** The structure of an interneuron. Interneurons are found in the spinal cord.

Key term

Interneuron – a type of nerve cell found inside the central nervous system that acts as a link between sensory neurons and motor neurons.

Impulses will travel along the spinal cord to the brain. This is why you become aware of the reflex action shortly after it happens.

The receptors in the reflex shown in Figure 9.10 are thermoreceptors in the dermis of the finger, which generates the sense of pain. The effectors are muscle fibres in the hand.

The thermoreceptors initiate nerve impulses that travel along the **afferent pathway**, which is along the sensory neuron to the spinal cord. The sensory neuron enters the spinal cord and forms a synapse with an interneuron located in the grey matter of the spinal cord.

The interneuron forms a synapse with a motor neuron. The impulse leaves the spinal cord via the **efferent pathway**, which is along the motor neuron to the effector, the muscles of the hand and arm. The muscles contract to move the finger away from the hot surface.

The neuromuscular junction

A neuromuscular junction is a synapse between a motor neuron and a muscle. The structure and function is similar to that of a synapse between two neurons.

When the axon reaches a muscle, it forms branches and loses its myelin sheath. The axon branches to make contact with different fibres in the muscle in a plate-like structure called the neuromuscular junction or motor end plate.

The motor end plates consist of folds of the muscle fibre surface and are located opposite the axon terminal knob. There is a small gap between the membrane of the neuron and the muscle fibre called the synaptic cleft.

A neuromuscular junction functions in a similar way to the synapse between two neurons. The following is a summary of transmission at the neuromuscular junction.

▸ The action potential arrives at the neuromuscular junction.

▸ Calcium ion channel proteins open and calcium ions diffuse into the synaptic cleft.

▸ The diffusion of calcium ions causes the synaptic vesicles to move to the junction membrane.

▸ The vesicles fuse with the junction membrane and release acetylcholine (neurotransmitter) into the synaptic cleft.

▸ Acetylcholine diffuses across the cleft and attaches to the receptor molecules on the muscle fibre.

▸ Sodium ion channels open in the muscle fibre membrane.

▸ The movement of sodium into the cytoplasm of the muscle fibre causes depolarisation.

▸ An action potential is generated across the muscle fibre.

▸ The muscle contracts.

A neuroglandular junction is where a neuron and a gland interact.

(II) PAUSE POINT   Can you explain how an impulse is generated and transmitted from neuron to neuron and at a neuromuscular junction?

Hint   Close the book and draw a flow diagram to show the stages involved for each type of synaptic transmission.

Extend   Some poisons have an antagonistic effect on synaptic transmission. Find out how curare, hemlock and botulin act on the synapse, and their resulting effects on the nervous system and the human body.

Stimuli detection by receptor cells and sense organs

The human body needs to detect and respond to changes in its surroundings. Sense organs are specialised organs, such as the eye, ear and skin, where sensory neurons are concentrated to form receptors. Receptors detect specific changes in the environment, which are called stimuli.

Receptor cells act by converting stimuli into electrical responses in neurons. The process of converting one type of energy into the electrochemical energy of an action impulse is called **transduction** (or signal transduction).

Receptors are only able to respond to specific stimuli. A summary is shown in Table 9.3.

▶ **Table 9.3:** Examples of receptors in the human body and their stimuli

Receptor	Stimuli detected	Examples
Chemoreceptors	Chemical stimuli	Nose and mouth
Photoreceptors	Light energy	Eyes
Thermoreceptors	Temperature changes	Skin
Mechanoreceptors	Changes in movement, pressure or vibrations	Pacinian receptor in the dermis
Electroreceptors	Electrical fields	Mainly found in fish

Receptor cells act as transducers. This means they convert the energy of the stimulus into the electrical energy of a nerve impulse which is transmitted along a sensory neuron to the **central nervous system**.

The frequency of the impulse sends messages to the brain about the strength of the stimulus, which enables the body to respond in an appropriate way. Receptor cells can act individually or in a group in a sense organ.

What do receptor cells do?

A receptor cell responds to a specific stimulus by initiating an action potential in a sensory neuron, which carries an impulse to the central nervous system where it is interpreted and a response is coordinated.

When a receptor cell is stimulated, sodium ions move across the cell membrane in a similar way to that which takes place when an action potential is generated in a neuron.

Figure 9.12 shows how the generator potential is developed in the receptor cell.

> **Key terms**
>
> **Transduction** – the conversion of a signal from outside the cell to a functional change within the cell, e.g. odour to electrochemical signals.
>
> **Central nervous system** (CNS) – consists of the brain and spinal cord.

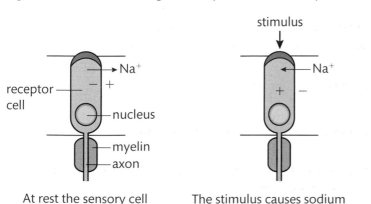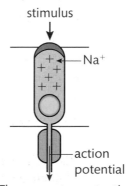

At rest the sensory cell inside is negative relative to the outside.

The stimulus causes sodium ions to enter the cell. The cell inside becomes less negative. The generator potential is initiated.

The generator potential builds up and initiates an action potential.

▶ **Figure 9.12:** Simple illustration of how a generator potential and action potential are developed by a receptor cell

As Figure 9.12 shows, the function of the receptor cell is to produce a generator potential which initiates an action potential. The way that this happens is specific to the type of receptor. In mechanoreceptors, receptors that detect movement or changes in pressure, it is physical changes to the cell caused by pressure or movement of a tiny hair that causes the sodium channels in the cell membrane to open up and cause depolarisation of the cell membrane.

In other receptors, stimuli may cause a series of chemical reactions to take place which then lead to the sodium channels in the cell membrane opening and causing depolarisation.

Neurological disorders

Motor neuron disease

The 2014 film *The Theory of Everything* was a biographical account of the life of the world-famous physicist, Stephen Hawking, who developed motor neuron disease (MND) as a university student. MND is a fatal disease, which arises from the degeneration of motor neurons in the spinal cord.

MND is characterised by:

▸ impairment of the use of the limbs

▸ twitching and cramping of muscles in the hands and feet

▸ difficulty in speaking and projecting the voice

▸ difficulty in breathing and swallowing.

▸ Stephen Hawking developed motor neuron disease when he was a university student

Parkinson's disease

Parkinson's disease develops as a result of a deficiency of the neurotransmitter, dopamine, which is caused by a loss of nerve cells in part of the brain called the substantia nigra. Nerve cells in this part of the brain produce dopamine, which acts as a messenger between the brain and **peripheral nervous system** to control and coordinate body movements. Loss of the nerve cells is a slow process. The symptoms of Parkinson's disease only start to develop when 80% of the nerve cells in the substantia nigra have been lost.

The symptoms of Parkinson's disease are:

▸ involuntary shaking

▸ slow movement

▸ stiff and inflexible muscles.

Multiple sclerosis

Multiple sclerosis (MS) is an autoimmune condition. This means that the body's immune system has begun to attack body tissues. In MS, the immune system mistakes the myelin for a foreign substance and starts to attack it. This disrupts the impulses travelling along the neurons, causing the impulses to be slowed, jumbled and sent down another neuron or stopped altogether.

There are many different symptoms of MS. The most common ones are:

▶ fatigue
▶ numbness and tingling in the limbs
▶ blurring of the vision

▶ mobility difficulties
▶ problems with balance
▶ muscle weakness.

Ⅱ PAUSE POINT — Can you explain how the three neurological disorders are caused? Cover the section about the nervous system disorders and write a summary of each of the three diseases discussed in this section.

Hint — Produce a large diagram to show the main structures of the nervous system. Annotate the diagram to show the structures and functions of the nervous system that are affected by each disease, and why each symptom occurs.

Extend — Mercury poisoning was common among hat makers in the 1800s, when a mercury solution would be applied to animal fur to make felt. Research how mercury affects the nervous system and produce a poster to show your findings.

Cardiovascular system regulation and control

Receptors in the cardiovascular system

The **cardiovascular system** is controlled by a number of reflex actions which are initiated by receptors that detect changes in blood pressure and blood pH levels.

Chemoreceptors in the cardiovascular system

If you have felt short of breath, it is because chemoreceptors in in your cardiovascular system have detected that your blood oxygen levels are low or that your blood carbon dioxide levels are too high. These chemoreceptors are located in the walls of the aorta and carotid arteries, as shown in Figure 9.13.

Key term

Cardiovascular system – the heart and blood vessels.

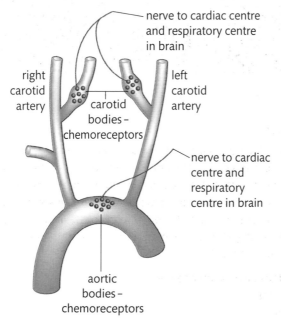

nerve to cardiac centre and respiratory centre in brain

right carotid artery

left carotid artery

carotid bodies – chemoreceptors

nerve to cardiac centre and respiratory centre in brain

aortic bodies – chemoreceptors

▶ **Figure 9.13:** Chemoreceptors in the walls of aorta and carotid arteries

These receptors are sensitive to the levels of carbon dioxide in the blood. As carbon dioxide levels rise, the pH of the blood decreases (the blood becomes more acidic) and this is detected by the aortic and carotid chemoreceptors.

When a chemoreceptor detects a fall in the blood pH, a small depolarisation occurs in its cell membrane and an action potential is produced. This generates an electrical impulse.

The impulse travels along sensory neurones to the cardiac control centre in the medulla of the brain. This increases the impulses travelling down the sympathetic nerve to the heart. As a result, the heart rate increases, and there is an increased blood flow to the lungs. The effect of this increased blood flow is that carbon dioxide is removed from the blood.

As blood carbon dioxide levels fall, the blood pH rises. The chemoreceptors respond to this by reducing the number of impulses to the cardiac centre. This reduces the number of impulses in the sympathetic nerve to the heart and reduces the acceleration of the heart rate, so it returns to the intrinsic rhythm.

The chemoreceptors are also involved in the control of the breathing rate.

Baroreceptors in the cardiovascular system

Baroreceptors are pressure receptors located in the walls of the aorta and carotid artery. Their function is to detect changes in blood pressure and send this information to the cardiovascular centre of the brain. It is important that the blood pressure remains at an adequate level so that blood can reach all of the tissues and organs.

If the blood pressure drops too low, then blood will not reach all of the tissues. If the blood pressure rises too high, then it may cause damage to the blood vessels and eventually lead to heart disease or stroke.

If the blood pressure rises, the walls of the blood vessels will stretch more. This stimulates the baroreceptors in the blood vessel walls. The baroreceptors generate a greater number of action potentials to the cardiovascular centre which then initiates responses to cause a decrease in blood pressure.

If blood pressure falls, there will be a decrease in the number of action potentials sent from the baroreceptors to the cardiovascular centre, which initiates responses to increase blood pressure.

<table>
<tr><td>

Key term

Baroreceptor – stretch receptors found in the blood vessels that respond to changes in blood pressure in the blood vessels.

</td></tr>
</table>

PAUSE POINT Explain how chemoreceptors and baroreceptors control the cardiovascular system.

Hint Make a simple flow diagram to show what happens when carbon dioxide levels in the blood increase.

Extend Find out how adrenaline affects the heart rate.

Gas exchange

In order to stay healthy, all of the cells in your body require a constant supply of energy. Energy is provided by **respiration**, which is a series of oxidation reactions that occur within the cells. Respiration therefore requires a supply of oxygen, which is brought to the cell by the blood. It also requires a supply of glucose.

Respiration produces energy as a useful product and waste products: carbon dioxide and water.

As respiration is a constant process in the body's cells, there is a constant need for oxygen to be brought to the cells and for carbon dioxide to be removed.

Gas exchange is the process where oxygen is supplied to the cells and carbon dioxide is removed.

The control of gas exchange

Oxygen is constantly taken up by cells and carbon dioxide is constantly released. This is called gas exchange and occurs by diffusion.

As organisms increase in size, their surface area to volume ratio decreases and diffusion alone is an insufficient mechanism for efficient gas exchange. A large fluctuation of respiratory gases can have harmful effects on the body.

A deficiency of oxygen (hypoxia) deprives cells of the vital requirement of **metabolism**. A build-up of carbon dioxide in the tissues leads to increased acidity of the blood and tissues, which inhibits enzymes, stops metabolism and would quickly prove fatal.

In the human body, breathing enables a constant supply of oxygen and constant removal of carbon dioxide. The cardiovascular system provides a transport mechanism to carry oxygen, nutrients, carbon dioxide, hormones and waste products to and from exchange surfaces.

The lungs and breathing

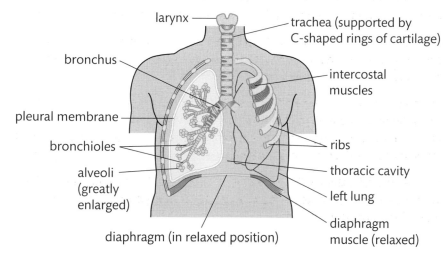

larynx — trachea (supported by C-shaped rings of cartilage)

bronchus

intercostal muscles

pleural membrane

bronchioles

ribs

alveoli (greatly enlarged)

thoracic cavity

left lung

diaphragm muscle (relaxed)

diaphragm (in relaxed position)

▶ **Figure 9.14:** The lungs and associated structures of the thoracic cavity

The human lung is an efficient structure which enables maximum gas exchange to take place with minimum heat loss. The structure of the lungs are shown in Figure 9.14.

The regular breathing pattern is an automatic action controlled by nerve impulses from the **ventilation centre** in the brain. However, as impulses are also received from higher centres in the brain, the breathing rate can be brought under voluntary control. This allows a person to hold their breath while diving, for example.

Breathing in (inhalation)

▶ The diaphragm muscles contract.

▶ The diaphragm flattens.

▶ The intercostal muscles between the ribs contract, lifting the rib cage upwards and outwards.

Key terms

Respiration – a series of oxidation reactions that take place in all living cells to produce ATP, carbon dioxide and water from organic compounds such as glucose.

Gas exchange – the diffusion of oxygen into cells and the diffusion of carbon dioxide out of the cells to enable respiration to take place.

Metabolism – the chemical reactions that occur within the body to maintain life.

Key term

Ventilation centre – groups of nerve cells located in the brain that control the pattern and rate of breathing.

- The volume of the thoracic cavity increases.
- The air pressure inside the thorax becomes lower than the external environment.
- Air moves down the concentration gradient into the lungs and into the alveoli where gas exchange takes place.

Breathing out (exhalation)
- The rib cage drops downwards and inwards.
- The diaphragm relaxes and domes upwards.
- The volume of the thoracic cavity decreases.
- The elasticity of the lung tissues means the lungs recoil to their original size.
- Air pressure inside the thorax becomes greater than the external environment and the air moves out of the lungs.

PAUSE POINT   What is gas exchange, where does it occur and why is it necessary?

Hint   Explain how oxygen moves from the lungs to a muscle cell and how carbon dioxide moves from the same cell to the lungs.

Extend   Find out about Fick's Law. How do the lungs follow this law?

Case study

The Hering-Breuer reflex

In 1868, two scientists, Josef Breuer and Ewald Hering, discovered a reflex action that prevents over inflation of the lungs. They discovered that stretch receptors in the smooth muscle of the airways respond to excessive stretching of the lung during large inhalations.

As the lungs inflate, the frequency of nerve impulses from stretch receptors in the bronchi to the ventilation centre increases until a point is reached where inhalation is inhibited. Tissues that were stretched during inhalation recoil and air is forced out of the lungs. This is the Hering-Breuer reflex.

When someone is placed on a ventilator because he or she is having problems breathing, care must be taken by hospital staff to avoid over-inflating the lungs. As the ventilator is doing the breathing for the patient, his/her Hering-Breuer reflex cannot initiate to regulate the size of their breaths.

The ventilator has to be programmed to adjust the volume of air pushed into the patient's lungs and the frequency of breaths. This ensures that the patient receives the right amount of oxygen and the lungs are not damaged.

Check your knowledge

1 Try taking a deep breath. Why can you only breathe in a limited amount of air before you have to breathe it out again?

2 Why is it important for anaesthetists to understand the Hering-Breuer reflex in order to do their job safely?

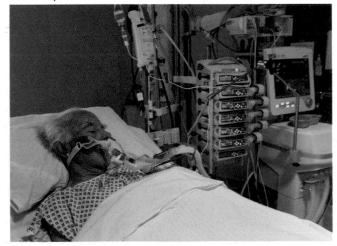

▶ When a patient is connected to a ventilator, hospital staff need to ensure that the lungs are not over-inflated

The ventilation centre sends nerve impulses to the intercostal muscles and the diaphragm.

When the intercostal muscles and the diaphragm contract, the space inside the thorax increases. This causes a decrease in the air pressure relative to the external environment. The result is that air moves into the lungs and you experience this as breathing in, inhalation.

Exhalation occurs because the intercostal muscles and the diaphragm relax. This causes the volume of the thoracic cavity to decrease and the air pressure inside to increase relative to the external environment.

Gaseous exchange in the alveoli

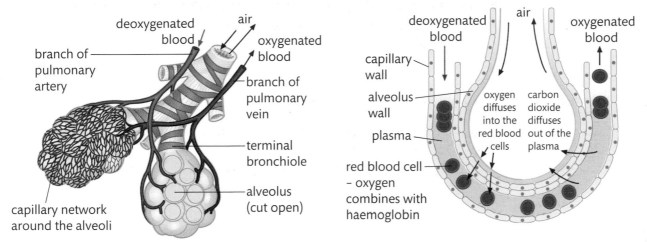

▶ **Figure 9.15:** Gaseous exchange in the alveoli

The lungs have a number of adaptations which make them highly efficient in the process of gas exchange.

▶ They contain millions of air sacs (alveoli) which creates a large surface area to enable rapid diffusion of oxygen and carbon dioxide.

▶ The alveoli have walls that are one cell thick creating a short diffusion pathway.

▶ The lungs have a rich blood supply enabling each alveolus to be close to a capillary. Capillary walls are one cell thick, which allows gases to pass rapidly from the alveolus into the blood and vice versa.

▶ The flow of blood through the capillaries means that a steep concentration gradient is maintained, which ensures rapid diffusion of gases (as described in the section below).

Diffusion of oxygen

The air in the alveolus is rich in oxygen. Blood arriving from the body in the capillaries of the alveoli is low in oxygen. Oxygen diffuses down the concentration gradient from the alveolus to the capillary where it combines with haemoglobin in the red blood cells and is transported to the rest of the body.

Diffusion of carbon dioxide

Carbon dioxide from cellular respiration in the body is transported in the blood to the lungs. Blood arriving at the alveolus is high in carbon dioxide and the air in the alveolus is low in carbon dioxide. Carbon dioxide diffuses down the concentration gradient into the alveolus, where it is exhaled into the external environment.

The constant pumping of blood through the capillary and ventilation of the lungs ensures that a steep concentration gradient is maintained which, when combined with short diffusion pathway, ensures that a rapid rate of diffusion is maintained.

Chemoreceptors

During exercise, carbon dioxide levels in the blood increase due to increased levels of respiration taking place in body cells. Carbon dioxide in the blood forms carbonic acid, which leads to a fall in blood pH levels.

When the pH of the blood decreases, chemoreceptors in the carotid artery and aorta are stimulated and send impulses to the ventilation centre. The ventilation centre responds by sending impulses to the external intercostal muscles and the diaphragm to increase the breathing rate. This is a function of the sympathetic nervous system.

Circulation of the blood

Figure 9.16 shows the general layout of the cardiovascular system and the direction the blood flows around it.

The cardiovascular system comprises a muscular pump, the heart, which pumps blood into a system of blood vessels. Blood is first pumped into arteries which divide into smaller vessels called arterioles. Arterioles divide into networks of tiny blood vessels in the tissues called capillaries, where exchange of materials between the tissues and blood takes place. From the capillaries the blood is carried into larger vessels called venules, which join to form veins, larger vessels that carry the blood back to the heart.

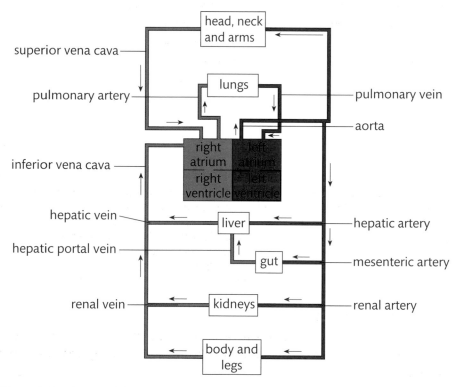

▶ **Figure 9.16:** The main blood vessels and direction of blood flow around the cardiovascular system

The structure and function of blood vessels

A closed circulation

Blood vessels form a closed system that begins and ends with the heart. The blood is always enclosed by arteries, arterioles, capillaries, venules or veins (see Table 9.4), which vary in diameter.

Arteries carry blood away from the heart and divide to form smaller arteries and arterioles. Arterioles subdivide to form capillaries which form networks of tiny blood vessels in the tissues. Capillaries join up to form venules, which join up to form veins. Veins carry blood back to the heart.

▶ **Table 9.4:** Structure and function of arteries, veins and capillaries

Arteries	Veins	Capillaries
artery lumen — endothelium — elastic fibres — smooth muscle — collagen fibres	**vein** lumen — endothelium — elastic fibres — smooth muscle — collagen fibres	**capillary** lumen — endothelium
• Carry blood away from the heart • Thick muscular walls • Large amount of elastin in walls • Small lumen (inner open space within the vessel) • High blood pressure • Rapid blood flow • Pulse • No valves	• Carry blood back to the heart • Thin muscular walls • Small amount of elastin in walls • Large lumen • Low pressure • Slow blood flow • No pulse • Valves to prevent backflow of blood	• Form networks in the tissues of the body • Link arterioles and venules • Walls are made up of a single layer of endothelium cells • No elastin fibres or muscle • Small lumen, just enough to allow blood cell to pass through • Little pressure • Slow blood flow • No pulse • No valves
Function Carry fast-flowing blood under high pressure away from the heart. Elastic walls enable the vessel to stretch and recoil to keep the blood flowing.	**Function** Carry slow-flowing blood under low pressure back to the heart. There is sufficient pressure to force valves in the veins to open, and backflow of blood causes the valves to close, therefore keeping blood flow in one direction.	**Function** Networks of tiny, thin blood vessels in the tissues of the body that supply blood to all tissues and cells of the body. Thin walls create a short diffusion pathway to enable rapid diffusion of substances between the tissues and the blood.

The structure of the heart

The human heart is made up of cardiac muscle. The cells in the muscle fibres of cardiac muscle are interconnected, which enables impulses to spread rapidly from muscle cell to muscle cell.

Heart muscle is myogenic, meaning it can contract and relax rhythmically without fatigue and of its own accord.

Figure 9.17 shows the external and internal structure of the heart. The heart is a double pump, which means it comprises of two pumps side by side. The right side of the heart pumps blood to the lungs and the left side of the heart pumps blood to the rest of the body. Each side of the heart is separated from the other so that oxygenated and deoxygenated blood are kept separate.

Each side of the heart is comprised two chambers: the atrium (upper chamber) and the ventricle (lower chamber). These are separated by a valve, which ensures that blood flows in only one direction through the heart.

Step-by-step: Dissection of a mammalian heart

1 Place the heart on the dissection board and locate the tip of the heart or apex. Only the left ventricle extends all the way to the apex.

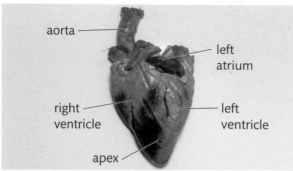

▶ Whole heart - Ventral view

2 Turn the heart around so that the front of the heart, or ventral side, is facing you. You can recognise the ventral side because it has a groove that extends from the right side at the broad end of the heart diagonally to a point to the left and above the apex.

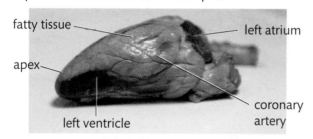

▶ Lateral view of the left side of the heart, showing the line of the coronary artery running through fatty tissue covering the heart.

3 Locate the upper chambers, the atria, and the lower chambers, the ventricles, and the blood vessels. The arteries have thick, rubbery walls and the veins have thinner walls.

▶ The pulmonary artery (showing the branches taking deoxygenated blood to the left and right lungs).

4 Using a scalpel, cut through the side of the pulmonary artery (curves out of the right ventricle) and continue cutting down the wall of the right ventricle. Take care to only cut deep enough to go through the wall of the heart chamber.

▶ The first incision into the right atrium.

5 With your fingers, push open the heart at the cut so that you can see the internal structure. Locate the ventricles, the atria, the septum and the valves (tough, stringy structures). Look at the blood vessels and note the differences between them.

▶ The interior wall of the right atrium showing the comb-like structure of the pectinate muscles.

6 Using scissors, make a further cut from the outside of the left atrium downwards through the left ventricle to the apex. Push open the heart with your fingers and locate the ventricles, atria and valves. Note the differences in thickness of the two sides of the heart. Clear away following the instructions of your tutor and wash your hands thoroughly when you have finished.

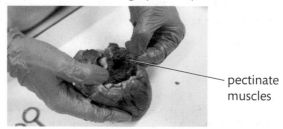

▶ Cusp (leaflet) of the right atrio-ventricular valve held in place by tendon-like cords (chordae tendineae or heart strings).

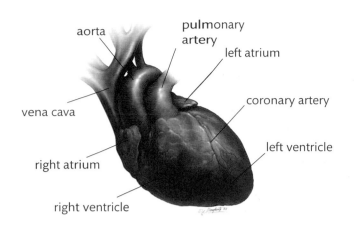 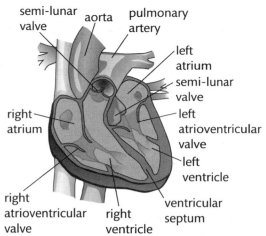

▶ **Figure 9.17:** External and internal structure of the heart

The cardiac cycle

The cardiac cycle describes the sequence of events in one complete heartbeat.

1   Both atria relax and fill with blood from the vena cava or pulmonary vein.

2   The atria contract and force open the atrioventricular (AV) valves.

3   Blood flows into the ventricles.

4   The pressure of the blood filling the ventricles makes the AV valves close.

5   The ventricle walls contract, causing increased pressure inside the ventricles.

6   When the pressures inside the ventricles exceed the pressure in the adjoining blood vessels, the semi-lunar valves are forced open.

7   Blood enters the pulmonary artery or the aorta.

8   The semi-lunar valves close and prevent the back flow of blood into the ventricles.

Control of the cardiac cycle

The heart does not need impulses from the nervous system in order to contract and relax. Each cardiac cycle is started by specialised muscle cells in the right atrium called the **sinoatrial node** (SAN).

The SAN sends electrical impulses to the atria walls. This causes depolarisation to take place. The effect is that the electrical impulses spread across both atria as a wave and both atria contract at the same time.

Collagen fibres between the atria and ventricles prevent the impulse wave spreading to the ventricles. This is important as it ensures that the ventricles do not contract until the atria have finished contracting.

When the wave meets the junction between the atria and the ventricles, it causes the **atrioventricular node** (AVN) to generate its own electrical impulse. The AVN transmits an impulse down strands of fibres lying between the ventricles called the Bundle of His.

Key term

Sinoatrial node (SAN) – specialised muscle cells in the right atrium that start the cardiac cycle by sending impulses across the atria walls. This is often called the heart's pacemaker as these cells control the speed of the cardiac cycle.

Key term

Atrioventricular node (AVN) – specialised muscle cells in the junction of the atria and ventricles that receive impulses from the SAN and send impulses across the ventricle walls.

Getting ready for assessment

Scott is working towards a BTEC National in Applied Sciences. He was given an assignment with the following title 'Assess the role of the nervous system in coordinating the cardiovascular and respiratory systems'. He had to produce a booklet suitable for trainee health care assistants to use. He had to ensure that his booklet included:

▶ information about the structures in the nervous, cardiovascular and respiratory systems

▶ information about how the nervous system responds to stimuli

▶ information about the changes that occur in the cardiovascular and respiratory systems and what causes them

▶ explanations of how the nervous system coordinates the cardiovascular and respiratory systems.

How I got started

First I collected all my notes on this topic and put them together into a folder. I decided to sort my notes into three sections – the nervous system, the cardiovascular system and the respiratory system. I needed to make sure I included enough work in each section to achieve all the criteria.

I then drew a concept map so that I could see clearly the ways that each system worked and I could add how they are linked on to my concept map. I was then able to see clearly the role that the nervous system played in the control of the systems.

I attended a public lecture about the cardiovascular system at my local FE college.

How I brought it all together

I organised my booklet into three chapters to ensure that my work was in a clear coherent order. In each chapter I included:

- an introduction which included the main organs in each system
- clear annotated diagrams to support the explanations
- an assessment of the role of the nervous system in the coordination of the system.

I ended each chapter with a summary and references for further reading.

What I learned from the experience

I learned the importance of planning and being organised. When writing about the body systems, there are a lot of parts to include and a lot of detail about interactions to include, so it's vital that you plan how you are going to present your work so that it makes sense to the reader.

Think about it

▶ Have you written a plan with timings so you can complete your assignment by the agreed submission date?

▶ Do you have notes for each of the systems mentioned in the assignment title?

▶ Is your information written in your own words and referenced clearly where you have used diagrams from textbooks or the Internet?

Biological Molecules and Metabolic Pathways 10

Getting to know your unit

Every living organism is built from different chemical elements. When these elements are combined, they make molecules such as carbohydrates, proteins and lipids. This unit will look at the structure and function of these molecules. Many biological molecules are made from smaller units called monomers, which are bonded together to form large structures. They are made from smaller units called monomers that are bonded together. Water is another important biological molecule. It enables the trees around us to grow. In fact, water is needed for all plants to photosynthesise and produce carbohydrates, so it is vital for their survival. Animals rely on plants for food and so they too would not survive without water. Water also enables nutrients to be transported in our blood and to allow chemical reactions to take place in our cells.

Biological molecules and metabolic pathways are extremely important in the science industry, in particular the health, chemical and environmental industries. For example, optimising biochemical pathways can improve the efficiency of photosynthesis and to increase the yield of crops. They are also used to produce biological agents that can neutralise contaminants in polluted soil or water. Biological molecules and metabolic pathways are an area of science that underpins and overlaps with many other branches of science such as pharmacology, physiology, microbiology and clinical chemistry.

How you will be assessed

This unit will be assessed by a series of internally assessed tasks set by your tutor. Throughout this unit, you will find assessment activity activities that will help you work towards your assessment. Completing these activities will not mean that you have achieved a particular grade, but you will have carried out useful research or preparation that will be relevant when it comes to your final assignment. In order for you to achieve the tasks in your assignment, it is important to check that you have met all of the Pass grading criteria. You can do this as you work your way through the assignment. If you are hoping to gain a Merit or Distinction, you should also make sure that you present the information in your assignment in the style that is required by the relevant assessment criterion. For example, Merit criteria require you to analyse and explain, and Distinction criteria require you to evaluate.

Assessment criteria

This table shows what you must do in order to achieve a **Pass**, **Merit** or **Distinction** grade, and where you can find activities to help you.

Pass	Merit	Distinction
Learning aim A Understand the importance of biological molecules in living organisms and the effect of disruption on the structure and function		
A.P1 Explain the structure of biological molecules in living organisms Assessment practice 10.1	**A.M1** Explain the links between the structure and function of biological molecules and their role in living organisms Assessment practice 10.1	**A.D1** A.D1 Evaluate the effects of disruption on the structure and function of biological molecules in living organisms Assessment practice 10.1
Learning aim B Explore the effect of activity on respiration in humans and factors that can affect respiratory pathways		
B.P2 Explain the stages involved in the human respiratory pathway Assessment practice 10.2	**B.M2** Analyse primary and secondary data to explain the effect of activity on respiration Assessment practice 10.2	**B.D2** Evaluate the effects of harmful substances on the efficiency of respiration Assessment practice 10.2
B.P3 Carry out an investigation involving the effect of activity on respiration in humans Assessment practice 10.2	**B.M3** Explain the harmful effects of factors on respiration Assessment practice 10.2	
B.P4 Describe factors that can affect respiration Assessment practice 10.2		
Learning aim C Explore the factors that can affect the pathways and the rate of photosynthesis in plants		
C.P5 Explain the stages involved in photosynthesis in plants Assessment practice 10.3	**C.M4** Analyse primary and secondary data to explain the outcomes of an investigation into a factor that affects the rate of photosynthesis Assessment practice 10.3	**C.D3** Evaluate the effect of factors on photosynthetic efficiency Assessment practice 10.3
C.P6 Carry out an investigation into a factor that affects the rate of photosynthesis Assessment practice 10.3		

Getting started

Water is one of the most important biological molecules that we use. Water makes up approximately 78% of the human body. It covers 75% of the earth existing in three different states of matter: a solid, a liquid and a gas. Water is vital for survival. Make notes on the roles of water in living organisms. When you have completed this unit, add any more roles to your list. Write down the chemical symbol of water. When you have completed this unit, draw the chemical structure including partial charges.

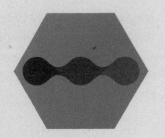

A Understand the importance of biological molecules in living organisms and the effect of disruption on the structure and function

Water structure and importance

Water structure

Link

Unit 1: Principles and Applications of Science 1 has more details about bonding.

The molecular structure of water (Figure 10.1) is what makes it unique. Water is a small molecule that consists of two hydrogen atoms that are **covalent bonds** to an oxygen atom. The electrons are not shared evenly; the oxygen atom pulls the electrons to it and away from the hydrogen atoms. In each covalent bond the electrons are not shared evenly (because oxygen is more electronegative than hydrogen), creating an uneven distribution across the molecule (see Figure 10.2). The oxygen pulls negatively charged electrons towards it and away from the hydrogen. The water molecule has regions of slight negative charge near to the oxygen and slight positive charges near to the hydrogens because of this water is described as a **polar molecule**.

$\delta-$

O

H   H

$\delta+$ $\delta+$

▶ **Figure 10.1:** Structural formula of water

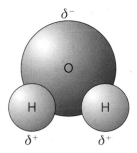

▶ **Figure 10.2:** A network of water molecules showing hydrogen bonds and partial positive charges and partial negative charges

Importance of water

Water is important, and has many functions, including:

▶ as a transport of molecules
▶ acting as a medium for chemical reactions
▶ regulating pH
▶ temperature regulator
▶ **electrolyte** balance.

Water is an excellent transport medium in living organisms because it stays in the liquid state over a large temperature range.

All metabolic processes in organisms rely on chemicals being able to react together in a solution. Water is a good **solvent** for chemical reactions, because molecules that are polar will dissolve in water. This is due to the fact that the solute (substance to be dissolved) also has an uneven charge distribution across the molecule. The slightly negative ends of the solute will be attracted to the slightly positive part of the water molecule, and vice versa. These interactions with the water molecule and the solute mean that the water molecule collects around the charged parts of the solute. The solute molecules are separated and they become dissolved. Once they are in the solution, molecules can move around and interact with other molecules.

Water has the ability to withstand temperature changes, as it takes a lot of energy to change the temperature. This is due to the **hydrogen bonds** between water molecules. When the temperature rises, water molecules gain more kinetic energy and vibrate more, so hydrogen bonds between the water molecules break. This means that water molecules make and break hydrogen bonds quicker so the make-break rate increases. It takes a relatively large amount of heat energy to break these bonds. When the temperature of water decreases, hydrogen bonds are able to be formed and the water molecules move less freely. This resistance to rapid temperature change means that water is an excellent habitat. Living organisms that live in water will not be exposed to potentially life-threatening temperature changes. Many organisms are mainly composed of water. The fact that water does not change temperature rapidly enables these organisms to regulate their internal body temperature. For example, your core body temperature does not drastically drop to the same temperature as the outside temperature while you are playing in the snow. It is very important that the body temperature remains stable to ensure the temperature is optimum for enzyme activity inside the body.

At pH 7, water contains equal concentrations of H^+ and OH^- **ions**. Living organisms are extremely sensitive to pH. They function best when internal conditions are closest to the optimum pH, so water plays an important role in regulating the pH in living organisms. **Buffer solutions**, such as that found in blood, stop the pH changing when hydrogen or hydroxide ions are added. Water has the ability to accept and donate H^+ where necessary, which means it plays an important role in keeping pH steady. Without water, solutions would not be able to keep the pH required.

Water also plays an important role in electrolyte balance. For instance, when **extracellular** electrolyte concentration rises, water diffuses out of the cell by **osmosis** into the extracellular space, diluting the extracellular fluid and raising the **intracellular** electrolyte concentration level.

Key terms

Electrolyte – a chemical compound that will conduct electricity in solutions.

Solvent – a substance that is able to dissolve other substances.

Hydrogen bond – a weak interaction that can occur between molecules that contain a slightly negatively charged atom and a slightly positively charged hydrogen.

Key terms

H^+ – positively charged hydrogen ion.

OH^- – oxygen and hydrogen atom held together by a covalent bond, carrying a negative electric charge.

Ions – electrically charged particles formed when atoms gain or lose electrons.

Buffer solution – a solution that resists changes in pH when small quantities of an acid or an alkali are added to it.

Extracellular – taking place outside a cell.

Osmosis – the movement of water from a region of high water potential to low water potential across a partially permeable membrane.

Intracellular – occurring within a cell.

Ⅱ PAUSE POINT   Explain the importance of water as a biological molecule.

> Hint   Draw a water molecule and list five functions of water.

> Extend   Explain (a) why water is a good solvent and (b) why it can regulate temperature.

Carbohydrate structure and importance

Carbohydrates are an essential part of your diet in order for you to gain energy. They are vital for all living organisms, as they act as an energy source and energy store, and are used for structure. Carbohydrates contain the elements:

▶ carbon ▶ hydrogen ▶ oxygen.

Carbohydrate means 'hydrated carbon'. The general formula for carbohydrates is $C_n(H_2O)_n$ (where n is the number of carbon atoms). Therefore for every carbon atom, there is an equivalent formula of water molecule.

Importance of carbohydrates

Carbohydrates are used by the body for:

▶ ATP production (the body's chemical energy)

▶ energy storage

▶ structural support

▶ lipid metabolism.

Using carbohydrates as an energy source also prevents important proteins from being broken down for energy in animals.

Monosaccharides (single sugars)

Monosaccharides are the simplest carbohydrate. They are **monomers**. When monosaccharides join together, they eventually make a more complex carbohydrate. Monosaccharides all have similar properties. For example, they:

▶ are soluble in water ▶ form crystals ▶ taste sweet.

Monosaccharides are classified according to the number of carbon atoms they have, as shown in Table 10.1.

▶ **Table 10.1:** Types of monosaccharide sugars

Number of carbons	Type of sugar
3-carbon	Triose
5-carbon	Pentose
6-carbon	Hexose

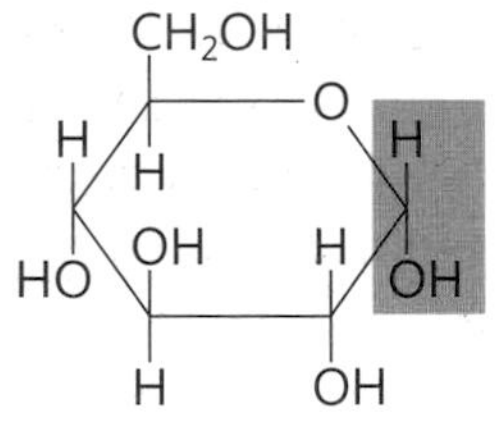

Glucose

Glucose is the main source of energy for many organisms. Glucose is described as a hexose monosaccharide, as it contains six carbons, and the formula of glucose is $C_6H_{12}O_6$.

In the molecular structure diagrams, the carbons are numbered clockwise. In alpha glucose, the hydroxyl group (OH) at carbon I is below the plane of the ring (Figure 10.3) and in beta glucose it is above the plane of the ring (Figure 10.4). The hydroxyl group (OH) on carbon 1 is in the opposite position on alpha and beta glucose. Both glucose molecules have different functions. Alpha glucose is used in respiration in plants and animals. This is because the **enzymes** used to speed up the respiration have active sites complementary to the shape of alpha glucose and not beta glucose.

Glucose molecules are polar and soluble in water, because hydrogen bonds can form in between the hydroxyl group and the water molecule.

Key terms

Monosaccharide – a single carbohydrate molecule.

Monomer – a single small molecule that can be joined with others to form a **polymer**.

Polymer – a single large molecule made from repeating units of monomers.

Enzyme – a biological catalyst.

▶ **Figure 10.3:** Structure of alpha glucose

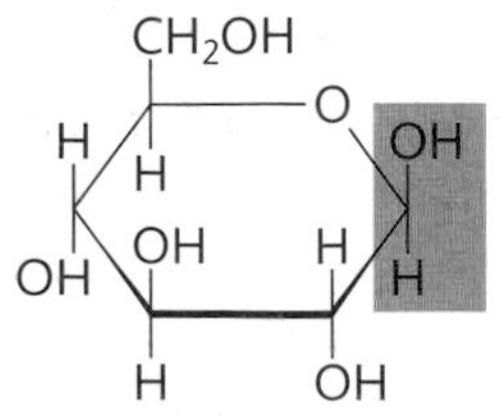

▶ **Figure 10.4:** Structure of beta glucose

Disaccharides

A **disaccharide** is formed (Figure 10.5) when two monosaccharides bond together in a **condensation reaction**. A new covalent bond is formed called a glycosidic bond and water is eliminated. The bond formed is called a 1,4 glycosidic bond. This bond can be broken by adding water. This is called a **hydrolysis** reaction and makes two monosaccharides again. Common disaccharides are:

▸ maltose   ▸ lactose   ▸ sucrose.

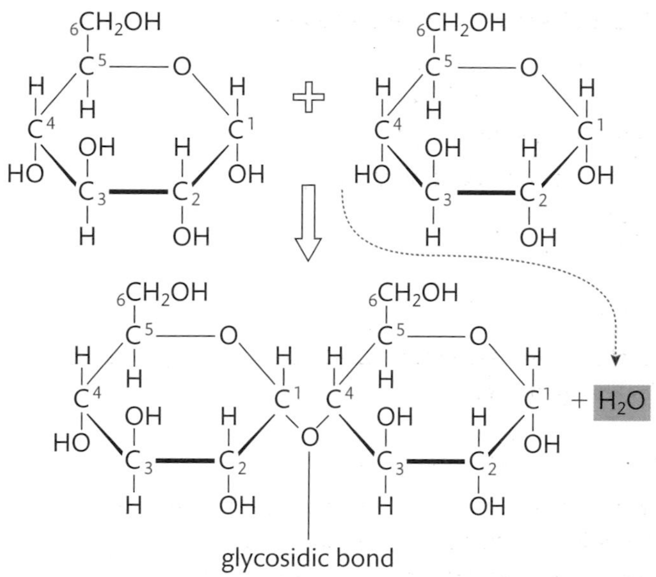

Figure 10.5: Disaccharide (maltose) forming when two monosaccharides bond together

> **Key terms**
>
> **Disaccharide (double sugars)** – two monosaccharides bonded together by a glycosidic bond.
>
> **Condensation reaction** – a chemical reaction involving the removal of a water molecule from two or more small molecules in order to form a larger molecule.
>
> **Hydrolysis** – a chemical reaction involving the addition of water molecules to break a covalent bond in order to break a larger molecule into smaller units.

Case study

Lactose intolerance

Jackson is lactose intolerant and has been since he was born. Lactose intolerance is a very common problem that affects the digestive system. Lactose is a disaccharide that is made from two monosaccharides: galactose and glucose. This sugar is usually found in milk and dairy products. The body is normally able to digest lactose because it produces an enzyme called lactase. This enzyme is essential in order for the lactose in dairy products to be broken down into its monomers so that they can be easily absorbed into the blood stream.

However, people who suffer from lactose intolerance are unable to produce enough lactase to break down lactose. The lactose stays in the digestive system and is fermented by bacteria. This produces lots of gas and causes symptoms such as stomach cramps, bloating and diarrhoea.

There is no cure for lactose intolerance, and it is normally controlled by making changes to diet and being aware of foods that contain high concentrations of lactose. There is also medication available in the form of drops or tablets. These can be taken just before or during a meal to help digest the lactose present in the meal, but they must be taken with all meals to have an effect.

Check your knowledge

1 What is lactose?

2 Where is lactose commonly found?

3 What does the body make to digest lactose?

4 What are some of the symptoms of lactose intolerance?

Polysaccharides

Polysaccharides are produced when a large number of monosaccharides form glycosidic bonds. Polysaccharides are called polymers because they are made from many repeating units of monosaccharides with each other. Amylose, amylopectin, glycogen and cellulose are important polysaccharides.

> **Key term**
>
> **Polysaccharides (multiple sugars)** – polymers of monosaccharides. They are made up of thousands of monosaccharide monomers bonded together to form a large molecule.

Energy storage and production

When thousands of alpha glucose monomers join together by alpha 1,4-glycosidic bonds, amylose is formed. Amylose (Figure 10.6) is a polysaccharide that is shaped like a coiled spring because of the position of the 1,4 glycosidic bonds.

— α-1,4-glycosidic bond

— alpha glucose

▶ **Figure 10.6:** Amylose shape structure like a coiled spring

Plants store their energy in the form of starch. This is a mixture of amylose and another carbohydrate called amylopectin. Amylopectin (Figure 10.7) consists of 1-4 glycosidic bonds and 1-6 glycosidic bonds. This refers to a condensation reaction occurring between carbon 1 on one alpha glucose molecule and carbon 6 of another. This makes the molecule branched and so it is not soluble in water. Starch is stored in the plant's chloroplasts and in starch grains that are surrounded by a membrane in the plant cell. Plants use enzymes to break down starch into glucose monomers. These are used in respiration to release energy for the plant to grow.

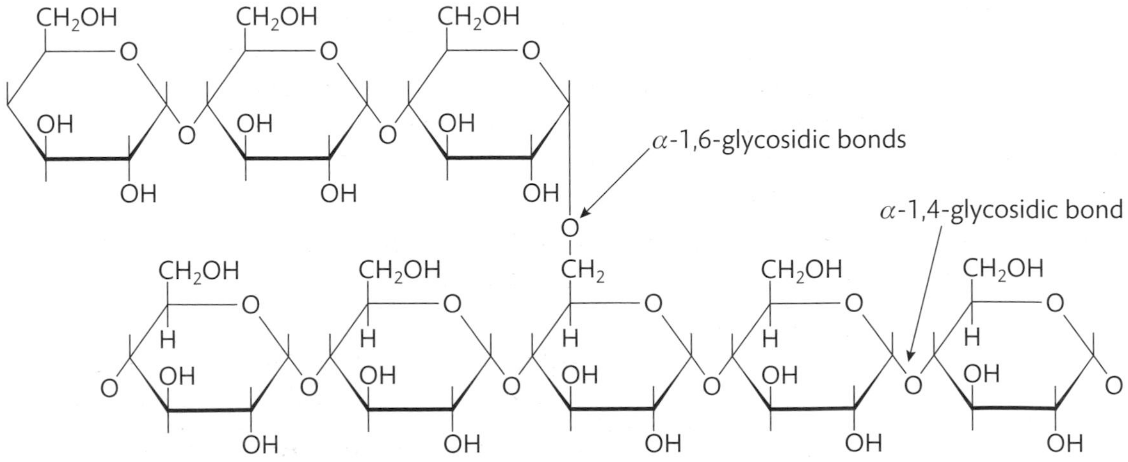

α-1,6-glycosidic bonds

α-1,4-glycosidic bond

▶ **Figure 10.7:** Amylopectin-branched carbohydrate with 1-4 glycosidic bonds and 1-6 glycosidic bonds

Animals store energy in the form of a polysaccharide called glycogen. Glycogen consists of alpha glucose monomers, and is a large, branched molecule. It forms more branches than amylopectin because a 1-6 glycosidic bond occurs every 10–15 glucose monomers. This makes glycogen more compact and ideal for storage.

This branching means that there are many free ends available to add or remove glucose molecules when necessary. Animals have a higher metabolic activity than plants so glucose needs to be released more readily.

Hydrolysis reactions

Glucose is stored as starch in plants and as glycogen in animals until it is needed for respiration. Biochemical energy in these stored molecules is converted into useable energy for the cell. A hydrolysis reaction is needed to release glucose molecules from these storage molecules. The reaction also needs enzymes in order to **catalyse** the reaction. The addition of water releases glucose molecules that are then free to be used in respiration.

> **Key term**
>
> **Catalyse** – to speed up or accelerate a reaction without itself being used up or changed.

> **Research**
>
> Research food sources that contain carbohydrates. Examples of food sources that contain carbohydrates are potatoes, pasta and bread. Find other food sources that contain carbohydrates. Put them in order from the food that contains the most carbohydrate to the food that contains the least carbohydrate.

Cellulose and structural support

Cellulose is a polysaccharide that is composed of repeating units of beta glucose molecules (Figure 10.8). When glycosidic bonds form in between these beta glucose molecules, they create a straight chain structure. Cellulose is not coiled like amylose. This is because of the difference in structure of the beta glucose compared to the alpha glucose molecules.

The hydroxyl (OH) groups of two beta glucose molecules are too far apart to react in a condensation reaction. The only way for them to bond together is for every other glucose molecule to be turned upside down. This creates a straight, unbranched, polysaccharide chain known as cellulose. These cellulose chains line up next to each other and form hydrogen bonds between adjacent hydrogen and oxygen atoms. This forms microfibrils (Figure 10.9), which combine to make fibres. These fibres are insoluble in water, strong, and are only found in plants cell walls. The strength of the molecules gives the whole plant a rigid structure.

Cellulose is an important part of our diet. It is extremely hard to break down and forms the fibre that we need to maintain a healthy digestive system.

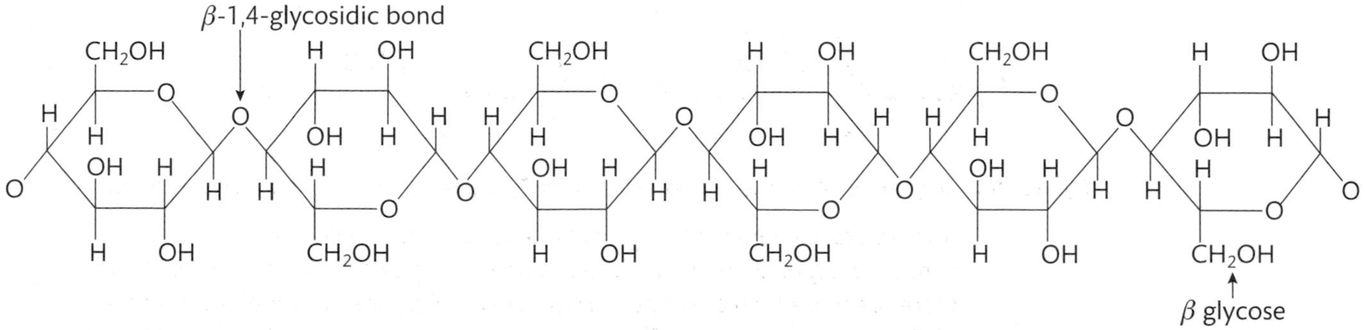

▶ **Figure 10.8:** Cellulose structure – straight chain, no branching

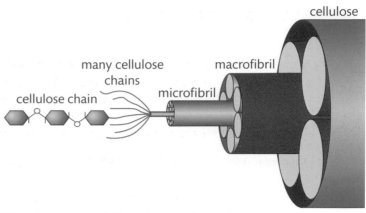

▶ **Figure 10.9:** Cellulose microfibril making up cellulose fibres

Case study

Diabetes mellitus

Insulin is a hormone produced in the pancreas. It is a protein made from 51 amino acids. It is needed in order to lower the blood glucose concentration in blood if it becomes too high, for example, after eating a meal that is rich in carbohydrates. If the production of insulin does not occur properly, it can cause medical problems.

Diabetes mellitus is the Latin name for diabetes. There are different types of diabetes, but the most common are type 1 and type 2.

Karlee suffers with type 1 *diabetes mellitus*, so she is unable to produce the insulin she needs in order to control her blood glucose levels. She developed diabetes at a young age and, in order to control her diabetes, she has to have daily insulin injections, and she must be very careful about what she eats.

Graham suffers from type 2 *diabetes mellitus*, which is much more common and occurs when the body cannot produce enough insulin, or the insulin is not working efficiently. Graham was diagnosed with diabetes later in life. Type 2 diabetes can sometimes be attributed to excess body weight and lack of exercise. Graham controls his diabetes by being careful about what he eats, but he may eventually require insulin.

1 What is insulin and why is it needed?
2 What are the most common types of diabetes?
3 What hormone is not produced if you suffer with diabetes?
4 How can people with diabetes control their symptoms?
5 How do type 1 and type 2 diabetes differ?
6 How does the control of each diabetes differ?

 PAUSE POINT Describe the structure and explain the function of carbohydrate molecules.

Hint Think about the different glucose molecules and their structure.

Extend Describe the difference in structure of glycogen and cellulose, and explain their different functions.

Theory into practice

You are a research scientist in the biochemistry department of a science laboratory. A local college has asked if you would provide materials for learners to show them the kinds of biological molecules they will be studying if they choose a career in biochemistry. Produce a scientific poster about the biological molecules water and carbohydrates, and include the structure and function of each. Include labelled diagrams and remember to include references to acknowledge the sources you have used.

Protein structure and importance

Proteins

Proteins are very important molecules in your cells, as they are involved with nearly all your cellular functions. The cells in your body are 50% protein. Proteins are extremely important for the growth and repair of your tissues. All proteins within the body have a specific job. The jobs that proteins do vary throughout the body. For example, they may be neurotransmitters, antibodies, hormones or enzymes.

Proteins are made from small molecules called amino acids. Amino acids are made from the elements below:

▶ carbon
▶ hydrogen

▶ oxygen
▶ nitrogen.

They are called amino acids (Figure 10.10) because they are made from an amino group and an acid group. The amino acid has a carbon in the centre with four carboxylic acid groups attached:

▶ a hydrogen
▶ an **amino group**

▶ a **carboxyl group**
▶ a variable R group.

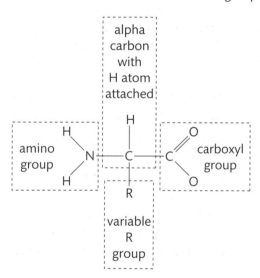

▶ **Figure 10.10:** Amino acid structure

Key terms

Carboxyl group – consist of a carbon atom double bonded to an oxygen atom and single bonded to a hydroxyl group (–COOH).

Amino group – the group –NH_2 present in amino acids.

There are 20 different amino acids, each with a different R group (see Figure 10.11). All the proteins in your body are made from different combinations of these amino acids. The structure of proteins is broken down into four levels:

▶ primary
▶ secondary

▶ tertiary
▶ quaternary.

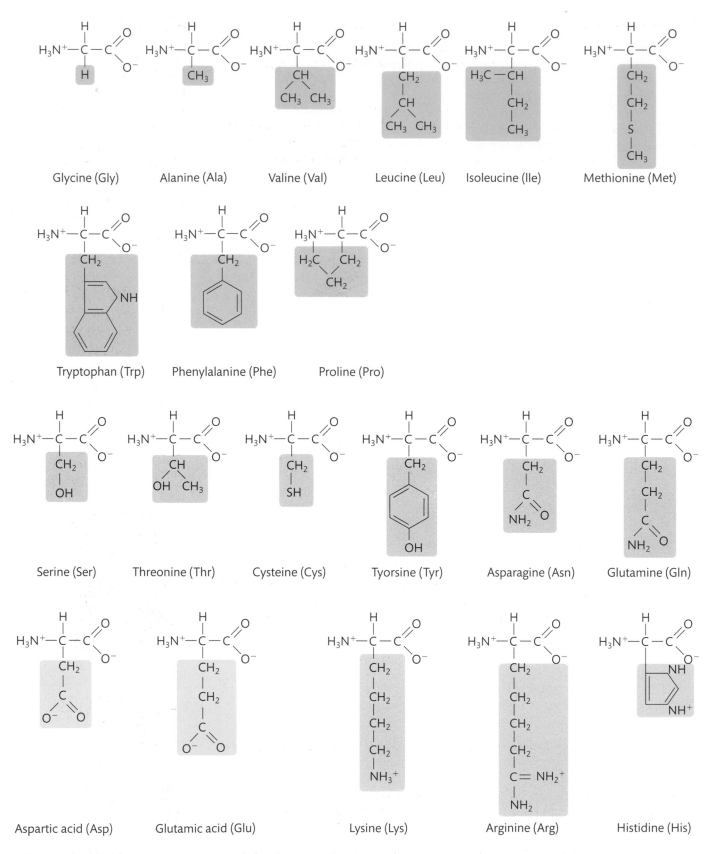

Glycine (Gly) Alanine (Ala) Valine (Val) Leucine (Leu) Isoleucine (Ile) Methionine (Met)

Tryptophan (Trp) Phenylalanine (Phe) Proline (Pro)

Serine (Ser) Threonine (Thr) Cysteine (Cys) Tyorsine (Tyr) Asparagine (Asn) Glutamine (Gln)

Aspartic acid (Asp) Glutamic acid (Glu) Lysine (Lys) Arginine (Arg) Histidine (His)

▶ **Figure 10.11:** Twenty different amino acids

Primary structure

The simplest level of the protein structure is called the primary structure. It consists of a unique sequence of amino acids that make up a **polypeptide**. The function of each protein depends on this unique sequence of amino acids.

Amino acids join together when a condensation reaction occurs between the carboxyl acid group of one amino acid and the amino group of another amino acid. A **covalent bond** is formed between the two amino acids, and a water molecule is produced. The bond that forms between the two amino acids is called a **peptide bond**, and a dipeptide molecule is produced.

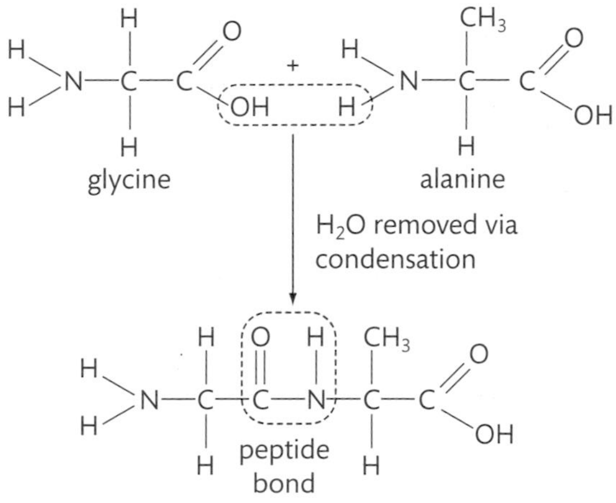

▶ **Figure 10.12:** Two amino acids undergoing a condensation reaction to form a dipeptide

As more amino acids form bonds with each other, the chain of amino acids gets bigger. A polypeptide is produced, along with many peptide bonds. Each peptide bond can be broken by the addition of a water molecule in a hydrolysis reaction.

Secondary structure

A protein's secondary structure forms because this unique chain of amino acids (primary structure) either coils to form an **alpha helix** or folds to form a **beta pleated sheets**. Regions of alpha helix and beta pleated sheets can exist within the same polypeptide chain. The secondary structures are held in shape by hydrogen bonds. Each bond is a weak force of attraction between a lone pair of electrons on an oxygen atom and a hydrogen atom attached to a nitrogen atom (Figure 10.13).

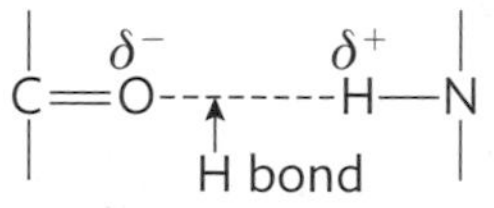

▶ **Figure 10.13:** Hydrogen bond between an oxygen atom and a hydrogen atom that is attached to a nitrogen atom

In the secondary structure, the hydrogen bonds are a type of intramolecular force. This means that they are forces of attraction between different parts of the same molecule.

An alpha helix (α helix) (Figure 10.14) is formed when the polypeptide chain is coiled into a spring shape. It is held together by hydrogen bonds, and although each bond is only a weak force, there are so many hydrogen bonds that their combined effect results in a very strong structure.

Key terms

Polypeptide – polymer consisting of a large number of amino acids bonded together to form a chain.

Covalent bond – a chemical bond formed by the sharing of one or more electrons between two atoms.

Peptide bond – a covalent bond formed between two amino acid molecules when the carboxyl group of one molecule reacts with the amino group of the other molecule.

Alpha helix – a right-hand coiled formation in proteins.

Beta pleated sheet – a flat flexible structure consisting of parallel polypeptide chains cross-linked found in proteins.

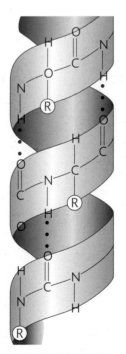

• • • hydrogen bond
Ⓡ = amino acid side chain

▶ **Figure 10.14:** An alpha helix in the secondary structure of a protein

In a beta (β) pleated sheet (Figure.10.15), the polypeptide chains are folded so that they run next to each other. This structure is also held together by hydrogen bonds.

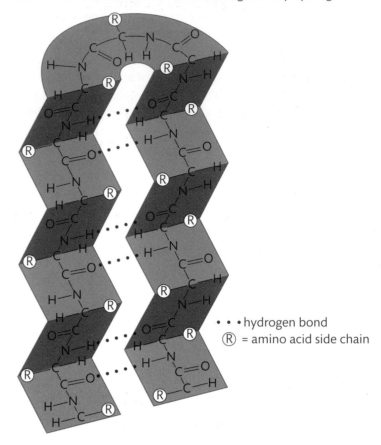

• • • hydrogen bond
Ⓡ = amino acid side chain

▶ **Figure 10.15:** A beta pleated sheet in the secondary structure of a protein

Tertiary structure

The tertiary structure is formed when the secondary structure coils and folds. The protein becomes a three-dimensional structure held in place by a number of different bonds and interactions into a 3D shape (see Figure 10.16) between the R groups of amino acids, which are now adjacent to each other because of the second structure.

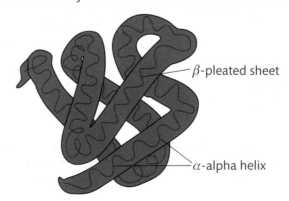

β-pleated sheet

α-alpha helix

▶ **Figure 10.16:** Tertiary structure of a protein

Disulfide bridges (Figure 10.17) (or S-S links) occur between two sulfur atoms. Sulfur atoms are present in the amino acid R group cysteine, so when there are two close together, a covalent bond forms between the two sulfur atoms. These bonds are the strongest intramolecular bonds in the tertiary structure.

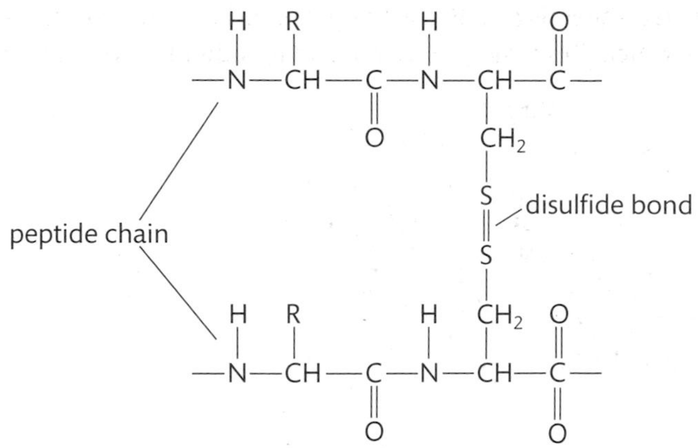

▶ **Figure 10.17:** Sulfur bridges

Ionic bonds (Figure 10.18) form between R groups if they carry opposite charges. For example, when proteins containing the amino acid aspartic acid and lysine are close together, an ionic bond will form. These are weaker than disulfide bonds but stronger than hydrogen bonds.

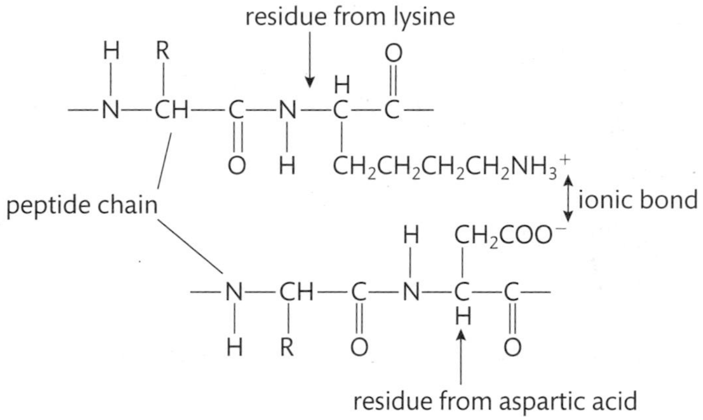

▶ **Figure 10.18:** Ionic bonds in the protein tertiary structure

Hydrogen bonds will occur where there are slightly positively charged hydrogen ions in R groups close to slightly negatively charged R groups.

Finally, **hydrophobic** and **hydrophilic** interactions occur. Amino acids with hydrophobic R groups tend to be found in the centre of the globular protein, and amino acids with hydrophilic R groups are found on the outside of the protein.

Van der Waals forces are weak interactions that also occur in the tertiary structure. They form between nearby atoms. They are not specific to any particular group or between any particular molecules.

> **Link**
>
> You met van der Waals forces in *Unit 1: Principles and Applications of Science 1*.

Quaternary structure

Some proteins are made from more than one polypeptide chain. This is the quaternary structure. These proteins sometimes contain essential **functional groups**, known as prosthetic groups. This is the non-protein part of the protein and it is essential for the functioning of the protein. An example of a protein with a quaternary structure and a prosthetic group is haemoglobin. Haemoglobin is found in red blood cells.

> **Key terms**
>
> **Hydrophobic** – does not mix with water.
> **Hydrophilic** – has a tendency to mix with water.

> **Key term**
>
> **Functional group** – specific part of a molecule that is responsible for particular characteristic chemical reactions.

Globular proteins

Haemoglobin is a soluble globular protein that consists of four globular sub-units arranged in a roughly spherical structure, each with a prosthetic group called haem. Haem contains an Iron (Fe) ion (Figure 10.19). Oxygen binds to the prosthetic group, which is bonded within the quaternary structure. The prosthetic group is essential for haemoglobin to perform its function of carrying oxygen. Prosthetic groups are organic groups that are bonded to proteins and allow the proteins to carry out their biological role.

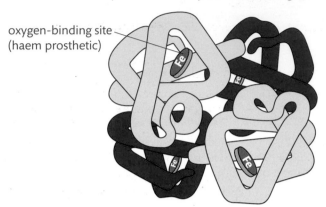

oxygen-binding site
(haem prosthetic)

▶ **Figure 10.19:** Haemoglobin showing a haem prosthetic group

Case study

Porphyria disorder

Laura was suffering with abdominal pain and skin problems. Upon further examination, Dr Hazel Butler diagnosed her with porphyria disorder.

Dr Butler explained that this disorder means that there is a problem with the production of haem in the body. Haem is required by the body to make the globular protein haemoglobin in red blood cells (erythrocytes). Porphyria is inherited in most cases, and therefore is passed on through different generations. There are seven essential enzymes needed to synthesise haem. People suffering with porphyria have a problem producing one of these essential proteins (enzymes).

When Lucy's body tries to synthesise haem, substances will be produced. However, because there is not enough of one of these essential enzymes, the reaction is unable to continue and substances called porphyrins build up in the body and cause symptoms.

1   Why is the fact that porphyria disorder is inherited important?
2   Why may the body not produce an essential protein properly?
3   Why is haem important?
4   Explain why the absence of one enzyme would affect the body.

Fibrous proteins

Fibrous insoluble proteins (Figure 10.20), like collagen, consist of different protein strands coiled around each other. They normally have a structural function. Collagen is found in the body's tendons supporting organs and bones. Collagen is made up of three polypeptide chains wound around each other. Each of the three chains is a coil made of around 1000 amino acids. It is very strong as hydrogen bonds form between the chains.

▶ **Figure 10.20:** Fibrous protein structure

PAUSE POINT   Describe the structure of polypeptides.

    Think about the number of different levels that make up the structure of a protein.

Extend   Explain the bonds that may be made in the tertiary structure, using specific examples of amino acids that may be involved in the bonding.

Importance of proteins

Table 10.2 explains the importance of proteins in the human body.

▶ **Table 10.2:** Proteins

Importance of protein	Description
Neurotransmitter	Neurotransmitters are chemical messengers that allow the communication between nerve cells. A neurotransmitter is released into the synaptic cleft. Neurotransmitters bind to receptor proteins within the postsynaptic cell membrane, and the message continues along the nerve cell.
Enabling vitamins and minerals to be used in the human body	Proteins such as transferrin and metallothionine are transport proteins that bind to vitamins and minerals and transport them around the body.
Antibodies	Antibodies consist of four polypeptide chains held together with disulfide bonds. The function of antibodies is to bind to specific antigens on the surface of invading organisms to provoke an immune response.
Hormones	Hormones are chemicals substances produced in the body to control and regulate activity of cells or organs. For example, insulin is a hormone needed to lower the blood glucose concentration.
Transport of other components	Various proteins transport different substances around the body. For example, haemoglobin is a globular protein that transports oxygen around the body via the circulatory system.
Growth, repair of body tissue	Body cells become worn out and they need replacing. Proteins repair damaged tissue and replace worn-out tissue.
Connective tissue	Collagen is used for structure in the human body. Within connective tissue there are thick collagen fibres.
Muscle contraction	Muscle contraction depends on interactions between actin and myosin. These are both filamentous proteins that slide past each other.
Blood clotting	Coagulation is the formation of a blood clot whereby blood turns from a liquid to a gel. In order to form a clot, platelets and proteins are needed.
Enzymes	Enzymes are proteins with a tertiary structure and they are used within the body to catalyse chemical reactions. Amylase is an enzyme present in saliva made in salivary glands to break down starch.

Nucleic acids

Nucleic acids are large molecules that are found inside the nucleus of a cell. Deoxyribonucleic acid (DNA) and ribonucleic acid (RNA) are both nucleic acids. They are responsible for storing your genetic information and for the synthesis of proteins. It is your secret code that is stored in your DNA that provides instructions in your cells to build the polypeptides which make up the structure and carry out most of the functions in your body.

Nucleotides

A **nucleotide** consists of three components (see Figure 10.21):

▶ pentose monosaccharide (a 5-carbon sugar)

▶ a phosphate group

▶ a nitrogenous base.

A condensation reaction occurs between the hydroxyl functional group (OH) of the sugar and the hydrogen atom of hydroxyl group from the phosphate, so water is expelled. The nitrogenous base also undergoes a condensation reaction. The hydroxyl functional group of carbon 1 on the sugar reacts with a hydrogen atom on the base.

> **Key term**
>
> **Nucleotide** – the basic structural unit of nucleic acids.

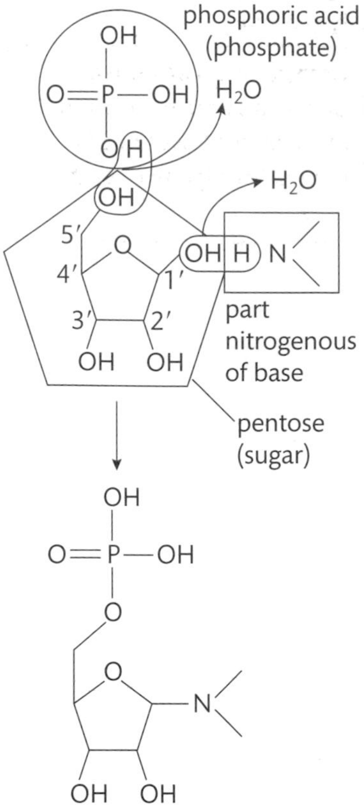

Figure 10.21: The formation of a nucleotide. Condensation reactions occurring and two molecules of water being expelled.

The five organic nitrogenous bases that make up the structure of nucleotides all contain carbon, hydrogen, oxygen and nitrogen. They are:

▶ adenine (A)

▶ thymine (T)

▶ guanine (G)

▶ cytosine (C)

▶ uracil (U).

Thymine is found only in DNA and uracil is only found in RNA. Figure 10.22 shows the five organic nitrogenous bases. Two are called purines and three are called pyrimidines. The two purine nitrogenous bases are adenine and guanine. They consist of a double ring structure. The three pyrimidine nitrogenous bases are thymine, cytosine and uracil, and they consist of a simple ring structure.

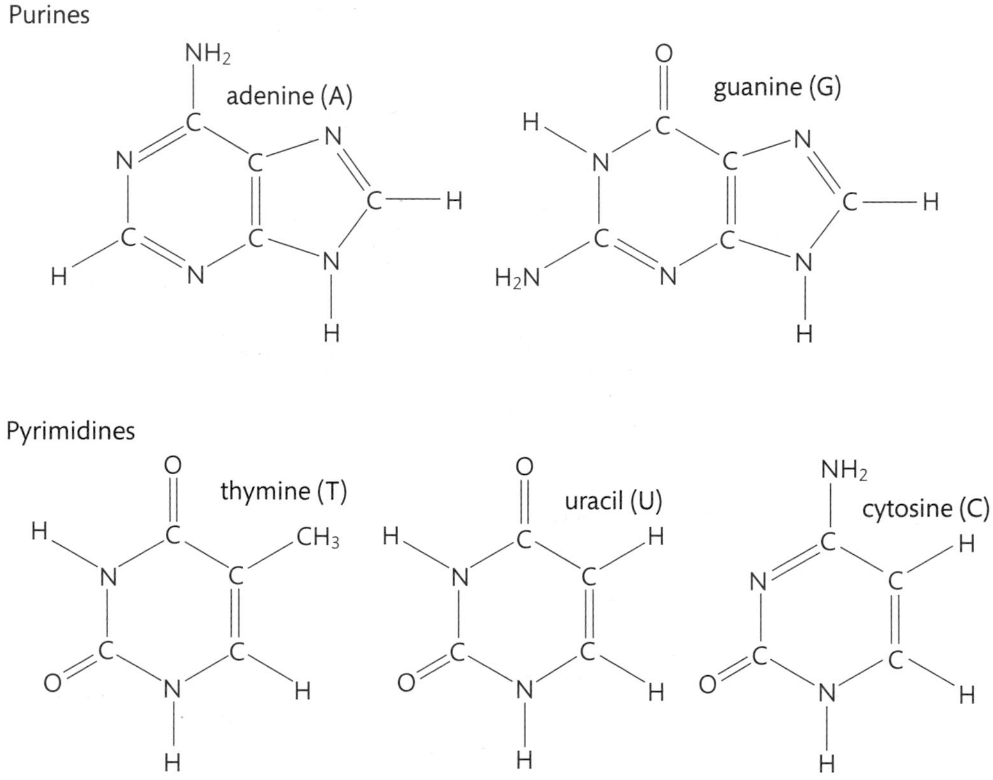

Purines

Pyrimidines

Figure 10.22: The five organic nucleotide bases: adenine, guanine, thymine, uracil and cytosine

Many nucleotides joined together make a polynucleotide. The phosphate group from one nucleotide forms a covalent bond with the hydroxyl (OH) group attached to the carbon 3 sugar of the next nucleotide. These bonds are called phosphodiester bonds. They produce a very strong sugar-phosphate backbone. Note that the nitrogenous bases do not take part in polymerisation (see Figure 10.23) and they extend out from the polynucleotide structure.

DNA is a double-stranded polynucleotide. Its individual nucleotides contain deoxyribose sugars and nitrogenous bases A, G, C and T. It is made up of two polynucleotide chains alongside each other. The sugars in the polynucleotide chains run in opposite directions, so the two chains are described as antiparallel. The two strands are joined together because hydrogen bonds form between the nitrogenous bases. The bases form base pairs, two hydrogen bonds form between the bases A and T, and three hydrogen bonds form between the bases C and G. This is known as **complementary base pairing**. The two strands form a double helix (see Figure 10.24) as they twist around each other. It is in this double-stranded polynucleotide chain that your genetic information is stored. DNA stores the genetic code that is used to build organisms and produce essential proteins.

Key term

Complementary base pair – the way in which the nitrogenous bases of DNA molecules align with each other.

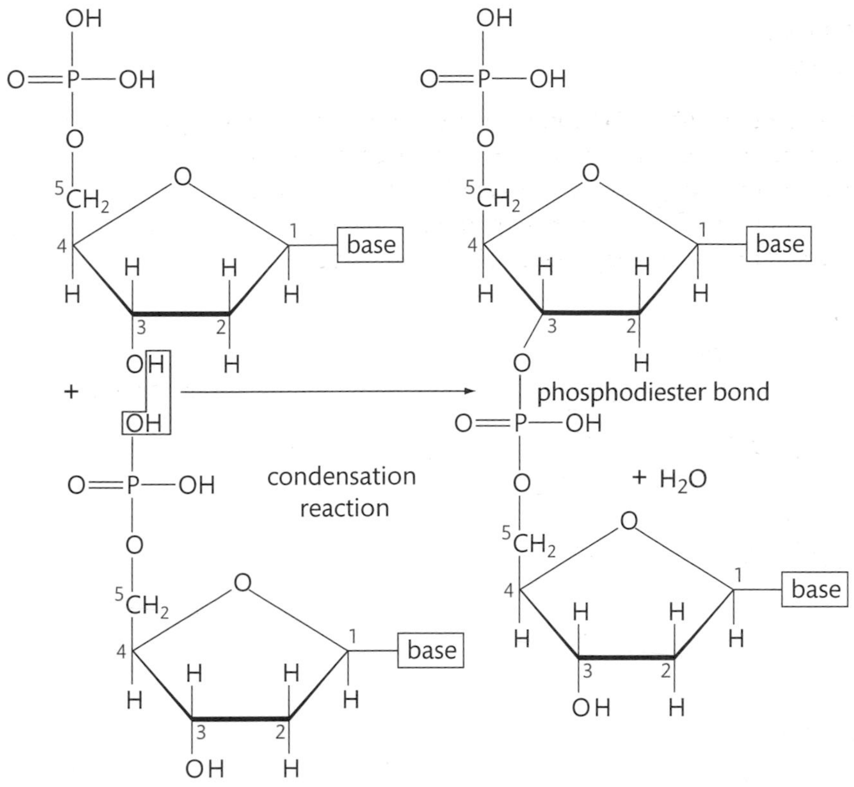

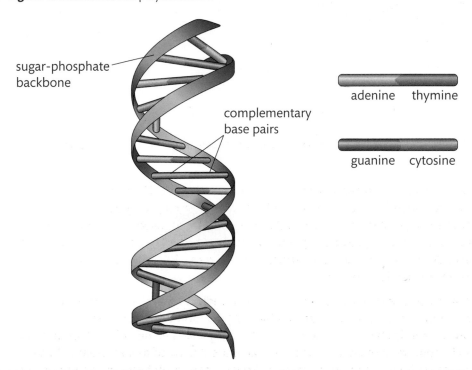

▶ **Figure 10.23:** Nucleotide polymerisation

▶ **Figure 10.24:** DNA double helix

RNA is a nucleic acid, but it is different than DNA. It is a single-stranded polynucleotide. It contains the sugar ribose, not deoxyribose, and it contains uracil, not thymine. RNA is used to read the genetic code in DNA after the DNA molecule unwinds and exposes the bases. RNA is used to produce a copy of the genetic code of DNA after the DNA molecule unwinds and exposes the bases. The code is read at the ribosomes and used as instructions to assemble amino acids to synthesise proteins.

PAUSE POINT

Describe the structure of mononucleotides and polynucleotides.

Hint   Draw or describe the structure of both.

Extend   Compare RNA and DNA, and produce a table of comparison.

Case study

Cystic fibrosis

Charley was diagnosed with cystic fibrosis when she was born. This is a genetic condition where the lungs and digestive system become blocked with thick sticky mucus. She suffers with a persistent cough, struggles to gain weight and she has recurring chest infections that make her very poorly.

When Charley was diagnosed, Dr Ken Tudor explained that cystic fibrosis is caused by a genetic mutation of the **cystic fibrosis transmembrane conductance regulator (CFTR) gene**. He explained that a genetic mutation is a permanent alteration of the nucleotide sequence in a person's DNA. The CFTR gene should provide instructions to produce a protein that regulates the level of sodium and chloride ions in a person's body cells. Charley has two copies of the defective gene, and so this protein is not produced. This results in the thick sticky mucus building up in the vessels on Charley's body, and on her airways, damaging her lungs and digestive system.

Check your knowledge

1 What parts of the body are affected by cystic fibrosis?

2 How is cystic fibrosis caused?

3 What is a genetic mutation?

4 What is the function of a healthy CFTR gene?

5 What symptoms do cystic fibrosis sufferers experience?

Lipid structure and importance

Lipids are mainly used as energy storage but they can also be used as an energy source. They are vital as a component of cell membranes and play an important role in insulating the body. Lipids are made from carbon, hydrogen and oxygen atoms. The common types of lipids are fats, waxes and oils.

Fats, waxes and oils are made from a molecule of glycerol and fatty acid chains. Glycerol contains three hydroxyl (OH) function groups. These groups can bond to three fatty acids chains by removing three molecules of water in a condensation reaction to form a lipid known as a triglyceride. The bonds that form are called **ester bonds**; three ester bonds are formed in a triglyceride (Figure 10.25). All fatty acids have a carboxyl functional group on one end and a large hydrocarbon chain at the other end.

Key term

Ester bond – the bond formed when the carboxyl group of a fatty acid combines with the hydroxyl group of glycerol.

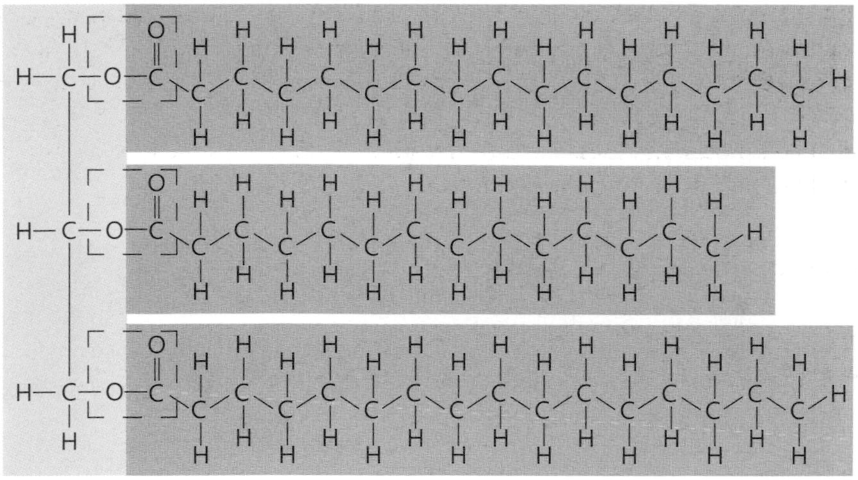

▶ **Figure 10.25:** The formation of a triglyceride after a condensation reaction

There are different types of fatty acids, but the main types are saturated and unsaturated. These have different types of hydrocarbon chains.

▶ Saturated contains just carbon-carbon single bonds (C-C) (see Figure 10.26).

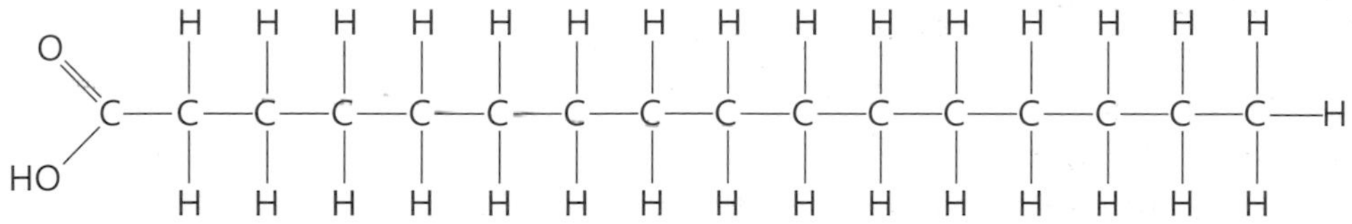

▶ **Figure 10.26:** Structure of palmitic acid, a saturated fat

▶ Unsaturated contains carbon-carbon double bonds (C=C) (see Figure 10.27).

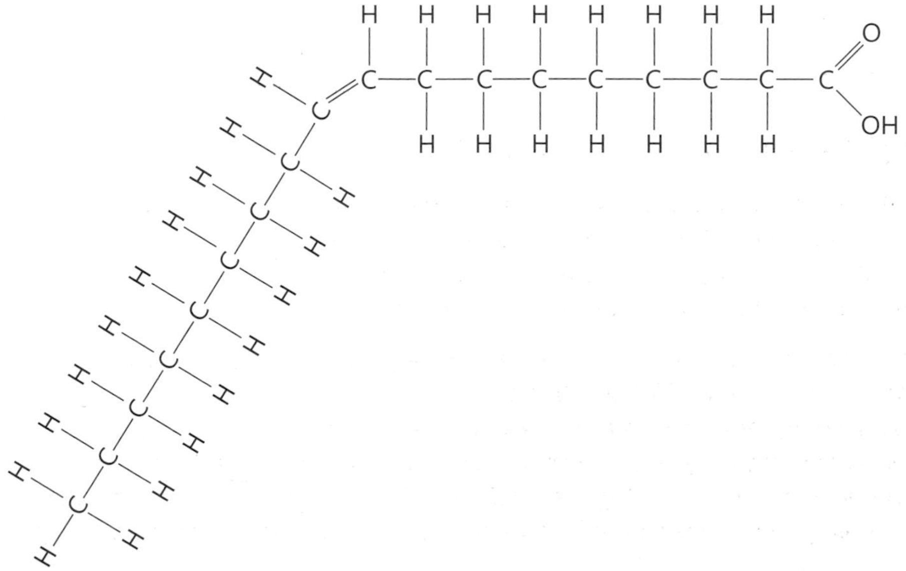

▶ **Figure 10.27:** The structure of oleic acid, an unsaturated fat

Saturated fats

These form straight chains. They can line up against each other more easily than molecules that have branched chains and all the molecules form attractions. A lot of energy is required to overcome these attractions. This means that saturated fatty acids have high melting points and tend to be solid at room temperature. Any lipids containing saturated fatty acids will have high melting points, e.g. waxes.

Unsaturated fats

These form chains with kinks in due to the double bonds present. Their shape means that they push away from each other and attractions are not formed as strongly between the molecules as in saturated fats. Because of this, they have low melting points and tend to be liquid at room temperature.

Phospholipids

Phospholipids (Figure 10.28) are a major component of all cell membranes. They can form lipid bilayers, made of two layers of lipid molecules due to their hydrophilic and hydrophobic properties (see Figure 10.29). Phospholipids are made up of two hydrophobic fatty acid chains and a phosphate group which has a hydrophilic head. The head and the fatty acid chains are joined together by a glycerol molecule.

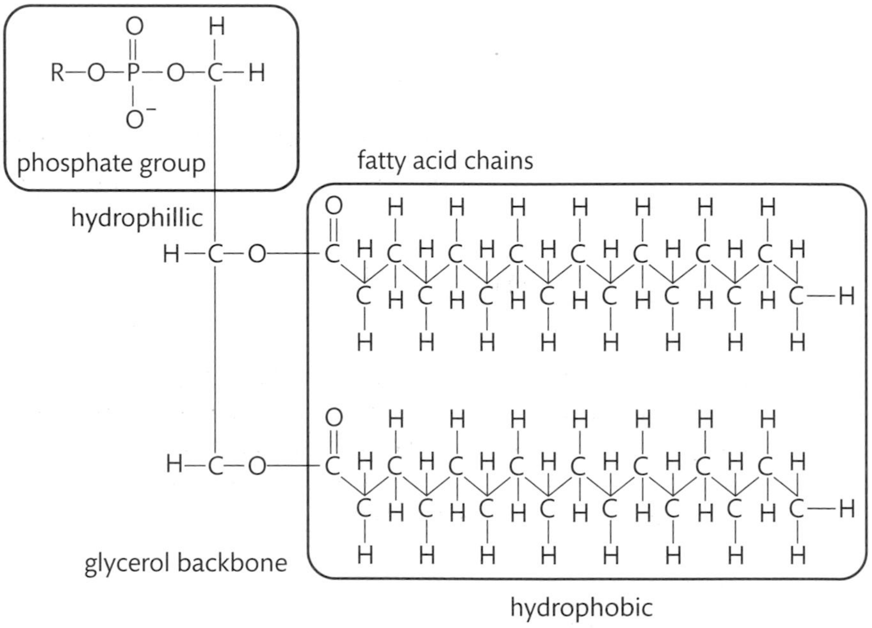

▶ **Figure 10.28:** Structure of a phospholipid

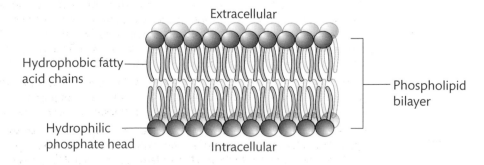

▶ **Figure 10.29:** A phospholipid bilayer

Importance of lipids

Table 10.3 explains the importance of lipids and the roles they play in the human body.

▶ **Table 10.3:** Roles of lipids

Importance of lipid	Role of lipid
Energy source	Triglycerides are a lipid used for energy storage, insulation and protection. They are found in fatty tissue under your skin and surrounding your organs. They are a good source of energy as they have a high carbon and hydrogen content.
In membranes	Phospholipids are found in cell membranes. They consist of one glycerol molecule bonded to two fatty acid chains and one phosphate group. They have a negatively charged phosphate head which is hydrophilic and is soluble in water, and uncharged hydrocarbon chains that are hydrophobic (insoluble in water).
Cholesterol	Cholesterol does not have the same structure as triglycerides or phospholipids, but it is still an important lipid. Cholesterol is present in cell membranes to provide rigidity. It is also used to form steroid hormones (see below).
Production of vitamins	Vitamin D is produced in your skin when it is exposed to sunlight. Your liver and kidneys modify Vitamin D to produce a hormone, 1,25 dihydroxycholecalciferol. This contains a steroid molecule which is a lipid. This hormone controls calcium absorption in your gut, and bone development.
Bile acids	Bile acids are needed for normal digestion, and absorption of lipids and fat-soluble vitamins such as A and D. Bile acids are made from cholesterol in the liver and act like cleaners in your gut. They dissolve fat from your food. Without bile acids, fat is not digested properly. This can result in diarrhoea.
Steroids	Steroids include cholesterol, sex hormones (progesterone, oestrogen and testosterone) produced by gonads, and cortisone. Cortisone is a steroid that prevents the release of substances in the body that cause inflammation.

PAUSE POINT   Describe the structure of lipids.

> Hint   Do you know the structural difference between saturated and unsaturated fats?
>
> Extend   Explain four functions of lipids in the body.

Disruption of living organisms

Plant growth regulators and their disruption

Plant hormones, or plant growth regulators, are chemicals that regulate the growth in plants. Plant hormones are signalling molecules produced in the plant. These hormones regulate cellular processes in plant cells and can also determine development and formation of flowers, stems and leaves, for example. Table 10.4 describes the roles of different plant growth regulators.

▶ **Table 10.4:** Growth regulators

Growth regulator	Description
Auxins	Auxin is a plant hormone that is responsible for controlling the direction of growth of root tips and stem tips. Auxin is made at the tips of stems and roots, and moves to older parts of the stem and root, where it causes the elasticity of the cells to change. Elastic cells absorb more water, and therefore grow longer and bend. Light and gravity can interfere with auxin distribution and cause it to become uneven. For example, a houseplant grows towards the window. It does this because light coming from the window side of the plant destroys the auxin in that side of the stem, and growth on that side slows down. If auxin production is disrupted, the growth of the plant will be affected. Spraying auxins on tomatoes, for example, can increase the yield.
Gibberellins	Gibberellins are plant hormones that control various developmental processes, including stem elongation, germination, flowering and fruit ageing. If this hormone is not produced, then these processes may not occur as they should. This may have effects, for example, flowers are not produced, or fruit ages quickly and deteriorates.
Ethylene	Ethylene is a plant hormone that is responsible for the ripening of fruit. Some fruit will produce ethylene as it ripens. Examples are apples and pears, which produce ethylene when ripening. Changes in texture, softening and colour are all caused by ethylene. If this system becomes disrupted, it may result in fruit ripening too quickly.

Use the internet and research cytokinins, abscisic acid and synthetic auxin 2,4-Dplus agent orange, and how they work chemically in plants. In small groups, talk about their role and then about what would happen if this did not work as expected.

Assessment practice 10.1

A year 12 learner is thinking about studying biochemistry at university and wants to apply for a scholarship. She has been asked to produce a portfolio of the biology work that she has covered so far. She contacted the university for an information sheet that explains the importance of biological molecules in living organisms and the structure and function. She wants to go above and beyond to increase her chances.

Produce an information sheet that will explain the structure of water, carbohydrates, proteins and lipids. You should also explain how the structure links to function and the importance of the molecules in living organisms. You should evaluate the effects that disruption of biological molecules can cause and how this can affect living organisms.

Plan
- What is the task? What am I being asked to do?
- How confident do I feel in my own abilities to complete this task?
- Are there any areas I think I may struggle with?

Do
- I know what it is I am doing and what I want to achieve.
- I can identify when I have gone wrong and adjust my thinking/approach to get myself back on course.

Review
- I can explain what the task was and how I approached the task.
- I can explain how I would approach the hard elements differently next time (i.e. what I would do differently).

B Explore the effect of activity on respiration in humans and factors that can affect respiratory pathways

Stages involved in respiratory pathways

Your metabolism refers to the thousands of chemical reactions that are taking place in your body cells. These reactions can be **catabolic** (e.g. respiration) or **anabolic** (e.g. photosynthesis).

In one day, you will use the equivalent of your own body weight of chemical energy. Chemical energy must be produced all the time so it can be used all the time by cells. Respiration is an extremely important process, because the end product is chemical energy that we rely on to keep our cells alive. Adenosine Triphosphate (ATP) is the molecule produced in respiration that acts as a store of chemical energy and is able to be released when it is needed by cells. This energy is required in our cells for processes such as muscle contraction and enzyme-catalysed reactions.

ATP consists of adenine, ribose and three phosphate groups. The bonds between the two phosphate groups on the right-hand side (see Figure 10.30 and Figure 10.31) can be broken easily by an enzyme. This immediately releases energy. ATP becomes adenosine diphosphate (ADP) and an **inorganic** phosphate.

Key terms

Catabolic – reactions that involve the breakdown of a molecule.

Anabolic – reactions that produce a molecule.

Inorganic – does not contain carbon.

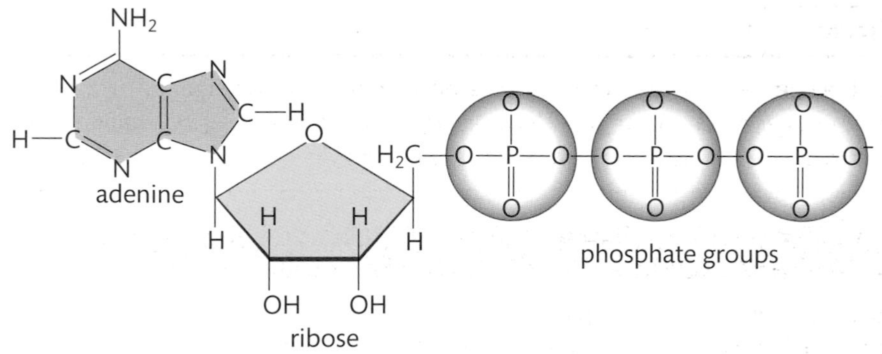

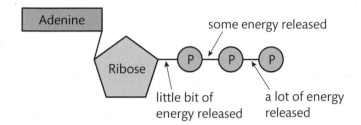

▶ **Figure 10.30:** Structure of ATP

some energy released

little bit of
energy released

a lot of energy
released

▶ **Figure 10.31:** Energy release from ATP

Respiration can be shown in a word equation.

glucose + oxygen ⟶ carbon dioxide + water (+ energy)

However, respiration is more than this. It is a series of very complex metabolic pathways that incorporates more than 30 different steps. It is important for biochemists to understand these pathways in order to know how to diagnose patients when they observe an abnormality.

Respiration is broken down into four stages. Stage 1 does not require oxygen and is described as **anaerobic**. Stages 2, 3 and 4 do require oxygen, so they are **aerobic**.

Key terms

Anaerobic – does not require oxygen.

Aerobic – requires oxygen.

Stage 1: Glycolysis

Glycolysis takes place inside the cytoplasm of cells. Glycolysis is a four-stage process and results in the conversion of glucose into two molecules of pyruvate. It is catalysed by enzymes to convert one molecule of glucose into two molecules of pyruvate. Glucose is a 6-carbon compound and pyruvate is a 3-carbon compound. This stage does not require any oxygen and can therefore occur during anaerobic conditions, providing the cell with some energy even when oxygen is not available. Within the four stages of glycolysis, there is a sequence of reactions, each catalysed by a different enzyme. Glycolysis only produces two molecules of ATP, but the pyruvate is used to form more ATP in stages 2 and 3.

Phosphorylation is the first stage of glycosis (see Figure 10.32). This refers to the addition of a phosphate group to activate a glucose molecule so that it can be split.

Key term

Phosphorylation – production of ATP from ADP and P_i.

- An ATP molecule is hydrolysed and a phosphate group is released. This phosphate group attaches to carbon 6 of a glucose molecule to produce glucose-6-phosphate. The enzyme hexokinase catalyses this reaction.

- Glucose 6-phosphate turns into fructose 6-phosphate. The enzyme glucose phosphate isomerase catalyses this reaction.

- Another ATP molecule is hydrolysed and the phosphate group that is released during this hydrolysis attaches to carbon 1 of fructose 6-phosphate. This becomes fructose 1,6-biphosphate (the phosphate groups attached to carbon 1 and 6). The enzyme phosphofructokinase catalyses this reaction.

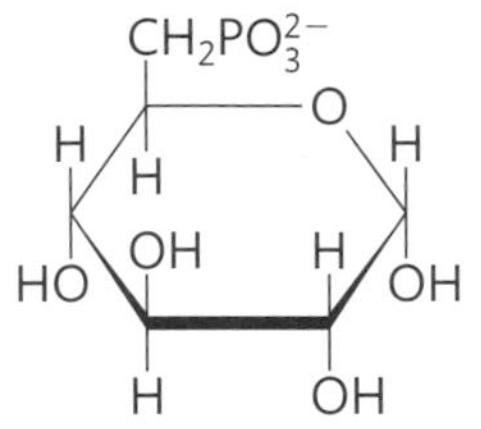

▶ **Figure 10.32:** Step 1 of phosphorylation

The second stage of glycosis involves splitting fructose 1,6-biphosphate to enable the products to form ATP.

- Fructose 1,6-biphosphate is split into two molecules. The enzyme fructose diphosphate aldolase catalyses this reaction.

- The enzyme triose phosphate isomerase produces two molecules of glyceraldehyde 3-phosphate.

By splitting fructose 1,6-biphosphate, two molecules of glyceraldehyde 3-phosphate are produced.

The third stage of glycolysis is oxidation, whereby glyceraldehyde 3-phosphate loses electrons.

- Two hydrogen atoms are removed from each glyceraldehyde 3-phosphate to produce two molecules of 1,3 bisphosphate glycerate. This reaction is catalysed by glyceraldehyde phosphate dehydrogenase, but this enzyme also needs another molecule, called a co-enzyme, to work. It requires two molecules of nicotinamide adenine dinucleotide (NAD). These molecules act as electron acceptors. They accept the hydrogen atoms and in this reaction, two NAD molecules become reduced NAD or NADH.

- Oxidation requires two NAD co-enzymes and produces two reduced NAD (NADH) and two 1,3 bisphosphate glycerate from two glyceraldehyde 3-phosphate.

The fourth stage of glycolysis is conversion of 1,3 bisphosphate glycerate to produce pyruvate.

- Four different enzymes are used to convert both 1,3 bisphosphate glycerate into pyruvate. This produces two ATP per molecule of 1,3 bisphosphate glycerate, so four overall by adding an inorganic phosphate to ADP during phosphorylation.

- The conversion of two molecules of 1,3 bisphosphate glycerate produces two pyruvate and four ATP.

- Overall, glycolysis produces four molecules of ATP per glucose. However, it uses two ATP during phosphorylation. So, for every glucose molecule, two molecules of ATP, two molecules of NADH and two molecules of pyruvate are produced (see Figure 10.33).

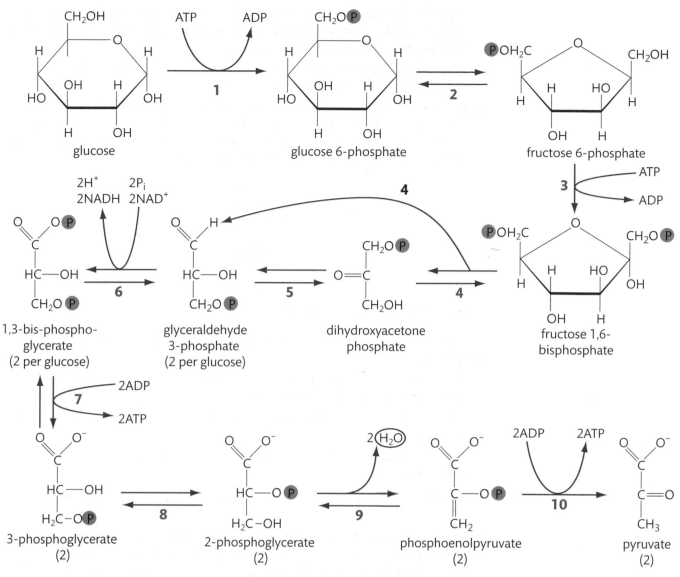

Enzymes:

1 hexokinase	**4** fructose diphosphate aldolase	**7** phosphoglycerate kinase
2 glucose phosphate isomerase	**5** triose phosphate isomerase	**8** phosphoglyceromutase
3 phosphofructokinase	**6** glyceraldehyde phosphate dehydrogenase	**9** enolase
		10 pyruvate kinase

▶ **Figure 10.33:** Ten-step enzyme-catalysed reactions to show the chemical changes taking place during glycolysis to produce pyruvate

Ⅱ **PAUSE POINT** State the products of glycolysis.

　　　　　　　Hint　　Do you know the four distinct stages?

　　　　　　　Extend　　Explain the four stages, and draw small diagrams to help you explain each stage.

Stage 2: The link reaction

This stage of respiration takes place inside the matrix of the mitochondria (see Figure 10.34).

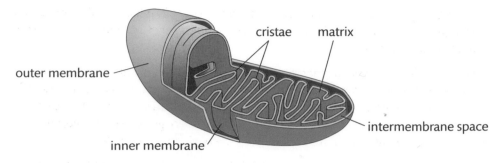

▶ **Figure 10.34:** Structure of mitochondria

If oxygen is present, the pyruvate that has been produced during glycolysis is changed into acetate during the link reaction.

Pyruvate is a 3-carbon molecule. Both molecules of pyruvate from glycolysis are decarboxylated (carbon dioxide, CO_2, is removed) and it is dehydrogenated (two hydrogen atoms are removed). Enzymes are needed for this to occur. The carbon dioxide is a product of respiration and diffuses into the bloodstream, where it is breathed out of the lungs. NAD accepts the hydrogen atoms and becomes reduced NAD (NADH). A 2-carbon compound called acetate is produced. The acetate joins to a co-enzyme called co-enzyme A (coA) and forms acetyl co-enzyme A.

For every two molecules of pyruvate, two NADH are made. No ATP is made during this reaction, but the two molecules of acetyl co-enzyme A and the NADH are used in stage 3 (see Figure 10.35).

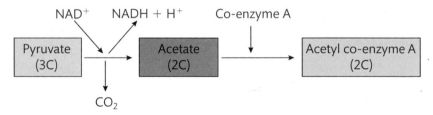

▶ **Figure 10.35:** The link reaction, showing one molecule of pyruvate being converted into acetyl co-enzyme A

Stage 3: Krebs cycle

The next stage of respiration also occurs in the mitochondrial matrix and is classed as aerobic respiration. This stage consists of five enzyme-catalysed reactions.

▶ The acetyl co-enzyme A from the link reaction releases the acetate. This reacts with oxaloacetic acid to form citric acid. Citric acid is a 6-carbon compound.

▶ Citric acid is decarboxylated (CO_2 is removed) and it is dehydrogenated (two hydrogen atoms are removed). NADH and a hydrogen ion, H^+, is produced. A 5-carbon compound is then produced.

▶ Two 4-carbon compounds are made in succession. The 5-carbon compound from step 2 is decarboxylated and it is dehydrogenated. NADH and a hydrogen ion, H^+, is produced, producing the first 4-carbon compound. This is turned into another 4-carbon compound and one molecule of ADP is phosphorylated (inorganic phosphate is added) to produce one molecule of ATP.

▶ The second 4-carbon compound from step 3 is changed into another 4-carbon compound. This new compound is dehydrogenated. This time a different co-enzyme, called flavin adenine dinucleotide (FAD), accepts the hydrogen atoms and becomes reduced FAD or FADH2.

▶ The 4-carbon compound from step 4 is dehydrogenated, reduced NAD is produced again and oxaloacetic acid is reformed. This goes on to accept another acetate from another link reaction.

The Krebs cycle produces one molecule of ATP per acetate molecule, one glucose molecule produces two pyruvate molecules, and each pyruvate molecule goes on to produce one acetate molecule (so two acetate molecules are produced for each molecule of glucose). So this cycle happens twice for every glucose molecule, and therefore two ATP molecules are produced for every glucose molecule.

The Krebs cycle also produces six reduced NAD molecules per glucose (three for every acetate molecule) and two reduced FAD molecules per glucose (one from each acetate molecule) (see Figure 10.36).

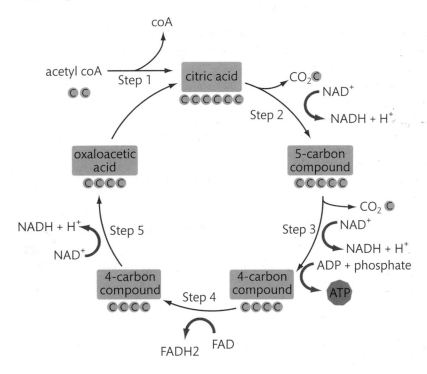

▶ **Figure 10.36:** Krebs cycle

Stage 4: Oxidative phosphorylation

Oxidative phosphorylation is the final stage of respiration. It occurs across the inner mitochondrial membrane. The inner mitochondrial membrane contains four large protein complexes: I, II, III and IV (referred to as the cytochrome system of carriers in the cristae of the mitochondria). These are used throughout the stage of oxidative phosphorylation. During this stage (see Figure 10.33), hydrogen atoms from all the reduced NAD that has been produced from stages 1, 2 and 3 release their energy to produce ATP.

▶ Reduced NADH molecules bind to complex I and release their hydrogen atoms as protons (H^+) and electrons (e^-) into the matrix of the mitochondrial. The reduced NAD is recycled back to NADHR and can be used during glycosis, the link reaction and in the Krebs cycle.

▶ Reduced FAD molecules made during the Krebs cycle bind to complex II and release their hydrogen atoms as protons (H^+) and electrons (e^-) into the matrix.

▶ The H^+ ions stay in the matrix while the electrons pass along all the protein complexes. This is known as the electron transport chain (ETC).

Research

Produce a table to show the products of glycolysis, the link reaction and the Krebs cycle.

- In complexes I, II and IV in the membrane the electrons release some of their energy. This energy is used to pump the protons from the matrix across the inner mitochondrial membrane to the space between the inner membrane and the outer membrane of the mitochondrial. The protons are pumped by complexes I, III and IV.

- In complex IV, the electron combines with protons and oxygen to form water, another product of respiration. This is the only stage that uses oxygen, but without it respiration does not occur. Four electrons, four protons and one molecule of oxygen are needed to make two molecules of water. Oxygen is therefore known as the final electron acceptor.

- Because protons have moved from the matrix to the intermembrane space, this has created a proton gradient, where there are more protons in the intermembrane space than in the matrix. This is a store of potential energy and is used to generate ATP.

- The protons cannot move back across the inner mitochondrial membrane without being pumped. However, the ATP synthase enzyme has a channel for the protons to move through. This movement of protons across a membrane due to a proton gradient is called **chemiosmosis**.

> **Key term**
>
> **Chemiosmosis** – the movement of ions across a semi-permeable membrane, down an electrochemical gradient.

- When the protons move from the intermembrane space back to the matrix, through the ATP synthase enzyme, the energy physically spins part of the enzyme which, in turn, causes phosphorylation of ADP to produce ATP.

Therefore, the stage of oxidative phosphorylation uses all the reduced NAD and reduced FAD produced in stages 1 to 3 to produce ATP. All the reduced NAD and FAD come from glucose molecules, so the energy that is stored in the glucose molecules is used to produce ATP (see Figure 10.37).

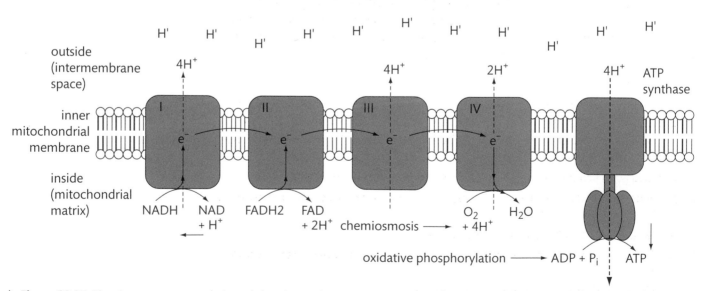

▶ **Figure 10.37:** The electron transport chain and chemiosmosis

Table 10.5 shows the estimated number of ATP molecules made for each molecule of glucose. It is approximately 2.5 molecules of ATP made for each molecule of reduced NAD and 1.5 molecules of ATO made for each molecule of reduced FAD.

▶ **Table 10.5:** ATP molecules made per molecule of glucose

Stage of respiration	Molecules produced	Total ATP produced after oxidative phosphorylation
Glycolysis	+4 ATP (it uses 2 ATP)	2
	+2 reduced NAD	5
Link reaction	+2 reduced NAD	5
Krebs cycle	+2 ATP	2
	+6 reduced NAD	15
	+2 reduced FAD	3

The total **yield** of ATP for each glucose molecule of glucose respired is approximately 32. However, this is rarely achieved, because:

▶ protons can leak across the inner mitochondrial membrane, reducing the number that remain in the cytoplasm to move through the ATP synthase enzyme and cause phosphorylation of ADP to produce ATP

▶ some ATP produced is used to active transport the pyruvate into the mitochondria as it is made in the cytoplasm

▶ some ATP is also used to shuttle the reduced NAD made during glycolysis in the cytoplasm into the mitochondria.

> **Key term**
>
> **Yield** – the amount produced.

⏸ PAUSE POINT   Explain oxidative phosphorylation.

> **Hint**   Can you explain why the proton gradient is a source of potential energy?
>
> **Extend**   Produce a table to show what molecules are needed in each stage of respiration and what is produced in each stage of respiration.

Anaerobic respiration

As you have just seen, oxygen acts as the final electron acceptor in oxidative phosphorylation. However, if oxygen is not present, the electron transport chain stops, and so do the Krebs cycle and the link reaction.

This means that the only source of ATP is through glycolysis, as this stage does not require oxygen. The reduced NAD produced during glycolysis needs to be re-oxidised so that glycolysis continues to occur. No more than two ATP are produced during this stage per glucose, but it does mean that glycolysis can continue.

Lactate fermentation

This is the process that happens in human tissue during vigorous activity such as sprinting, to sustain muscle contraction. Lactate fermentation (see Figure 10.38) relies on the enzyme lactate dehydrogenase, as it is responsible for both the oxidation of NAD and reduction of pyruvate. During lactate fermentation, the following must happen:

▶ reduced NAD produced during glycolysis must be reoxidised to NAD+

▶ pyruvate (the product of glycolysis) accepts as the hydrogen acceptor from reduced NAD

▶ pyruvate becomes reduced and forms lactate

▶ NAD is now re-oxidised and available again to accept more hydrogen atoms from glucose

▶ glycolysis can therefore continue generating two ATP.

The lactate that is produced is carried away from the muscles in the blood and is transported to the liver for each glucose until oxygen becomes available again. When oxygen becomes more readily available, it will convert the lactate back into pyruvate, which can then enter the Krebs cycle and continue to generate more ATP.

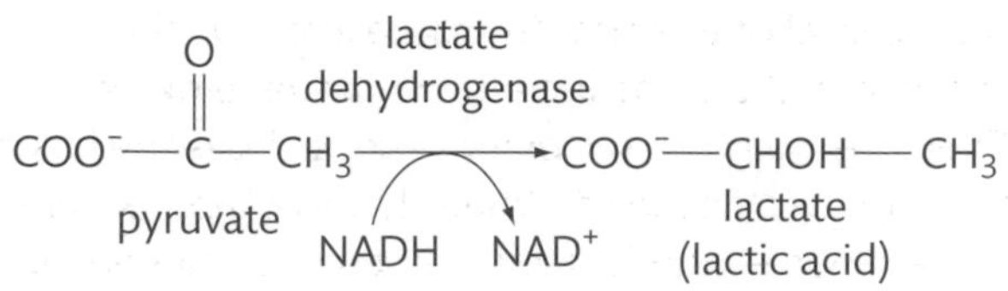

Figure 10.38: Lactate fermentation

❚❚ PAUSE POINT   Explain anaerobic respiration.

Hint   Draw the chemical word equation for lactate fermentation.

Extend   Explain why NAD needs to be re-oxidised, and why this is important for mammals.

Effect of activity on respiration

Exercise places demands on the body, its organs and organ systems. Can you think of ways that your body changes when you exercise?

All of the changes you think of challenge the human body because the body tries to respond by maintaining the body's original stable state. Your body responds by increasing:

▶ breathing and heart rate
▶ blood flow to the skin
▶ sweat secretion
▶ glycogen conversion to glucose.

Recovery rates

After exercise, the body returns to resting level gradually. The fitter you are, the quicker your body returns back to normal. The body needs to recharge the ATP/CP (Creatine Phosphate) system and remove lactate, both of which take time.

Alactacid oxygen debt component

You may know this as Excess Post-exercise Oxygen Debt (EPOC) from level 2. Oxygen debt occurs because the demand for oxygen is too high, so the body is not always able to deliver enough oxygen to the cells for respiration during exercise. After exercise, the body must recharge the ATP/CP system that is used to make ATP. CP is a chemical compound that is stored in muscles, which helps the manufacture of ATP. The combination of ADP and CP produces ATP.

Lactacid oxygen debt component

This part of the recovery process is the removal of the lactate ions produced during exercise. Lactate is acidic and lowers the pH of the blood. This eventually interferes with the enzymes that are needed during aerobic respiration, which in turn reduces the ATP supply to muscles. If lactate builds up, this can cause pain and muscle fatigue, which would make you stop exercising. The removal of lactate can be speeded up by gently exercising. This is known as the warm down, and keeps the capillaries dilated (opened) to allow oxygenated blood into the muscles.

Effect of exercise on carbon dioxide output

During exercise, your body breathing rate will vary to meet the demand for oxygen. More oxygen is used in the respiring muscles to produce more ATP and, as a consequence, more carbon dioxide is produced. The body is sensitive to an increase in carbon dioxide in the blood, so this indicates to the body that more oxygen is required. The level of oxygen can vary, but the levels of carbon dioxide change in direct proportion to the level of exercise. The more intensive the exercise, the greater the carbon dioxide concentration. The increase in carbon dioxide and lactate lowers the pH of blood. Chemoreceptors in the blood vessels above the heart are sensitive to the carbon dioxide content of the blood. When chemoreceptors detect this change, they send more frequent nerve impulses to the intercostal muscles and diaphragm to increase ventilation, making you breathe harder and faster. There are also chemoreceptors in the cardiovascular centre in the brain. They send more frequent nerve impulses down the accelerator nerve to speed up heart rate and deliver more oxygen to muscle cells.

Bromothymol blue is a chemical indicator that changes colour when the pH of a solution changes (see Investigation 10.1). It is blue in neutral or **basic solution** and turns yellow in the presence of acid. When carbon dioxide dissolves in water, carbonic acid is produced. Carbonic acid has a pH of 5.7. Cellular respiration produces carbon dioxide and this increases with exercise. As your cells produce CO_2 during respiration, it is carried by your blood to your lungs where it is breathed out.

> **Key term**
>
> **Basic solution** – solution containing a base that is a substance that react with acids and neutralise them.

Factors that can affect respiration

Cigarettes and nicotine

One of the biggest causes of death and illness in the UK is smoking cigarettes, causing approximately 100 000 deaths per year. Nicotine is the addictive drug in tobacco smoke and it causes smokers to continue to smoke. Smokers need enough nicotine to satisfy cravings or control their mood. Smokers also inhale over 4000 other chemicals when smoking a cigarette. Tobacco smoke contains over 60 known cancer-causing chemicals. The most damaging components of tobacco smoke are:

▶ tar   ▶ carbon monoxide   ▶ carcinogens.

Tar

Tar is a sticky brown substance which contains **carcinogenic** chemicals. It stains teeth, fingers and lung tissue. The chemicals settle in the lining of the lungs and increases the diffusion distance of the respiratory gases, oxygen and carbon dioxide. Tar also destroys the tiny cilia in the respiratory airways. It stimulates the cells to secrete more mucus. The mucus builds up and makes the airways smaller, which restricts the flow of oxygen and carbon dioxide into and out of the lungs.

> **Key term**
>
> **Carcinogenic** – causing cancer.

Carbon monoxide

Carbon monoxide is an odourless gas that binds to haemoglobin in the blood better than oxygen. It therefore reduces how much oxygen the blood can carry. More red blood cells are produced to carry the oxygen, which makes the blood thicker. During exercise, the demand for oxygen is higher. This is when smokers most notice the effects of smoking. Not as much oxygen can reach the brain and muscles compared to a non-smoker. This is because the blood is thicker and does not flow as well through the system due to the presence of carbon monoxide. Smokers are therefore unable to exercise effectively. If the concentration of carbon monoxide becomes large enough, it can cause death.

Carcinogens

Tobacco smoke contains carcinogenic chemicals. Benzopyrene is one of the most harmful. This is present in the tar that lies on the surface of the lungs. These chemicals enter the nucleus of the cells in the lung tissue and have a direct effect on the genetic material inside. The chemicals can cause mutations to DNA and affect genes. If it happens to affect the genes that control cell division, then cells divide uncontrollably and this causes a cancerous growth. This significantly reduces lung function and the ability to exchange the respiratory gases needed for, and produced by, cellular respiration.

Using bromothymol blue to investigate the effect of exercise on the rate of respiration

Bromothymol blue is a **chemical indicator** that changes colour from blue in a neutral solution to yellow-green in an acidic solution. You can use it to investigate how the rate of respiration PH changes as your physical activity increases. This is because the carbon dioxide that you breathe out, when dissolved in water, forms carbonic acid. The following investigation can be divided into two parts:

1 First you must establish a control, that is, you must find out the level of carbon dioxide that you produce before you start doing any exercise.

2 Then, keeping all other aspects of the investigation the same, you should carry out the same investigation having done some exercise.

For each step in the investigation, it's important that you understand the purpose of it, and what you need to pay particular attention to, in order that your results are as accurate as possible.

Steps in the investigation	Pay particular attention to...	Think about this...
1. Fill a 100 ml beaker with 40 ml of water and 10 ml bromothymol blue solution.	Make sure that you measure the volumes accurately, as you will have to repeat this step, and you must keep all the non-variable aspects the same throughout.	In order to find the effect of changing one 'variable' (condition that you can change), you must keep all the other conditions the same in each test you carry out. This is why it is crucial to measure out the volumes accurately each time you repeat the experiment.
2. Place a straw into the solution in the beaker.	**Safety tip**: for the next step, you must only exhale, i.e. breathe out through the straw. Never inhale, as you will swallow the liquid solution!	
3. Start the timer and start breathing out (only) through the straw into the bromothymol solution.	Ensure that you start the timer and start breathing out into the solution at exactly the same time.	
4. Continue to exhale through the straw, until the solution turns from blue to yellow-green. At this point stop the timer.	The problem here is deciding when the colour change has happened. Whatever point you decide to stop the timer, you must do the same in the next part of the investigation.	This is your control. It provides a reference for you to compare future results with.
5. Rinse the beaker out and refill with 40 ml of water and 10 ml bromothymol blue solution.	The accuracy of your results relies on keeping all the conditions the same, except the one condition you wish to change (the 'variable').	
6. Jump up and down on the spot for one minute, so you are panting when you finish.	**Safety tip**: remember to be careful when carrying out any sort of exercise in a laboratory: ensure that you have enough space and that there are no hazards around you.	The variable in this investigation is the amount of physical activity you do.
7. When you have finished exercising, start the timer and exhale into the solution through a straw and time how long it takes for the solution to turn yellow-green.		

8.	Stop the timer when you observe the colour change, and record the time.	The point at which you identify that the colour has changed should be the same as in the first test you carried out.	Think about how you might accurately measure the point at which the colour change occurs, rather than just doing this 'by eye'.
9.	Repeat this investigation a number of times.	Scientific investigations produce more accurate results if they are carried out a number of times.	
10.	Record your results in an appropriate way and write a report on your investigation.	Your report should inform a reader about how you carried out the investigation, and what you did to ensure accurate results. You should present your results in a way that shows your findings as clearly as possible.	Consider all the ways that you might present your results, e.g. as a table, a graph, a chart, etc. You should aim to make your findings as easy to understand as possible.

Drugs

Different classes of drugs have different effects on the body. They can affect ventilation and breathing rate, thus having a knock-on effect on cellular respiration.

▶ Depressants slow the brain activity down and slow breathing rate down. This affects the intake of oxygen, which will affect the body's ability to aerobically respire.

▶ Hallucinogens alter what we see and what we perceive as reality, although the drug itself does not affect breathing rate.

▶ Painkillers block nerve impulses.

▶ Stimulants increase brain activity.

Ketamine

Ketamine is a general anaesthetic that is used during operations on humans and animals, to stop the feeling of pain. It has become a very common drug on the 'clubbing' scene, but in high doses, especially when taken with other substances like alcohol or opiates, it can seriously slow down your breathing and heart rate.

Cocaine

Cocaine is a powerful stimulant and it speeds up the activity in the brain. Cocaine is commonly snorted, which can damage nasal passages. They can become ulcerated and perforated, resulting in infection as well as chronic sinusitis. These airways deliver the oxygen needed for aerobic respiration to take place, so if they are damaged, this may decrease their ability to perform their function efficiently. Taking cocaine can also cause increased levels of energy. This would increase the rate of respiration, as the body would work harder to produce more ATP in order to meet the demand for more energy. Cocaine causes anxiety and agitation which naturally increase breathing and heart rate, and would impact on the rate of respiration. Smoking crack cocaine would damage lung tissue in the same way that smoking nicotine-based products causes damage to cells.

Pollutants

Asbestosis is a disease caused by asbestos, which is the term for a group of minerals made of microscopic fibres. Asbestos materials were widely used in construction in the past, because they are strong, durable and fire-resistant. If you inhale materials containing asbestos fibres, **macrophages** that usually break down foreign particles release substances to destroy the asbestos fibres, but these substances actually cause damage to the alveoli in your lungs and cause permanent scarring. The alveoli are the

Safety tip

Do not inhale the solution.

Key term

Chemical indicator – any substance that gives a visible sign, usually by a colour change, of the presence or absence of a chemical, such as an acid or an alkali in a solution.

Key term

Macrophage – a large white blood cell that digests foreign material.

gas exchange surface in the lungs to deliver the oxygen needed for cellular respiration and to remove the carbon dioxide released during cellular respiration. If they become damaged, this will eventually have an effect on the rate of respiration. Asbestosis can lead to lung cancer and can be fatal.

Cyanide

Cyanide is a potentially lethal chemical that can be inhaled. It can enter the body by breathing air, drinking water, eating food or touching soil that contains cyanide. Cigarette smoke is a major source of cyanide exposure. Cyanide directly affects cellular respiration because it inhibits the last enzyme in the electron transport chain (in stage 4 of respiration). The electron transport chain produces a proton gradient in the matrix of mitochondria so that they can diffuse back through a channel which in turn synthesizes ATP, but cyanide stops this last step happening and therefore stops the synthesis of ATP.

Disease

Lung diseases such as asthma and emphysema can occur and affect the lungs and therefore the ability to breathe efficiently. Asthma is a common disease that can be present from birth or can become apparent in childhood or adulthood. People who suffer with asthma have difficulty breathing. It takes much more effort to deliver oxygen to the lungs because the smooth muscle in the bronchioles contracts, which in turn narrows the airways. Emphysema decreases lung function because the bronchi and bronchioles are continually inflamed and clogged. This causes the alveoli to swell, burst and merge together. This damage to the alveoli makes it more difficult for gas exchange to occur efficiently, therefore slowing the rate of respiration.

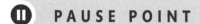

PAUSE POINT   Explain five factors that can affect respiration.

> **Hint**   Think about the gases needed for respiration to occur and the products of respiration.
> **Extend**   Explain the consequence of these factors on the rate of respiration.

Assessment practice 10.2

| B.P2 | B.P3 | B.P4 | B.M2 | B.M3 | B.D2 |

You are a junior research scientist working in a university. You have been asked to find out about aerobic and anaerobic respiration and present your findings to the department. You should:

- explain the stages involved in aerobic and anaerobic respiration
- use diagrams to help with your explanations
- identify where ATP is produced
- produce a table to compare the production of ATP at different stages
- carry out an investigation into the effect of activity on respiration in humans and collect and analyse primary data
- research, analyse and use secondary data to corroborate your findings
- explain the effect of activity on respiration.

You should also explain factors that may affect respiration and evaluate the effects the factors may have on the efficiency of respiration.

Plan
- What is the task? What am I being asked to do?
- How confident do I feel in my own abilities to complete this task? Are there any areas I think I may struggle with?

Do
- I know what it is I am doing and what I want to achieve.
- I can identify when I have gone wrong and adjust my thinking/approach to get myself back on course.

Review
- I can explain what the task was and how I approached the task.
- I can explain how I would approach the hard elements differently next time (i.e. what I would do differently).

Explore the factors that can affect the pathways and the rate of photosynthesis in plants

Pathways in photosynthesis

Photosynthesis is the process whereby light energy from the sun is transformed into chemical energy and used to make large **organic molecules** from inorganic substances. It also releases oxygen into the atmosphere. We all rely on photosynthesis to survive.

▶ We eat plants and products made from them.

▶ The animals we eat have also eaten plants.

▶ Plants release oxygen that we need for respiration.

> **Key term**
>
> **Organic molecule** – a molecule that contains carbon.

> **Reflect**
>
> Look outside and find a large tree. How long do you think it took to grow to that size? Why are trees so important for our existence?

Photosynthesis occurs in some bacteria and plant cells. Plant cells contain organelles called chloroplasts (Figure 10.39) and this is where photosynthesis takes place.

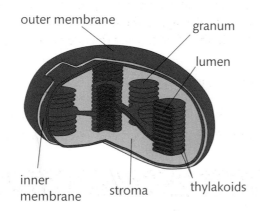

▶ **Figure 10.39:** Chloroplast

Chloroplasts

Chloroplasts vary in shape and size. They:

▶ are approximately 2–10 μm long

▶ are surrounded by a double membrane

▶ contain a stack of sacs (thylakoids) which hold a pigment called chlorophyll.

The outer membrane is **permeable** to small ions and the inner membrane is less permeable because it contains transport proteins, to have more control over entry and exit of molecules. There is an intermembrane space between this double membrane of about 10–20 **nm**.

> **Key terms**
>
> **Permeable** – allowing movement of substance through it.
>
> **nm** – a nanometre, one billionth of a metre.

The inner membrane is folded, and the folds are called lamellae. The lamellae are stacked up. They look like a pile of plates, and these are called grana (the plural of granum). Between the grana are intergranal lamellae. The inner membrane of the chloroplast contains protein carriers so it controls the entry and exit of substances between the cytoplasm and the stroma inside the choroplasts.

The chloroplast has two distinct regions: the grana and the stroma.

▶ The grana look like stacks of plates. Each individual membrane-bond sac (plate) is called a thylakoid. This is where the light is absorbed and ATP is synthesised during the light-dependent reaction.

▶ The stroma is a fluid-filled matrix where the enzymes needed for the light-independent stage are present. They also contain grains of starch and oil, DNA and ribosomes.

The thylakoid membranes inside the grana provide a large surface area for the photosynthetic pigments, electron carriers and ATP synthase enzymes, all of which are needed for the light-dependent reaction. Photosynthetic pigments are arranged into special structures called photosystems (see Figure 10.40) to allow maximum absorption of light energy. The grana have proteins embedded that hold the photosystems in place.

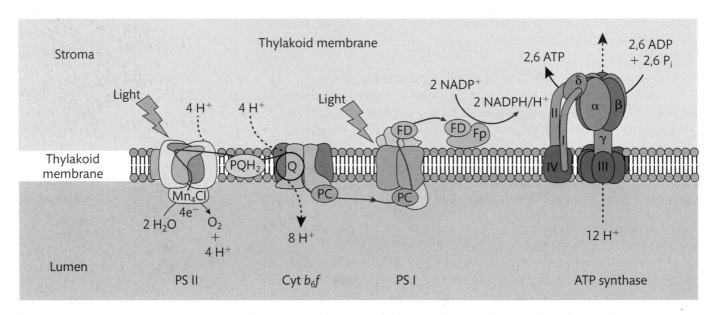

▶ **Figure 10.40:** Thylakoid membrane and photosystems

The fluid-filled stroma stores the enzymes needed to catalyse the light-independent reactions (LIRs). The stroma surrounds the grana. This allows all the products form the light-dependent reactions to pass easily into the stroma as they are needed for the light-independent reactions. Chloroplasts are also able to make their own proteins needed for photosynthesis by using the chloroplast DNA and ribosome.

Photosynthetic pigments

Photosynthetic pigments are substances that absorb different wavelengths of light and reflect others. We see them as the colour of the light wavelength that they are reflecting. Many of the pigments act together to capture as much light as possible. The pigments are arranged in photosystems. These are funnel-shaped structures in the thylakoid membrane which are held in place by proteins (see Figure 10.41).

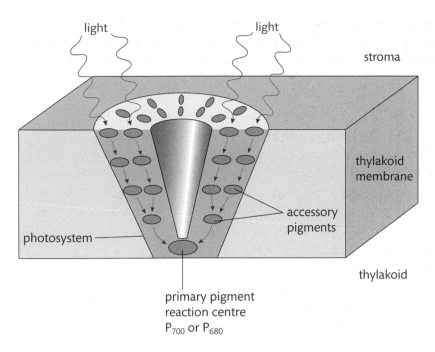

▶ **Figure 10.41:** Funnel-shaped photosystem

Chlorophyll

Chlorophyll is a green pigment found in green plants, but it actually contains a mixture of different pigments. The pigments all have long phytol (hydrocarbon) chains and a porphyrin group containing magnesium (this is similar to the functional group haem in haemoglobin). Chlorophyll works in conjunction with the metal magnesium. There are two forms of chlorophyll, a and b, which are very similar molecules that absorb light in the red and blue areas of the spectrum. They consist of a head that absorbs the light and a tail that anchors it to the thylakoid membrane. The head is a complex chemical ring structure with a magnesium ion in the middle (see Figure 10.42).

There are two types of chlorophyll a: P_{680} and P_{700}. Both appear yellow/green but they absorb red light at slightly different wavelengths. They are found in the primary pigment reaction centre, which is the centre of the photosystem. P_{680} is found in one type of photosystem called photosystem II and absorbs red light strongly at a wavelength of 680 nm. P_{700} is found in photosystem I and absorbs red light strongly at a wavelength of 700 nm. Chlorophyll a absorbs blue light to a lesser extent at a wavelength of 450 nm. Chlorophyll b absorbs light at wavelengths of 500 nm and 640 nm, and this appears blue-green.

Carotenoids (see Figure 10.43) are pigments that reflect yellow and orange light and absorb blue light. They do not contain porphyrin and are not directly involved in the light-dependent reaction. They absorb wavelengths that chlorophyll does not absorb and pass this energy onto the chlorophyll. Carotene is orange and xanthophyll is yellow. They are the main carotenoid pigments and they are shown as accessory pigments in Figure 10.41.

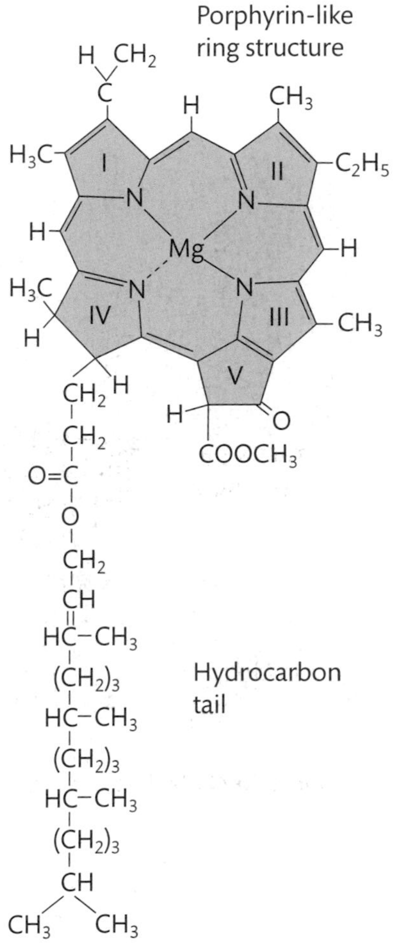

▶ **Figure 10.42:** Structure of chlorophyll with long phytol chain and porphyrin group containing magnesium

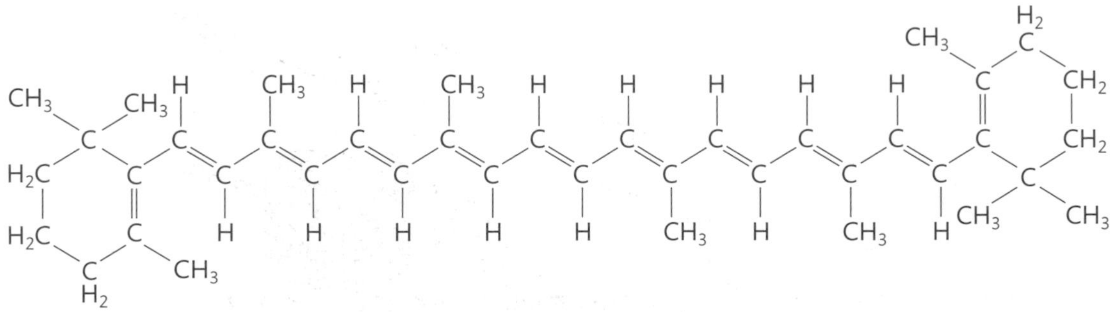

▶ **Figure 10.43:** Carotenoid structure

Photosynthesis can be divided into two steps:

▶ the light-dependent stage

▶ the light-independent stage.

Light-dependent stage

The light-dependent stage of photosynthesis takes place on the thylakoid membrane of the chloroplast. The photosystems with photosynthetic pigments are embedded in this membrane.

Photolysis

Photosystem II contains an enzyme that can split water into protons (H^+ ions), electrons and oxygen, when light is present (**photolysis**). Some of the oxygen is used in the plant for aerobic respiration, but most of it diffuses out of the leaf through the stomata. The H^+ ions are used during chemiosmosis to produce a proton gradient for the production of ATP, and they are used to reduce a co-enzyme called NADP which is used in the light-independent stage. Finally, the electrons are used to replace those lost by oxidised chlorophyll.

The chemical symbol equation for photolysis is as follows:

$$2H_2O \longrightarrow 4H^+ + 4e^- + O_2$$

Key term

Photolysis – splitting of water in the presence of light.

Photophosphorylation

When a particle of light called a photon hits chlorophyll in photosystem II (PSII), the electrons inside PSII absorb the energy from the light and they become excited (see Figure 10.44). The electrons are picked up by electron acceptors and pass along a series of electron carriers (proteins) embedded in the thylakoid membrane. Energy is released as the electrons pass along the chain. This energy pumps protons across the thylakoid membrane into the thylakoid spaces. This creates a proton gradient across the thylakoid membrane. The protons travel down the gradient through channels with ATP synthase enzymes that join ADP and an inorganic phosphate to make ATP. This flow of protons down a proton gradient is known as chemiosmosis. The kinetic energy from the protons moving is converted into chemical energy in ATP molecules. These ATP molecules are then used in the light-independent stage of photosynthesis. This is known as photophosphorylation because it uses light to make ATP.

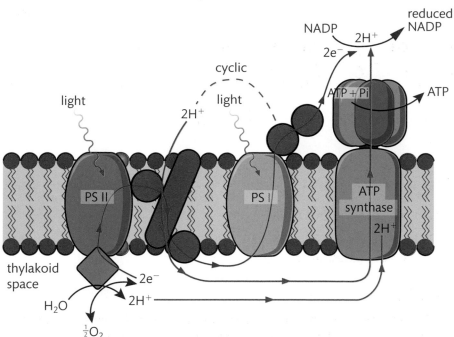

▶ **Figure 10.44:** Photophosphorylation when light hits PSII, from *OCR Biology A2 Heinemann*, Heinemann (Hocking, S., 2008) p.62, Pearson Education Limited

There are two types of photophosphorylation.

Cyclic photophosphorylation uses only photosystem I, with chlorophyll a (P_{700}) present. When light hits PSI the excited electrons pass to an electron acceptor and then back to the chlorophyll a molecule that they were lost from. Only a small amount of ATP is made and there is no photolysis of water and no reduced NADP is produced.

Non-cyclic photophosphorylation involves both photosystems I and II. Light hits photosystem II and the pair of excited electrons pass along the electron carrier and the energy released is used to make ATP. Light also strikes photosystem I and a pair of electrons here are excited too. These electrons and the protons produced during photolysis of water join to NADP and produce reduced NADP. The electrons from the oxidised photosystem II replace the electrons lost from photosystem I. Electrons lost by oxidised chlorophyll in PSII are replaced with some from photolysed water. Protons from photolysed water take part in chemiosmosis to make ATP.

Ⅱ **PAUSE POINT**   Describe the light-dependent stage of photosynthesis.

    Hint   Think about the structure of the chloroplast and the photosystems involved.

    Extend   Compare cyclic and non-cyclic photophosphorylation.

Light-independent stage

This stage is also known as the Calvin cycle and takes place in the stroma. Light is not used during this stage. However, products of the light-dependent stage are used, so the light-independent stage will eventually stop if light becomes a limiting factor.

Carbon dioxide is needed during this stage to provide carbon to produce large organic molecules. Carbon dioxide from the atmosphere diffuses into the open stomata in the leaf. It travels through the spongy mesophyll layer and reaches the palisade mesophyll layer (see Figure 10.45). It diffuses across the cell wall and membrane into the stroma.

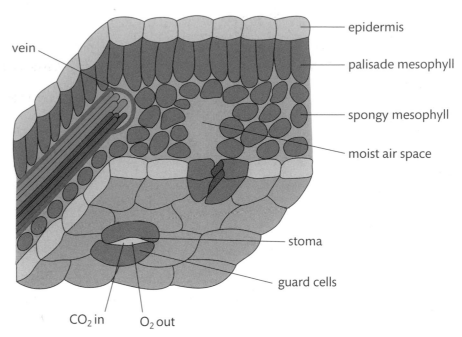

vein

epidermis

palisade mesophyll

spongy mesophyll

moist air space

stoma

guard cells

CO_2 in O_2 out

▶ **Figure 10.45:** Structure of a leaf

The Calvin cycle

1   The carbon dioxide combines with ribulose bisphosphate (RuBP), which is known as a carbon dioxide acceptor. The enzyme ribulose bisphosphate carboxylase-oxygenase, otherwise known as rubisco, is needed during this reaction.

2   RuBP is **carboxylated**. This fixes the carbon dioxide and produces two 3-carbon compound molecules known as glycerate 3-phosphate (GP).

3   GP is phosphorylated and reduced to a different 3-carbon compound, triose phosphate (TP).

4   ATP and NADP from the light-dependent stage are used during this process.

5   ATP from the light-dependent stage and five out of six molecules of TP are used during phosphorylation to regenerate three molecules of RuBP.

The other products of the Calvin cycle are used in different ways:

▶   GP is used to produce amino acids and fatty acids needed in the plant.

▶   Hexose sugars such as glucose are produced by the pairing of TP molecules.

▶   **Isomerisation** of glucose molecules can produce fructose.

▶   Glucose and fructose join during a condensation reaction, and form sucrose. Sucrose is a disaccharide.

▶   Glucose and fructose can be polymerised into other polysaccharides, such as starch and cellulose.

▶   Lipids can be produced by converting TP into glycerol and combining it with fatty acids which can be produced from GP.

> **Key term**
>
> **Carboxylated** – combined with carbon dioxide.

> **Key term**
>
> **Isomerisation** – the process of transforming one molecule into another which has exactly the same atoms, but they are arranged differently.

Figure 10.46 shows all the products of the Calvin cycle.

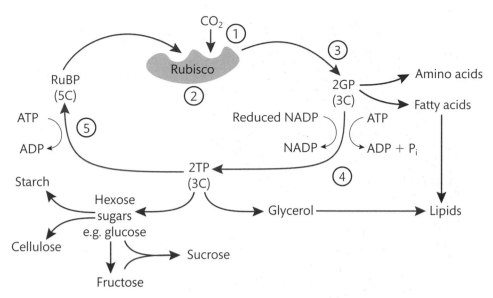

▸ **Figure 10.46:** The Calvin cycle, adapted from *OCR Biology A2*, Heinemann (Hocking, S., 2008) p.64, Pearson Education Limited

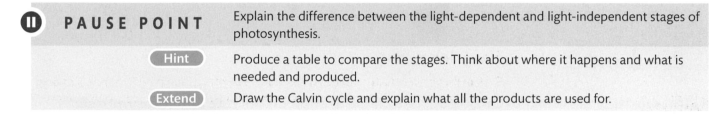

Ⅱ PAUSE POINT   Explain the difference between the light-dependent and light-independent stages of photosynthesis.

Hint   Produce a table to compare the stages. Think about where it happens and what is needed and produced.

Extend   Draw the Calvin cycle and explain what all the products are used for.

Factors that can affect the pathways in photosynthesis

When we summarise photosynthesis in a word equation, it enables us to see all the factors that affect it.

$$\text{carbon dioxide} + \text{water} \xrightarrow[\text{chlorophyll}]{\text{light}} \text{glucose} + \text{oxygen}$$

$$6CO_2 + 6H_2O \xrightarrow[\text{chlorophyll}]{\text{light}} C_6H_{12}O_6 + 6O_2$$

Carbon dioxide, light and water are all environmental factors that can influence the rate of photosynthesis. If one of the factors is less available than the others, then it can be described as a **limiting factor** because it could run out and therefore slow down the rate of photosynthesis.

Light intensity

Light can be a limiting factor during photosynthesis. The rate of photosynthesis is directly proportional to the light intensity. Light intensity changes throughout the day, so the rate of photosynthesis will also vary. As light intensity increases, the rate of photosynthesis increases. Light causes the stomata to open so that carbon dioxide can enter through the leaves. Light excites electrons in the chlorophyll and, with the help of enzymes, it splits water molecules.

Carbon dioxide concentration

Carbon dioxide makes up around 0.03–0.06% of the Earth's atmosphere. Increasing levels of carbon dioxide can increase the rate of photosynthesis, but only if there is no other limiting factor. Carbon dioxide is required in the light-independent reaction (LIR) so, by increasing CO_2, the rate of LIR would increase.

> **Key term**
>
> **Limiting factor** – a factor that may slow down the rate of a chemical reaction or process.

Water

Water is essential for photosynthesis to occur. However, a lack of water will probably cause the plant to wilt before the rate of photosynthesis is affected.

Temperature

Temperature does not have much of an effect on the light-dependent stage of photosynthesis. The enzymes in the Calvin cycle work at their best at 0–25 °C so the rate of photosynthesis doubles every 10 °C. However, when the temperature rises above 25 °C, the rate of photosynthesis starts to decrease because the enzymes stop working as efficiently as they become denatured. Oxygen competes with carbon dioxide for the active site of rubisco so carbon dioxide is less likely to be accepted by rubisco. Oxygen becomes a competitive inhibitor to carbon dioxide so CO_2 becomes less likely to be accepted by rubisco. An increase in temperature may also increase water loss from the stomata, so the stomata will close, reducing the amount of CO_2 that enters the plant.

Wavelength of light

The wavelength of light will affect the rate of photosynthesis because the pigments only absorb certain wavelengths of light. When these wavelengths are available, the rate of photosynthesis will increase. When these wavelengths are less available, the rate of photosynthesis will fall.

Investigation 10.2

Investigating the effect of carbon dioxide concentration on the rate of photosynthesis

A **photosynthometer** can be used to measure the rate of photosynthesis by collecting and measuring the volume of oxygen produced during a certain time. You can use it to investigate how the rate of photosynthesis changes as the concentration of carbon dioxide varies, by adding sodium hydrogen carbonate. This is because sodium hydrogen carbonate changes the concentration of carbon dioxide and this will in turn change the volume of oxygen produced. The following investigation can be divided into two parts.

▶ First you must establish a control, that is, you must set up the apparatus and leave it for the same amount of time as the test apparatus adding no sodium hydrogen carbonate.

▶ Then, keeping all other aspects of the investigation the same, you should carry out the same investigation adding 2, 3, 4, 5, 6, 7 and 8 drops of sodium hydrogen carbonate solution.

For each step in the investigation, it's important that you understand the purpose of it, and what you need to pay particular attention to, in order that your results are as accurate as possible.

Steps in the investigation	Pay particular attention to...	Think about this...
1. Set up the **photosynthometer** as in Figure 10.47.	Make sure that you the apparatus set up is air tight and there are no bubbles in the capillary tubing.	
2. Fill the barrel of the syringe and plastic tubing with water by removing the plunger from the syringe and adding a gentle stream of tap water in until it is full. Place the plunger back in and gently push water out of the capillary tube until the plunger is nearly at the end of the syringe.	**Safety tip**: apparatus is glassware so it should be set up in the centre of the table. If any glass were to smash, you should inform your assessor and it should be disposed of in a glass bin.	In order to find the effect of changing one 'variable' (condition that you can change), you must keep all the other conditions the same in each test you carry out. This is why it is crucial to measure out the volumes accurately each time you repeat the experiment.

3.	Cut a piece of **Elodea** about 7 cm long and place the cut end upwards in the test tube. Fill the test tube with the pond water that it has been kept in.	Make sure that you measure the Elodea accurately, as you will have to repeat this step, and you must keep all the non-variable aspects the same throughout.	Why is it crucial to keep the Elodea in the same pond water it has been kept in?
4.	Stand the test tube in a beaker of water that it approximately 20 °C. Use a thermometer to monitor the temperature and add cooled water to keep the water temperature constant when necessary.	This can be difficult especially if room temperature is significantly below or above. The accuracy of your results relies on keeping all the conditions the same, except the one condition you wish to change (the 'variable').	
5.	Place a light source as close to the beaker as possible. Measure this distance between the light source and the beaker. Keep it the same each time, e.g. 10 cm.	The accuracy of your results relies on keeping all the conditions the same, except the one condition you wish to change (the 'variable').	This is your control.
6.	Carry out steps 1 to 5 and add 2 drops of sodium hydrogen carbonate solution and measure the volume of gas produced in a fixed period of time.	Ensure that you start the timer and remove the elodea after the time agreed, to ensure this does not increase the volume of oxygen.	You can work out the volume if you know the radius of the capillary tube: volume of gas collected = length of bubble $\times \pi r^2$
7.	Repeat steps 1 to 5 and add 3, 4, 5, 6, 7 and 8 drops of sodium hydrogen carbonate solution.		The variable in this investigation is the concentration of carbon dioxide.
8.	Repeat the entire investigation three times and work out averages.	Ensure that you record to the same number of decimal points in each result and keep this consistent throughout your results.	Repeating the investigation increases reliability. Anomalous results should be identified and discussed.
9.	Record the length of the bubble in an appropriate way and write a report on your investigation.	Your report should inform a reader about how you carried out the investigation, and what you did to ensure accurate results. You should present your results in a way that shows your findings as clearly as possible.	Think about how you could investigate other factors that affect rate of reaction using the same apparatus.
10.	Analyse your results.		Consider all the ways that you might present your results, e.g. as a table, a graph, a chart, etc. You should aim to make your findings as easy to understand as possible.

Key terms

Photosynthometer – apparatus used to measure the rate of photosynthesis by collecting and measuring the volume of oxygen produced.

Elodea – aquatic plant.

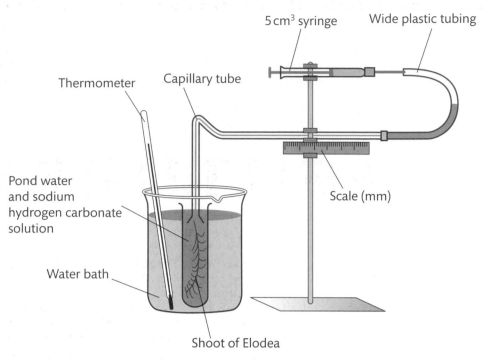

Labels on figure:
- 5 cm³ syringe
- Wide plastic tubing
- Thermometer
- Capillary tube
- Pond water and sodium hydrogen carbonate solution
- Scale (mm)
- Water bath
- Shoot of Elodea

▶ **Figure 10.47:** Photosynthometer

Assessment practice 10.3

C.P5 C.P6 C.M4 C.D3

You were so successful at presenting your research into respiration that your boss has asked you to present the stages involved with photosynthesis. You should:

- explain the stages involved in photosynthesis
- use diagrams to help with your explanations
- identify where ATP is produced
- compare the light-dependent and light-independent stages
- carry out an investigation into a factor (light intensity, carbon dioxide or temperature) that affects the rate of photosynthesis
- collect primary data
- research secondary data
- analyse how that factor affects the rate of photosynthesis.

You should also present an evaluation on the effect of factors on the efficiency of photosynthesis.

Plan
- What is the task? What am I being asked to do?
- How confident do I feel in my own abilities to complete this task? Are there any areas I think I may struggle with?

Do
- I know what it is I am doing and what I want to achieve.
- I can identify when I have gone wrong and adjust my thinking/approach to get myself back on course.

Review
- I can explain what the task was and how I approached the task.
- I can explain how I would approach the hard elements differently next time (i.e. what I would do differently).

Further reading and resources

Annets, F., Foale, S., Hartley, J., Hocking, S., Hudson, L., Kelly, T., Llewellyn, R., Musa, I. and Sorenson, J. (2010) *Applied Science Level 3*. Pearson.

Boyle, M. and Senior, K. (2008) *Biology* (3rd edition). Collins.

THINK ▶FUTURE

Nicki Sprung,
clinical biochemist
in a hospital

I have been working in the hospital for seven years, analysing samples taken from patients' blood, urine and other bodily fluids. This helps doctors to diagnose patients and decide on the correct treatment. People do not realise the level of responsibility we have, not only in carrying out the scientific tests with precision, but also interpreting the results and communicating effectively with other clinical staff on the correct use of tests and follow-up investigations. The results of tests will determine a patient's diagnosis and treatment, so it is imperative that the test is done properly.

Our laboratory in the hospital processes thousands of samples per day. If one of the samples is abnormal, that's when it must be scrutinised by myself or one of my colleagues. We are responsible for carrying out intricate analysis using incredibly complex equipment, for example, spectrophotometry, mass spectrometry, HPLC and electrophoresis. As we use biohazards such as blood and urine, risk assessing is essential and is a primary responsibility to ensure the safety of all staff. We do many other administration jobs such as report writing and bidding for funding, but our main aim is to process the biological samples as quickly as possible to provide a result to doctors.

Focusing your skills

Risk-assessing a scientific practical

- Identify the equipment and chemicals that are a potential hazard. When working in the laboratory, you must be able to identify equipment that can cause harm.

- Describe the risk that they could cause by thinking about how you are going to use it and the potential harm that could result from misuse. You should think about specific body parts that may be affected.

- Explain what you could do to reduce the risk from occurring (control). It is important that you pre-empt potential problems and think in advance what actions you could take in order to limit the risk from occurring.

- Research and explain how you would manage the risk if it did occur (emergency procedure). Use COSHH regulations. Many chemicals come with data safety sheets that explain what should be done in the event of an accident.

Interpreting results

- State the result. It may be a colour indication. We call this qualitative results. It may be a numerical measurement. We call this a quantitative result. This must be recorded accurately, including units where necessary.

- Use the result to interpret what this means. The presence of a specific colour may indicate the presence of a specific biological molecule. For example, a black/blue colour when adding iodine to potatoes indicates starch is present.

- You then use science to explain why the test produced that result.

Transferrable skills

- Communication skills are important to ensure that the correct information is conveyed to necessary people, for example, doctors and nurses. Working in the hospital means that you must always be aware of communicating in a professional manner with the public.

- Leadership skills are important in order to ensure that the laboratory runs efficiently and samples are analysed in a timely manner and to ensure prioritisation of samples.

Getting ready for assessment

Gemma is working towards a BTEC National in Applied Science. She was given an assignment with the title 'Why are biological molecules important?' for learning aim A. She had to write a professional-looking report on the disruption to the structure and function of biological molecules and the effect this can have on living organisms. The report had to:

▶ include information on the structure and function of water, carbohydrates, protein and lipids

▶ discuss the potential effects to living organisms if these molecules structure or function was disrupted.

Gemma shares her experience below.

How I got started

At the end of each lesson, I filed all my notes in a file, in date order with titles of each lesson/topic. I decided to divide my work into the different biological molecules. For each biological molecule, I had sub-divided my notes into structure, function and disruption. To help meet the grading criteria for structure, I described the structure in words and sourced a diagram from the internet. I referenced the source using Harvard-style referencing. After each structure I discussed the functions.

How I brought it all together

I decided to lay out my work in line with local university expectations, using size 12 Arial font, 1.5 line spacing and justifying text. I included a header with my name, date and unit number, and included page numbers for organisation. I also used references throughout my text and included a reference list. To start, I wrote a short introduction to introduce important biological molecules and then used side headings to indicate where I was going to concentrate on each biological molecule. For each biological molecule I included:

▶ a diagram of the structure with a description, including bonding

▶ an explanation of the importance of each biological molecule in living organisms

▶ a discussion on how the molecules structure can be disrupted, how this can change function and the effect this would have on the living organism.

I also tried to give specific examples. Finally, I wrote a short summary as a conclusion to the report.

What I learned from the experience

I wish I had organised my notes more in class to make it easier to find the information to go in each part. Next time I will highlight important titles and number the pages of my class notes.

I focused too much on the structure. I could have given a more detailed explanation on each function and their roles in living organisms. I struggled with the discussion and relaying the disruption to a change in function in the living organism.

Think about it

▶ Have you broken down the task into small manageable success criteria to ensure you cover all the unit content?

▶ Have you thought about timings so that you can complete your assignment by the agreed submission date?

▶ Is your information interpreted and written in your own words? Is it referenced clearly where you have used quotations or information from a book, journal or website?

Genetics and Genetic Engineering 11

Getting to know your unit

In 1859, when Charles Darwin published his book *On the Origin of Species by Natural Selection*, he did not know exactly *how* the characteristics that made some organisms better adapted than others for survival were passed from parent to offspring. Gregor Mendel published his own work on inheritance of characteristics in the 1860s, but it was not well understood or appreciated. In the early 1900s, Mendel's work was rediscovered and other scientists began exploring the branch of biology known as genetics. By 1953, scientists had established that DNA was the chemical that transmitted hereditary information. James Watson and Francis Crick, with Maurice Wilkins and Rosalind Franklin, had worked out its structure. During the next few decades, scientists determined how the coded information within DNA governed the synthesis of proteins in living cells. By the 1970s, restriction enzymes in some bacteria had been discovered. An increased knowledge about bacteria enabled scientists to begin recombinant DNA technology; this is genetic manipulation. Massive advances in DNA technology over the last 30 years have driven the study of genetics forward at an exponential rate, creating huge potential for many applications in food production and health care. Many genomes, including that of humans, have been sequenced. A way of gene editing has been developed that offers hope to many sufferers of genetic diseases.

This unit will allow you to develop an understanding of genetics and genetic engineering techniques and their uses. It will be of particular interest if you wish to follow a career in health, medical, veterinary, industrial, agricultural or forensic sciences, pharmacology or research. It will also enable you to understand and follow media coverage of medical advances such as gene editing, pre-implantation genetic screening and stem cell surgery, and to participate in informed debates surrounding the use of GM crops.

You will investigate the mechanisms of cell division and research how the behaviour of chromosomes during cell division to make gametes contributes to genetic variation and evolution. You will learn more about how genes control characteristics in living organisms by directing the process of protein synthesis via the information encoded in them. You will learn how the principles of Mendelian genetics can be used to explain patterns of inheritance. You will also explore modern genetic engineering techniques and their uses and have opportunities to extract and work with DNA and to genetically modify the bacterium *E. coli*.

How you will be assessed

This unit will be assessed by a series of internally assessed tasks set by your tutor, involving portfolios of evidence. Throughout this unit, you will find assessment activities that may help you work towards your assessment. Completing these activities will not mean you have achieved a particular grade, but the research you carry out for them will be relevant and useful when you come to carry out your final assessment.

It is important to check that you have met all the Pass grading criteria as you work your way through the assignments.

To achieve a Merit or Distinction, you need to present your work in such a way that you meet the criteria for those grades.

To achieve Merit, you need to analyse and discuss relevant issues and demonstrate skills, and for Distinction you need to assess and evaluate.

The assignments set by your tutor will consist of a number of tasks designed to meet the criteria in the table below. Some tasks will be written and some lab-based practical exercises. Tasks may also involve reviewing and analysing case studies.

Assessment criteria

This table shows what you must do in order to achieve a **Pass**, **Merit** or **Distinction** grade, and where you can find activities to help you.

Pass	Merit	Distinction
Learning aim A Understand the structure and function of nucleic acids in order to describe the process of protein synthesis		
A.P1 Explain the structure and function of DNA and various nucleic acids **Assessment practice 11.1**	**A.M1** Discuss the functional role of nucleic acids in DNA, in the stages of protein synthesis **Assessment practice 11.1**	**A.D1** Assess the impact of errors in the stages of protein synthesis **Assessment practice 11.1**
Learning aim B Explore how the process of cell division in eukaryotic cells contributes to genetic variation		
B.P2 Prepare microscope slides to observe and draw the stages of mitosis and meiosis **Assessment practice 11.2**	**B.M2** Demonstrate skilful preparation of microscope slides to observe and draw the stages of mitosis and meiosis **Assessment practice 11.2**	**B.D2** Evaluate how the behaviour of the chromosomes leads to variation **Assessment practice 11.2**
B.P3 Explain the structure and function of human chromosomes **Assessment practice 11.2**	**B.M3** Discuss the behaviour of the chromosomes during the cell cycle stages of mitosis and meiosis **Assessment practice 11.2**	
Learning aim C Explore the principles of inheritance and their application in predicting genetic traits		
C.P4 Carry out investigations to collect and record data for mono and dihybrid phenotypic ratios **Assessment practice 11.3**	**C.M4** Analyse data to explain the correlation between observed pattern of monohybrid and dihybrid inheritance **Assessment practice 11.3**	**C.D3** Make valid predictions on patterns of monohybrid and dihybrid inheritance and variation using principles of inheritance **Assessment practice 11.3**
C.P5 Explain genetic crosses between non-affected, affected and symptomless carriers of genetic conditions **Assessment practice 11.3**	**C.M5** Apply Mendel's laws of inheritance to the results of genetic crosses **Assessment practice 11.3**	
Learning aim D Explore basic DNA techniques and use genetic engineering technologies		
D.P6 Extract, separate and amplify DNA **Assessment practice 11.4**	**D.M6** Analyse the uses of genetic engineering technologies in industry and medicine **Assessment practice 11.4**	**D.D4** Evaluate possible future uses of genetic engineering technologies **Assessment practice 11.4**
D.P7 Explain the use of genetic engineering technologies in industry and medicine **Assessment practice 11.4**		

Getting started

Can you name any genetic diseases? Do you know if they have a recessive or dominant inheritance pattern? What comes into your mind when you hear the word 'mutation'? Do you think all mutations are harmful or are some useful, or neutral? Do you understand what genes, alleles and chromosomes are? What are your views on genetic modification? Do you understand what stem cells are and how they can be used in medicine?

Write down your answers to these questions and look at them again when you have completed this unit to see if your understanding has increased and if any of your views have changed.

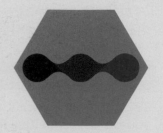

A Understand the structure and function of nucleic acids in order to describe gene expression and protein synthesis

Although by the late nineteenth century, scientists knew about the chemical called DNA (deoxyribonucleic acid) they did not understand its role, or just how important it was for living organisms. Scientists knew it occurred in all cells, but because it appeared to do nothing, they called it 'the stupid molecule'. In this section you will learn that this was not a good description of it. You will learn about the structure of nucleic acids (DNA and RNA), the features of the genetic code and the process of protein synthesis.

Nucleic acids

Nucleic acids are universal. They occur in all living organisms on earth – in prokaryotic and eukaryotic cells as well as in viruses (akaryotes). Both DNA (deoxyribonucleic acid) and RNA (ribonucleic acid) are **polymers**. In this case the monomers are **nucleotides**. Figure 11.1 shows the structures of deoxyribose and ribose sugars. Nucleic acids are polynucleotides. Figure 11.2 shows three nucleotides joined together.

Nucleotides are joined together by **condensation reactions**.

> **Key terms**
>
> **Polymers** – large molecules consisting of repeating smaller subunits (monomers).
>
> **Nucleotide** – a nucleotide consists of a pentose (5-carbon) sugar, a phosphate and a nitrogenous base (adenine, thymine, cytosine, guanine or uracil). Nucleotides are the monomers of nucleic acids.
>
> **Condensation reaction** – chemical reaction where two molecules are joined, with the elimination of water.

▸ **Figure 11.1:** Ribose and deoxyribose sugars

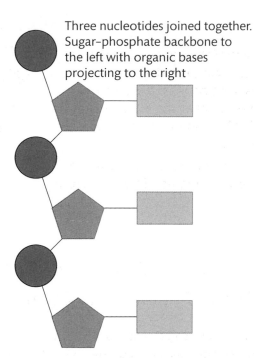

Three nucleotides joined together. Sugar-phosphate backbone to the left with organic bases projecting to the right

▶ **Figure 11.2:** Three nucleotides joined together. The sugars (orange) and phosphate groups (red) form the backbone. The organic nitrogenous bases are shown in green.

There are five different nucleotide bases.
▶ Adenine (A) and guanine (G) have a double ring structure. These are purines.
▶ Cytosine (C), thymine (T) and uracil (U) have a single ring structure. These are pyrimidines.

DNA contains the bases A, T, G and C; RNA contains the bases A, U, G and C.

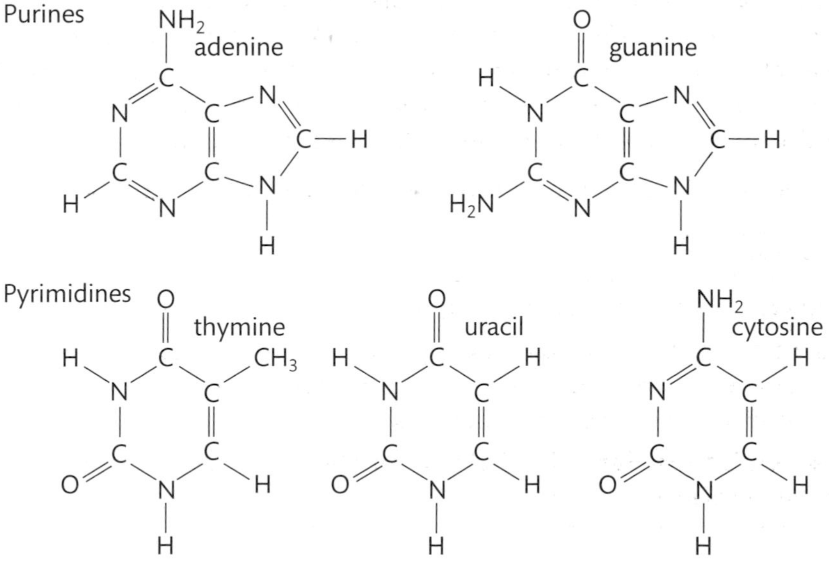

▶ **Figure 11.3:** Purine and pyrimidine nucleotide bases

DNA is a double helix

▶ Specific types of viruses contain either DNA or RNA, surrounded by a protein coat.
▶ In prokaryotic cells (bacteria), their DNA is found as one circular chromosome lying freely in the cytoplasm. They also have some smaller loops of DNA, called **plasmids**.

▶ In eukaryotic cells, most of their DNA is enclosed in the nucleus and organised into several linear chromosomes. Each chromosome contains one huge molecule of DNA and there are also **histone proteins** associated with that DNA. Eukaryotic cells also have DNA in their mitochondria and, in the case of some plant and algal cells, in chloroplasts.

Link

See *Unit 1B: Structure and functions of cells and tissues* for information about the structure and functions of chloroplasts and mitochondria, both of which originated from bacteria.

Each DNA molecule consists of the following.

▶ Two backbone chains of deoxyribose sugars and phosphate groups. These backbones are described as antiparallel as they each run in opposite directions.

▶ Pairs of nitrogenous bases joining the backbones together. The base pairs join to each other by hydrogen bonds (H bonds). A purine base always joins with a pyrimidine base so the 'rungs of the ladder' are always the same size. A joins, by two H bonds, with T; G joins, by three H bonds, with C. This type of specific base pairing is called **complementary base pairing**. See Figure 11.4.

Nucleotides with adenine as the base can make two hydrogen bonds with nucleotides with thymine as the base.

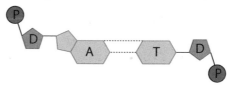

Nucleotides with guanine as the base can make three hydrogen bonds with nucleotides with cytosine as the base.

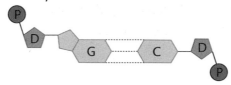

▶ **Figure 11.4:** Hydrogen bonding between purine and pyrimidine nucleotide bases

The hydrogen bonds make each molecule of DNA very strong and stable, while enabling it to unzip in order to copy itself before cell division, or allowing part of it (a gene) to unzip before being used as a template to make messenger RNA before the assembly of a new protein.

The sequence of base pairs form the coded information and this is protected from corruption by these base pairs being inside the backbones.

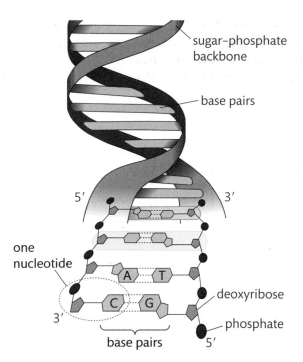

▶ **Figure 11.5:** The DNA double helix. The antiparallel chains, with the base pairs, are twisted into a helix. The directions 3′ to 5′ relate to the position of the carbon atoms on the sugars.

Worked example

A scientist analysed a sample of DNA. She found that 23% of the bases consisted of adenine. She could then work out the percentage of the other bases in that sample.

Because A always pairs with T, 23% were thymine bases.

A and T together make up 46% of the bases

So C and G make up 54% of the bases

Because C always pairs with G, the percentage of bases that are cytosine is 27% and 27% are guanine.

Another scientist analysed another sample of DNA. He found that 30% of the nitrogenous bases consisted of guanine.

(a) What percentage of this DNA sample consisted of thymine?

(b) In a length of DNA 250 base pairs long, 100 pairs were adenine and thymine. How many bases were cytosine?

(a) T = 20% (If 30% is C, then 30% is G, therefore 60% for C and G is G, so (100 – 60) = 40% for T + A)

(b) 150 base pairs are G and C, so C = 75 bases.

Ⅱ **P A U S E P O I N T**   Explain why DNA is described as 'a macromolecule' and 'a polynucleotide'.

> **Hint**   Think about what macromolecule and polynucleotide mean and say how the structure of DNA fits these descriptions.

> **Extend**   Explain how purines and pyrimidines differ from each other.

DNA replication

Key term

Interphase – the phase of the cell cycle in which most cells spend most of their time. They synthesise molecules, grow and the organelles and DNA replicate prior to mitosis.

Before a cell divides, during a part of the cell cycle known as the S phase of **interphase**, each DNA molecule replicates.

1 The double helix of the whole DNA molecule unwinds, a bit at a time and the hydrogen bonds between complementary base pairs break. This is catalysed by the enzyme helicase.

2 This exposes the nucleotide bases.

3 Free DNA nucleotides within the nucleoplasm of the nucleus, bond onto the exposed nucleotide bases, following complementary base-pairing rules, A bonds with T and C bonds with G. This is catalysed by the enzyme DNA polymerase.

4 Covalent bonds form between the sugar of one nucleotide and the phosphate group of the adjacent nucleotide, forming the new backbones. This is catalysed by the enzyme ligase.

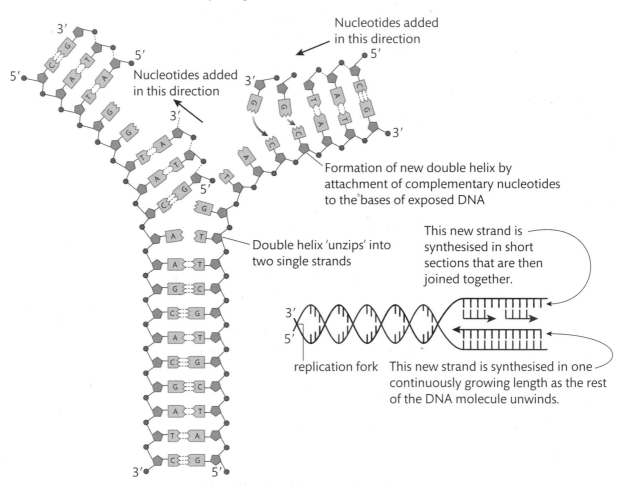

▶ **Figure 11.6:** How DNA replicates. The addition of the new DNA nucleotides is catalysed by the enzyme DNA polymerase.

At the end of replication, two new molecules of DNA, both identical to each other and to the parent molecule, are made. Each new molecule contains one old strand and one new strand, so this type of replication is called **semi-conservative replication**.

Key term

Semi-conservative replication – mode of replication where two new molecules are formed, each identical to the other and to the parent molecules and each consisting of one old strand and one new strand of DNA.

Differences between RNA and DNA

RNA is structurally different from DNA.

In RNA:

▶ the pentose sugar is ribose

▶ the nitrogenous bases present are A, U, C and G (there is no thymine)

▶ the polynucleotide chain is usually single-stranded, not double-stranded (however, some viruses contain double-stranded RNA and short sections of some RNA molecules, such as transfer RNA, in eukaryotes are also double-stranded).

In DNA:

▶ the pentose sugar is deoxyribose

▶ the nitrogenous bases present are A, T, C and G

▶ the polynucleotide chain is almost always double-stranded.

In eukaryotic cells:

▶ RNA occurs in the nucleus (including the nucleolus) and cytoplasm, as well as in ribosomes

▶ DNA occurs in the nucleus, chloroplasts and mitochondria.

There is more than one type of RNA.

▶ Messenger RNA (mRNA) carries the genetic code of a gene from the nucleus into the cytoplasm to ribosomes where the information is translated and proteins are assembled from amino acids.

▶ ncRNAs; these are non-coding RNA molecules. They do not convey information for protein synthesis but some are involved in regulating genes. Transfer RNA and ribosomal RNA are also non-coding RNAs.

▶ Transfer RNAs (tRNA) each carry a specific amino acid to the ribosomes for assembly into proteins.

▶ Ribosomal RNA (rRNA): each ribosome is made of RNA and protein which together form nucleoproteins. The RNA is the catalyst in ribosomes, enabling protein assembly.

▶ siRNA: small interfering RNA molecules. These help regulate gene activity by breaking down or blocking mRNA or by causing certain genes to become methylated (adding a CH_3 group to some nucleotides).

▶ Micro RNAs may help break down mRNA, block translation of mRNA or cause genes to become methylated, and so help regulate **gene expression**.

▶ Antisense RNA can regulate gene expression; some form complementary base pairing to a length of mRNA, forming double-stranded RNA that can then be broken down by enzymes.

▶ CRISPR RNA is found in prokaryotic cells and can regulate gene expression.

> **Key term**
>
> **Gene expression** – the production of the product encoded by a gene; may be a protein or a length of RNA that regulates expression of another gene.

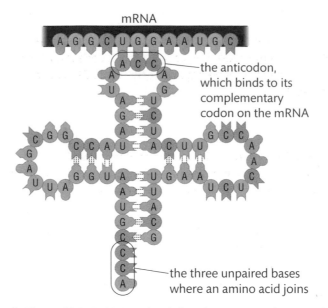

mRNA

the anticodon, which binds to its complementary codon on the mRNA

the three unpaired bases where an amino acid joins

▶ **Figure 11.7:** A tRNA molecule bonding temporarily to a codon on a section of an mRNA molecule

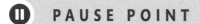 **PAUSE POINT**   Make a table to compare DNA and RNA.

Hint   When comparing, remember to think of both similarities and differences.

Extend   Compare prokaryotic and eukaryotic (a) DNA (b) RNA.

The basis of the genetic code

On each chromosome, there are specific lengths of DNA called **genes**. Genes carry the code for making proteins, many of which form the structure of an organism or cellular tools such as enzymes and antibodies, and some of which are **transcription factors** that activate or suppress the expression of other genes.

Proteins are polymers of amino acids and many of them are very large molecules. Their **primary structure**, the sequence of amino acids in the chain, determines how the protein pleats and coils (secondary structure) and finally folds into its **tertiary structure**. The tertiary structure (3D shape) is crucial for the protein to carry out its task (think of enzymes and their active sites that must be complementary in shape to their substrate molecules; antibodies that must fit to specific antigens, or receptors, on cell membranes that have a complementary shape to a specific signal molecule); if the primary structure of the protein is wrong then the tertiary structure will be wrong and the protein will not function well, or may not function at all.

Link

See *Unit 10: Biological Molecules and Metabolic Pathways* for more about protein structure.

Key terms

Gene – length of DNA that codes for one (or more) proteins/polypeptides or codes for one (or more) length(s) of RNA that may regulate the expression of another gene/other genes.

Transcription factors – proteins that activate or suppress the expression of genes.

Primary structure (of a protein) – the sequence of amino acids within the protein chain.

Tertiary structures (of a protein) – the 3D shape of a protein, caused by its folding.

Triplet codes

The sequence of amino acids in the protein is determined by the sequence of base triplets on the coding strand of the DNA molecule. Hence the genetic code is a triplet code. It is near universal because in almost all living organisms the same DNA base triplet codes for the same amino acid.

Theory into practice

There are 20 amino acids involved in protein synthesis. Hence at least 20 different base triplets are needed to code for these amino acids. However, there must also be one or more triplets that signal 'stop' and cause termination of the building of the protein or length of RNA.

If the four DNA bases were read in groups of two, there would only be 4^2 different combinations. This would be 16 different base pair codes – not enough.

Because the DNA bases are read in triplets, there are $4^3 = 64$ different combinations. This is more than enough. As a result, some amino acids have more than one base triplet code.

Imagine that on another planet the life forms contain proteins made from 100 different amino acids. Their DNA also contains four different bases. In what size groups would the DNA bases be read? Explain your answer.

Degenerate code

The genetic code is also degenerate because for all amino acids except methionine and tryptophan, there is more than one base triplet. This may reduce the effect of **point mutations**, because a change of one base triplet could produce another base triplet that still codes for the same amino acid.

Key term

Point mutation – change in base sequence of DNA caused by a substitution of one base for another, e.g. CGA becomes CCA.

Non-overlapping code

The genetic code is also non-overlapping. It is read from a fixed point in groups of three bases (triplets) that occur one after the other and do not overlap. If a mutation occurs where a base is added or deleted, then it causes a frame shift as every base triplet after that. This means that every amino acid coded for after that is changed.

Codons

DNA is in the cell nucleus but proteins are assembled at ribosomes in the cell cytoplasm. Hence the genetic code has to be copied and carried out of the nucleus to the ribosomes. This process is transcription and a length of messenger RNA (mRNA) is made that corresponds to the coding strand of the gene (length of DNA). The base triplets on the mRNA are called codons. Each is a copy of the base triplet on the DNA coding strand but in place of the base thymine there is the base uracil.

First position	Second position				Third position
	T	C	A	G	
T	Phe	Ser	Tyr	Cys	T
	Phe	Ser	Tyr	Cys	C
	Leu	Ser	STOP	STOP	A
	Leu	Ser	STOP	Trp	G
C	Leu	Pro	His	Arg	T
	Leu	Pro	His	Arg	C
	Leu	Pro	Gln	Arg	A
	Leu	Pro	Gln	Arg	G
A	Ile	Thr	Asn	Ser	T
	Ile	Thr	Asn	Ser	C
	Ile	Thr	Lys	Arg	A
	Met	Thr	Lys	Arg	G
G	Val	Ala	Asp	Gly	T
	Val	Ala	Asp	Gly	C
	Val	Ala	Glu	Gly	A
	Val	Ala	Glu	Gly	G

Key:

- Asp Aspartic acid
- Glu Glutamic acid
- His Histidine
- Ile Isoleucine
- Arg Arginine
- Thr Threonine
- Ser Serine
- Lys Lysine
- Gly Glycine
- Asn Asparagine
- Gln Glutamine
- Trp Tryptophan
- Tyr Tyrosine
- Ala Alanine
- Cys Cysteine
- Phe Phenylalanine
- Leu Leucine
- Met Methionine
- Pro Proline
- Val Valine

▶ **Figure 11.8:** The DNA triplet codes and the corresponding amino acids. Three base triplets do not code for an amino acid but act as stop codes.

Anticodons

When the mRNA has passed out through a pore in the nuclear envelope and arrived at a ribosome, amino acids are brought by tRNA molecules. Each of these is specific to a particular amino acid and also has a triplet of bases, called an anticodon, complementary to the mRNA codon (see Figure 11.8). See pages 457–458 for more about codons and anticodons and their importance in protein assembly (transcription and translation).

Protein synthesis

Protein synthesis, the assembly of amino acids into proteins, takes place in two main stages:

▶ transcription

▶ translation.

Transcription

Transcription is the process where the instructions on the coding strand of the length of DNA are copied on to a messenger molecule – a length of mRNA. Transcription occurs in the cell nucleus in eukaryotes.

1 The gene (length of DNA) unwinds and unzips as the hydrogen bonds between the nitrogenous bases break due to the action of the enzyme RNA polymerase.

2 This exposes the bases of the DNA nucleotides that make up the gene.

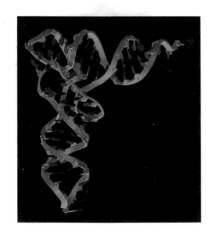

▶ A tRNA molecule

3 Free RNA nucleotides line up along the template strand of the DNA and make temporary hydrogen bonds with their complementary bases. Adenine, from an RNA nucleotide, pairs with thymine on the DNA template strand. Uracil from an RNA nucleotide pairs with the base adenine on the DNA template strand. The enzyme RNA polymerase catalyses these reactions.

4 Sugars and phosphate groups of adjacent RNA nucleotides bond together.

5 This forms a single polynucleotide chain of RNA that is complementary to the DNA template strand of the gene. It is therefore a copy of the coding strand of the gene.

6 Each codon on the mRNA codes for a specific amino acid.

7 The mRNA can now break away from the gene, which winds up to form double-stranded DNA again.

Before this can act as mRNA during the next stage of protein synthesis (translation), it has to be edited. Therefore, the mRNA is known as pre-mRNA at this stage.

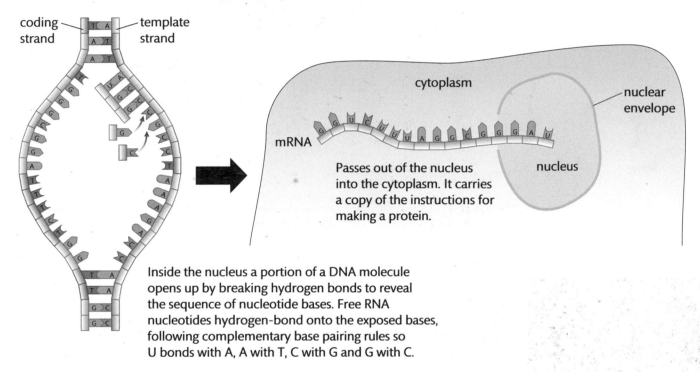

coding strand — template strand

cytoplasm

nuclear envelope

mRNA

Passes out of the nucleus into the cytoplasm. It carries a copy of the instructions for making a protein.

nucleus

Inside the nucleus a portion of a DNA molecule opens up by breaking hydrogen bonds to reveal the sequence of nucleotide bases. Free RNA nucleotides hydrogen-bond onto the exposed bases, following complementary base pairing rules so U bonds with A, A with T, C with G and G with C.

▶ **Figure 11.9:** Transcription of a gene

Introns, exons and splicing

Within a length of DNA that forms a gene, there is a specific sequence of base triplets that determines the sequence of amino acids in the protein encoded by that gene. However, within a gene there are non-coding regions of DNA called introns. These are not expressed. They separate the coding or expressed regions of the gene, which are called exons.

1 All the DNA of a gene, both introns and exons are transcribed and the resulting mRNA is called pre-mRNA.

2 This pre-mRNA is then edited and the RNA introns, lengths corresponding to the DNA introns, are removed and the remaining mRNA exons, corresponding to the DNA exons, are joined together.

3 Endonuclease enzymes may be involved in the editing and splicing (joining) processes.

4 Some introns may become short non-coding lengths of RNA involved in gene regulation.

5 Some genes can be spliced in different ways so that a length of DNA with its introns and exons can, according to how it is spliced, encode more than one protein.

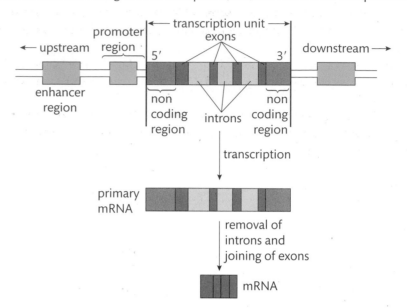

▶ **Figure 11.10:** Removal of introns and joining of exons during splicing of pre-mRNA to produce mRNA that will be translated into a protein

Translation

Transfer RNA molecules

Transfer RNA molecules (tRNAs) are made in the nucleolus and then pass out of the nucleus into the cytoplasm. Each is a single-stranded polynucleotide but can twist into hairpin shapes. At one end is a trio of nucleotide bases that recognises and attaches to a specific amino acid. At the loop of the hairpin is another triplet of bases called an anticodon that is complementary to a specific codon on the mRNA.

Amino acid activation

There is a pool of amino acids in the cell cytoplasm.

Before they can be used to form proteins, amino acids have to be attached to their corresponding tRNAs. These coupling reactions are catalysed by enzymes called aminoacyl-tRNA synthetases.

The following equations summarise the reaction. Note that 'aa' is an abbreviation of amino acid.

$$\text{Amino acid} + \text{ATP} \rightarrow \text{aa-AMP} + \text{PP}_i \text{ (pyrophosphate)}$$
$$\text{aa-AMP} + \text{tRNA} \rightarrow \text{aa-tRNA} + \text{AMP}$$

Each of the 20 amino acids is recognised by its specific aminoacyl-tRNA-synthetase enzyme. Energy from the breakdown of ATP is stored within the amino acid-tRNA complex and during translation is released and used to form a peptide bond between adjacent amino acids being assembled into a protein.

Translation at the ribosome

Ribosomes catalyse the synthesis of polypeptides (proteins). Ribosomes are made of two sub-units (a larger and a smaller) that are made within the nucleus and pass out into the cytoplasm where they join together.

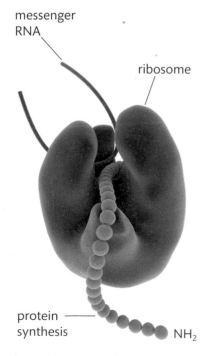

messenger RNA

ribosome

protein synthesis

NH₂

▶ Translation at a ribosome

1 The length of mRNA binds to a ribosome so that two of its codons are attached to the small ribosomal subunit.
2 The first exposed codon is always AUG. A tRNA with the corresponding anticodon, UAC, and holding the amino acid methionine, forms hydrogen bonds with this codon. Energy from ATP and a catalyst (ribosomal RNA) allow this reaction to happen.
3 A second tRNA molecule brings the next amino acid coded for and binds to the second codon.
4 Now two amino acids are side by side and a peptide bond forms between them.
5 The ribosome now moves along the mRNA so that codons 2 and 3 are exposed to the sub-unit.
6 A third tRNA brings a third amino acid and binds to its codon. The first tRNA molecule leaves and is free to collect and bring another amino acid of the same type to the ribosome.
7 This continues until an mRNA codon that does not code for an amino acid is reached. This codon is effectively a stop codon.
8 The chain of amino acids leaves the ribosome and can be modified into its functioning form.

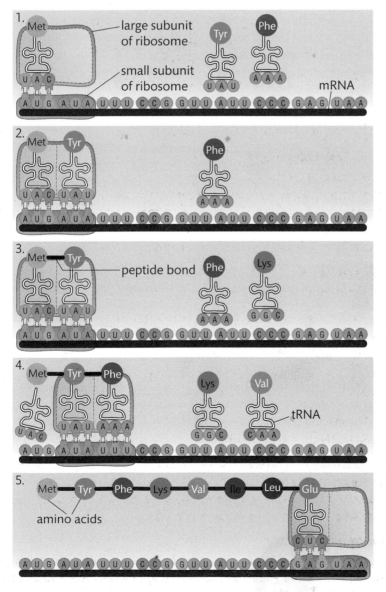

▶ **Figure 11.11:** Translation – stages of assembly of a protein at a ribosome

Chaperone proteins

Chaperone proteins assist the synthesised protein chain to fold into its correct shape. This tertiary structure is held by disulfide bonds, hydrogen bonds, ionic bonds and weak intermolecular forces called van der Waals forces, provided that when folded, certain amino acids are brought close together so these bonds can form. Hence the sequence of amino acids is very important for protein function. Mutations – changes to the genetic material of a cell – can alter the sequence of amino acids in a protein.

(II) PAUSE POINT   Make a table to compare transcription with DNA replication.

Hint   Remember to look at ways in which the two are similar and also how they differ.

Extend   Explain the difference between transcription and translation.

Mutations

Despite the structure of the DNA molecule making it stable and fairly resistant to corruption of the genetic information stored within it, errors may occur during the replication of a DNA molecule.

▶ Mutations associated with DNA replication prior to mitotic division are somatic mutations and are not passed to offspring. They may, however, be associated with the development of cancerous tumours.

▶ Mutations associated with DNA replication prior to meiosis and gamete formation may be inherited by offspring.

There are three main classes of DNA mutations.

▶ Point mutations, in which one base pair replaces (is substituted for) another.

▶ Insertion or deletion (indel) mutations, where one or more nucleotide pairs are inserted or deleted from a length of DNA. These may cause a frameshift where every base triplet after the insertion or deletion is altered, altering every amino acid in the chain after that.

▶ Expanding triple nucleotide repeats, where a base triplet is repeated many more times.

Point mutations

There are three types of point mutation:

▶ silent
▶ missense
▶ nonsense.

Silent mutations

You can see from Figure 11.8 that most amino acids have more than one triplet of bases coding for it. A point mutation where one base is substituted for another but the base triplet still codes for the same amino acids is a silent mutation. The primary structure, and therefore the secondary and tertiary structure of the protein, is not altered.

Missense mutations

A point mutation that changes the base triplet so it now codes for a different amino acid is a missense mutation.

Nonsense mutations

If a point mutation alters a base triplet so it becomes a stop triplet, then the assembly of the protein will stop too soon, resulting in a truncated (very short protein) that is unable to function.

Link

See page 487 about sickle cell anaemia, which is the result of a missense mutation.

Phenylketonuria

Sheila gave birth to her son, Mark, two years ago. Mark is very fair skinned and has pale blue eyes and very blond hair. When Mark was six days old, the midwife took some of his blood from a heel prick to test for a genetic disorder called phenylketonuria, PKU. This condition is caused by a nonsense mutation resulting in a truncated (very short) protein – the enzyme phenylalanine hydroxylase. This truncated enzyme does not function and the amino acid phenylalanine accumulates, as it cannot be used to make the pigment melanin, causing irreversible brain damage. Mark tested positive and Sheila was told that she would need to feed him a special diet, low in phenylalanine, to prevent the brain damage and ensure he will develop normally.

In 1921 Pearl S. Buck gave birth to a daughter, Carol, who became severely mentally retarded and was placed in special care. To pay for that care Pearl wrote

The Good Earth in 1931, followed by other novels and biographies about her life in China, for which she was awarded the Nobel and Pulitzer Prizes. In the 1960s doctors confirmed that Carol suffered from PKU but unfortunately this knowledge had not been available 40 years earlier and Pearl had not been advised about the dietary treatment that could have saved Carol.

Although this is a rare genetic condition (about 1 in 10 000 births), it has devastating consequences for those affected so all newborns are screened.

Check your knowledge

1 Explain why a truncated enzyme protein chain cannot function.

2 There are genetic conditions, such as having shorter than normal fingers, which are far more common than is PKU. Suggest why newborns are not screened for these disorders.

Indel mutations

If nucleotide base pairs, not in triplets, are inserted into the gene or deleted from the gene, because the code is non-overlapping and read in groups of three bases, all the subsequent base triplets are altered.

▶ This is a frameshift.

▶ All the subsequent amino acids in the protein are altered.

▶ The protein will be very abnormal and unable to function.

Some forms of thalassaemia, a haemoglobin disorder, result from frameshifts due to deletions of nucleotide bases.

Discussion

Not all deletions lead to a frameshift. If a whole triplet of bases is lost, as is the case with 70% of cases of cystic fibrosis, then what is the effect on the protein chain? In what ways do you think the protein will be different from the normal one and why can it not carry out its function? (See Learning aim C for more about cystic fibrosis.)

Expanding triple nucleotide repeats

Some genes contain a repeating triplet such as -CAG CAG CAG-. Huntington's disease results from an expanding triple nucleotide repeat – the number of CAG triplets increasing at DNA replication prior to meiosis and increasing from generation to generation. If the number of repeating CAG sequences goes above a certain critical number, then the person with that **genotype** will develop the symptoms of Huntington disease later in life.

Key term

Genotype – genes/alleles present in an individual/ cell; may refer to just one characteristic.

Link

There is more about the inheritance pattern of Huntington's disease on page 484.

Discussion

In small groups, discuss the following:

1 Where do you think the amino acids, in your cell cytoplasm, to be used for protein synthesis, have come from?

2 How are genes able to direct the synthesis of non-protein molecules such as triglycerides and phospholipids?

Share your ideas with others in your class.

PAUSE POINT Explain how the degenerate nature of the code reduces the effect of point mutations.

Hint Look back to see what degenerate code means.

Extend Sometimes during translation a tRNA molecule attaches to the wrong amino acid. How may this affect the protein being assembled at the ribosome?

Assessment practice 11.1   A.P1 A.M1 A.D1

You have been asked to help a secondary school teacher make some visual aids to help her teach about nucleic acids and protein synthesis to her class.

Make simple 3D models to show the structure of DNA and RNA. You can use pipe cleaners, wire, modelling clay or any other suitable material of your choice. Make sure that you can explain your model to someone else.

Produce a 2D or 3D annotated poster(s) to show how DNA and RNA are involved in protein synthesis.

Add a section to assess the impact of errors in various stages of protein synthesis.

Plan
- What is the task?
- How confident do I feel in my own abilities to complete the task?
- Are there any areas I think I may struggle with?

Do
- I know what it is I am doing and what I want to achieve.
- I can identify where I have gone wrong and can adjust my approach to get back on course.

Review
- I can explain what the task was and how I approached it.
- I can explain how I might approach the difficult parts differently next time.

B   Explore how the process of cell division in eukaryotic cells contributes to genetic variation

Human chromosomes

Humans are eukaryotic organisms and the majority of our DNA is packaged into linear chromosomes housed within each cell nucleus. Every somatic (body) cell, with the exception of red blood cells that do not have a nucleus, contains 46 (23 pairs of) chromosomes.

▸ Each chromosome contains one molecule of DNA. Within that molecule of DNA are specific lengths of DNA, each of which codes for a protein or a regulatory length of RNA; these are called genes.

▸ When the cell is not preparing to divide or dividing, this DNA is diffuse and spread out and is called chromatin.

Chromatids and centromeres

Before the cell divides (see page 467 on the cell cycle) each molecule of DNA duplicates and the chromatin condenses and coils very tightly (supercoiling). Each molecule of DNA is complexed with histone proteins and some non-histone proteins, to form a linear chromosome. The histone proteins associated with DNA in eukaryotic cells were once regarded as simply for packaging but scientists now know that these proteins may help regulate gene expression.

These chromosomes will take up stains and can be seen with a light microscope. Because each chromosome now consists of two identical molecules of DNA, ready to be split apart during cell division, each consists of two sister chromatids, still joined at the centromere.

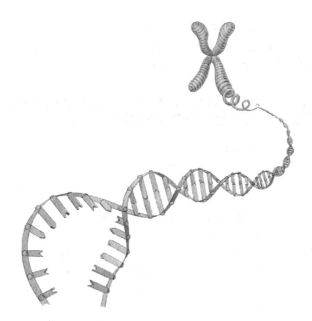

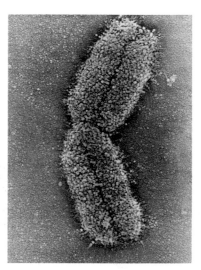

▶ Coloured scanning electron micrograph of human chromosome number 1, the largest human chromosome. The DNA has replicated so this chromosome consists of two identical chromatids joined at the centromere. Mag × 16 800.

▶ **Figure 11.12:** Organisation of DNA into a chromosome

Worked example

An average-sized human chromosome contains about 5 cm of DNA (as chromatin) which is condensed to a chromosome filament 0.5 cm long and 10 nm wide.

When this condenses into a sister chromatid it becomes 10 μm long and 1 μm wide.

A cytology screener who often examines karyotypes (pictures of sets of human chromosomes) wants to calculate the order of magnitude changes to the dimensions of the chromosome as it condenses into a sister chromatid.

Length:

Changes from 0.5 cm (=5 mm = 5000 μm) to 10 μm. Order of magnitude change is 5000/10 = 500. It has become 500 times shorter.

Calculate the order of magnitude change:

(a) in width as a chromosome condenses into a sister chromatid

(b) in length as a molecule of chromatin changes into a sister chromatid.

(a) 10 nm to 1 μm; 10 nm to 1000 nm, therefore order of magnitude of 1000/10 = 100.

(b) 0.5 cm = 5 mm = 5000 μm to 10 μm; order of magnitude of 5000/10 = 500.

The 46 chromosomes in every one of your cell nuclei are derived from your parents. Your mother's egg nucleus and your father's sperm nucleus both contained 23 chromosomes.

In your gonads (testes and ovaries) before meiotic cell division happens to make ova (eggs) and spermatozoa, these chromosomes synapse (pair up) into their homologous pairs (see below).

Homologous and non-homologous chromosomes

Homologous chromosomes

Twenty-two of those pairs of chromosomes are matching or homologous. This means they are the same size and contain the same genes at the same positions, gene loci, on the two chromosomes. However, they may contain different versions of these genes. A version of a gene is called an allele.

Homologous chromosomes synapse (pair up) at meiosis.

Non-homologous chromosomes

Non-homologous chromosomes do not synapse at meiosis. For example, chromosome 1 is not homologous with any of the other chromosomes 2–23.

Autosomes

These 22 pairs of homologous chromosomes, pairs 1–22, are called autosomes. Between them they contain many thousands of genes that govern many of our characteristics and aspects of your metabolism, but they do not play any part in determining your sex (whether you are male or female).

The other pair of chromosomes does not entirely match. One is the larger X chromosome and the other is the smaller Y chromosome. A small part of each matches to a small part of the other so that before meiosis these chromosomes can form a pair.

Sex chromosomes

This pair of chromosomes is also referred to as the sex chromosomes. If you inherit two X chromosomes you develop as female. If you inherit one X and one Y chromosome you develop as male. However, there are other genes on these chromosomes that govern body characteristics that are not to do with your sex. Such characteristics are described as sex-linked. You will learn more about this in learning aim C.

❚❚ PAUSE POINT	Make yourself a small glossary to define the following terms: chromosome, chromatid, centromere, sex chromosomes.
Hint	Don't just copy from the book; try to write a definition in your own words and then check your answers.
Extend	Explain why part of the X and Y chromosomes are homologous to each other.

Chromosome number and karyotyping

You have seen that in each of your cell nuclei you have 46 chromosomes (23 pairs of chromosomes). Although in your body cell nuclei, these chromosomes do not pair up unless the cell is preparing for meiosis, each member of a pair still match for size and gene loci.

Chromosome number

Each of your body cells contains two sets of chromosomes, 23 derived from your mother and 23 derived from your father. Because they contain two sets of chromosomes these cell nuclei are described as diploid ($2n$). Ova and sperm nuclei contain only one set of chromosomes and are described as haploid (n).

Karyotyping

Karyotyping is a method of finding the number, size and shapes of the chromosomes from a cell of an individual or species. A karyotype refers to a **photomicrograph** of chromosome preparations, made when the cell is in metaphase of mitosis.

If some cells, such as skin cells, that divide frequently are cultured (grown in a special medium) and then placed in distilled water, as water enters by osmosis, the cells swell and burst. Any dividing cells will not have an intact nuclear envelope so, when they burst, their chromosomes will spread out. Because the chromosomes, each consisting of two sister chromatids, can be seen under a light microscope, they can be photographed under the microscope. The photograph can be enlarged and the chromosomes cut out and arranged in their matching pairs – 22 homologous pairs of autosomes and one pair of non-homologous sex chromosomes. This process is called karyotyping. It can be used to diagnose any abnormalities with chromosome number in a fetus.

> **Key term**
>
> **Photomicrograph** – photograph of an image seen using a light microscope.

(a)
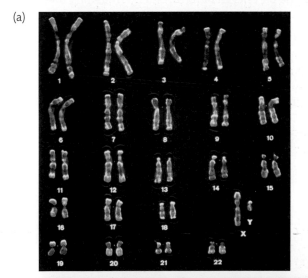

(b)
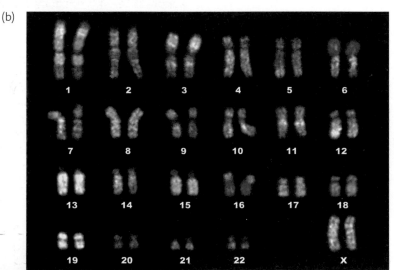

(c)
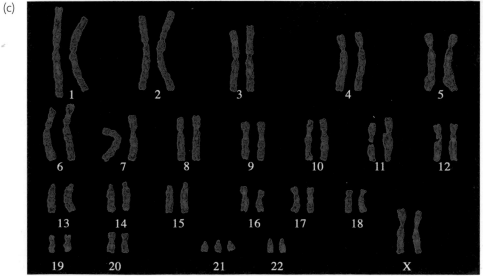

▶ Karyotypes: (a) normal male (b) normal female and (c) trisomy 21

Ben and Amira are expecting a baby. They are both over the age of 40 and have spoken to a genetic counsellor who tells them that their risk of having a baby with a chromosome abnormality, such as Down's syndrome, is higher than in a younger couple.

The genetic counsellor has informed than that there is a test available. It involves obtaining cells of the fetus, culturing them for several days and preparing a karyotype.

However, obtaining the fetal cells is invasive. Chorionic villus sampling (CVS) can be carried out in the first 13 weeks of the pregnancy and a needle inserted through the vagina and cervix can obtain fetal cells from the placenta.

Amniocentesis involves inserting a needle through the mother's abdominal wall, guided by an ultrasound scan, and removing fetal cells from the amniotic fluid.

CVS is slightly less accurate than amniocentesis because sometimes maternal placental cells may be obtained rather than fetal cells. There are also risks, such as infection or miscarriage, with the procedure. CVS can be carried out earlier than amniocentesis and this is better if the parents want to consider terminating the pregnancy.

Amniocentesis also carries a risk of miscarriage. It is carried out around week 15 or later but as the results take about three weeks, the couple have to bear in mind the latest date, 24 weeks, for termination.

Some couples decide that even if the fetus has a chromosome abnormality they will not terminate the pregnancy. Some couples say they will not terminate the pregnancy but want to be prepared and know as much as possible about caring for a Down's syndrome baby.

Both CVS and amniocentesis have a risk of inducing a miscarriage and that could cause the loss of a healthy fetus.

There is a blood test available that indicates an increased risk of fetal abnormality. This can be carried out in the early stages of pregnancy and can inform the couple whether they should take a CVS or amniocentesis test.

Role-play

In small groups, play the parts of Ben and Amira and the genetic counsellor for one of the following scenarios.

A Amira has had a blood test but the midwife lost the results and by the time they had repeated the test, the pregnancy is in its 21st week.

B Amira has had a blood test and it does not indicate a higher than normal risk of the baby having Down's syndrome. The pregnancy is in its 12th week.

C Amira has had a blood test that shows she has an increased risk of the baby having Down's syndrome. The pregnancy is in its 14th week.

Watch the role play of other groups. You may think about how you might have carried out each role play, including your own, differently and reached different conclusions.

Cell division and its role in variation

Cells reproduce by duplicating their contents and then splitting into two daughter cells. Body cells divide by **mitosis** and each new daughter cell receives genetic material identical to that of the parent cell and to each other. These daughter cells are diploid.

Meiosis in certain cells of the ovaries and testes produces haploid daughter cells – ova or spermatozoa. Each contains half the genetic material of the parent cell. Because of certain things that happen during meiosis the resulting haploid sex cells are genetically different from each other and from the parent cell. During sexual reproduction, two unrelated haploid cells fuse at fertilisation and this also contributes to genetic variation.

Cells spend only some of their time dividing, so mitosis or meiosis and cytokinesis occupy only a small part of the **cell cycle**. For the rest of its cycle, each cell is growing and carrying out its metabolism. The cell then prepares for division and the genetic material replicates.

The cell cycle

Early researchers observing cell division under the microscope could easily see the behaviour of chromosomes during mitosis, which is nuclear division, followed by cytokinesis or cytoplasmic division, with the resulting two daughter cells.

These events, called the M phase, occupy only a small part of the cell cycle. Between each M phase is an interphase. Interphase, when studied under the microscope, appears to be uneventful. However, more sophisticated techniques have enabled scientists to learn that during interphase there are elaborate preparations being made for cell division, in a carefully ordered and controlled sequence, with checkpoints.

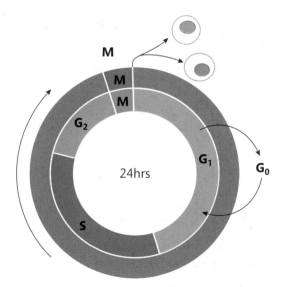

▶ **Figure 11.13:** The cell cycle in eukaryotic cells. M indicates the division phase. Interphase is divided into G_1, S and G_2. Cells may also enter a G_0 phase where they may undergo differentiation, apoptosis (programmed cell death) or enter senescence – where they can no longer divide.

▶ **Table 11.1:** The events that take place during different phases of the eukaryotic cell cycle

Stage of cell cycle	Events within the cell
M phase	• Cell growth stops. • A checkpoint chemical triggers condensation of chromatin. • Nuclear division consisting of stages: prophase, metaphase, anaphase and telophase. • Halfway through the cycle, the metaphase checkpoint ensures that the cell is ready to complete mitosis. • Cytokinesis (cytoplasmic division).
G_0 (Gap 0) phase	• A resting phase triggered during early G_1 at the restriction point, by a checkpoint chemical. • Some types of cells (e.g. neurones) remain in this phase for a very long time or indefinitely. • Some cells, e.g. epithelial cells lining the gut, do not have this phase. • In this phase cells may undergo apoptosis (programmed cell death), differentiation or senescence (ageing).
G_1 (gap 1) phase – also called the growth phase	• Cells grow and increase in size. • Transcription of genes to make RNA. • Biosynthesis, e.g. protein synthesis, including making the enzymes needed for DNA replication in the S phase. • Organelles, e.g. ribosomes, mitochondria and chloroplasts, duplicate. Mitochondria and chloroplasts are derived from bacteria and they contain DNA. These organelles divide by binary fission and their DNA duplicates. • The p53 (tumour suppressor) gene helps control this phase. • A G_1 checkpoint control mechanism (cyclin-CDK promotes the expression of transcription factors that promote the expression of S-cyclins and the enzymes needed for DNA replication) ensures that the cell is ready to enter the S phase and begin DNA synthesis.
S (synthesis) phase of interphase	• Once the cell has entered this phase it is committed to completing the cell cycle. • DNA replicates. • Because the chromosomes are unwound and the DNA is diffuse, every molecule of DNA is replicated. There is a specific sequence to the replication of genes. Housekeeping genes – those which are active in all types of cells, are duplicated first. Genes that are normally inactive in specific types of cells are replicated last. • When all chromosomes have been duplicated each one consists of a pair of identical sister chromatids. • This phase is rapid as the exposed DNA base pairs are more susceptible to mutagenic agents so rapidity reduces risk of mutation.
G_2 (gap 2) phase of interphase	• Cells grow. • Cyclin-CDK complexes ensure the cell is ready for mitosis by stimulating proteins that will be involved in condensing chromosomes and making the spindle for mitosis, in the M phase.

Regulation of the eukaryotic cell cycle

Cell cycle checkpoints

These monitor and regulate the progress of the eukaryotic cell cycle. There are two main checkpoints, the G_1/S checkpoint, also called the restriction point, and the G_2/M checkpoint. There are other checkpoints, for example, there is one halfway through mitosis and one in early G_1.

The purpose of the checkpoints is to:
▶ prevent uncontrolled division that would lead to tumours (cancer)
▶ detect and repair damage (for example caused by UV light) to DNA.

Because the molecular events that control the cell cycle happen in a specific sequence, they also ensure that:
▶ the cycle cannot be reversed
▶ the DNA is only duplicated once during each cell cycle.

The regulatory chemicals

The p53 gene, also known as the tumour-suppressor gene, is important in triggering both checkpoints.

The regulatory proteins are cyclins and cyclin-dependent kinases (CDKs).

▸ Each has no effect without the other.

▸ Cyclins are the regulatory sub-units and kinases are the catalytic sub-units.

▸ When they are bound together they phosphorylate (add phosphate groups to) target proteins to activate or deactivate them and this leads into the next phase of the cell cycle.

▸ The cyclins are synthesised in response to various signalling molecules, or growth factors, that attach to receptors in the plasma membrane of the cell.

When regulation goes wrong

There are also proto-oncogenes, which help regulate cell division by coding for proteins that help regulate cell growth and differentiation. If proto-oncogenes undergo mutations and become oncogenes, they can cause cells that should undergo apoptosis to keep dividing.

If the p53 gene undergoes a mutation, it may not be able to regulate cell division and uncontrolled cell division leads to tumour formation.

Cells should normally only undergo a certain number of cycles or divisions. This is about 50 and is known as the Hayflick constant. If cell division becomes uncontrolled and cells divide more than 50 times, a tumour can form which may become cancerous.

In 2001, Sir Paul Nurse, Sir Tim Hunt and Leland Hartwell shared the Nobel Prize for Physiology or Medicine for discovering the proteins, cyclin and CDK that regulate cell division. In 1982, Tim Hunt had discovered cyclin in sea urchin eggs and vertebrate cells. He had showed that cyclin is synthesised during interphase, with its levels dropping suddenly to zero during mitosis. He named the molecule cyclin as he loved cycling.

It is important to know how the cell cycle is regulated as medical treatments for cancer aim to correct unregulated cell cycles.

Ⅱ PAUSE POINT   Prepare a poster with a large annotated diagram of the eukaryotic cell cycle.

(Hint)   Include checkpoints and their purpose and indicate at which phase each of the following events occur.
(a) replication of organelles (b) condensation of chromatin (c) replication of DNA (d) differentiation (e) apoptosis (f) growth of the cell.

(Extend)   Epithelial cells lining your intestine divide two or three times a day whereas liver cells divide about once a year. During which part(s) of their cell cycle will liver cells differ from intestinal cells?

Worked example

If a cytology screener fixes and stain a sample of a population of cells and examines them under a microscope, then he can see cells in various stages of the cell cycle.

If the duration of the cell cycle is 24 hours and 1/12 of the cells are in mitosis, then the duration of mitosis is approximately 1/12 of 24 hours = 2 hours.

In the same population of 12 cells, 4 were in S phase. What is the approximate duration of S phase for these cells?

4/12 of 24 hours = 8 hours.

The role of centrioles

Centrioles occur in most animal cells but are absent from many plant and fungal cells.

▶ They are microtubule-organising centres composed mainly of a protein called tubulin.

▶ A pair of centrioles is situated within an area of dense material called a centrosome.

▶ The position of the centriole in a cell determines the position of that cell's nucleus and the spatial arrangement of organelles within the cell.

▶ Centrioles duplicate during the part of the cell cycle when DNA duplicates.

▶ Each centriole is made of nine sets of microtubule triplets arranged in a cylinder.

▶ During prophase of both mitosis and meiosis, centrioles separate and organise the microtubule protein threads into a spindle, to which the chromosomes attach by their centromeres. These tubulin threads move the separating chromatids apart from each other during anaphase.

▶ In eukaryotic cells that do not have centrioles, the spindle threads form from the cytoskeleton.

Many animal cells have cilia on their surface and at the base of each cilium is a centriole (see Figure 11.14).

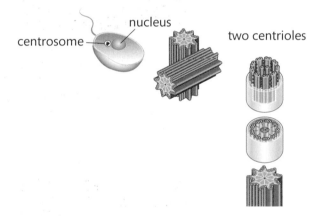

▶ **Figure 11.14:** Centrioles

Mitosis

Some eukaryotic organisms (some plants, fungi, protists and animals) reproduce asexually by mitosis. Humans, like other multicellular organisms, use mitosis for:

▶ growth – humans grow by producing more genetically identical cells by mitosis

▶ tissue repair – wounds heal when growth factors secreted by platelets and macrophages (white blood cells) and damaged cells of the blood vessel walls, stimulate the proliferation of endothelial and smooth muscle cells to repair damaged blood vessels

▶ replacement – adult stem cells can divide to give rise to differentiated cells, for example, stem cells in bone marrow divide to replace blood cells.

The stages of mitosis

There are four stages to mitosis (nuclear division) – prophase, metaphase, anaphase and telophase. Cytokinesis (cytoplasmic division) occurs immediately afterwards.

Interphase is not part of mitosis. Table 11.2 shows the events of the stages of mitosis.

▶ **Table 11.2:** The stages of mitosis

Stage of mitosis - diagram	Stage of mitosis - photo	Events occurring during the stage
Prophase spindle forming nuclear envelope breaking down sister chromatids · centromere	▶ Prophase of mitosis in white blood cells. The condensed chromatin is visible as chromosomes	• The chromosomes that have replicated during the S phase of interphase, and now consist of two identical sister chromatids, now shorten and thicken as the DNA supercoils. • The nuclear envelope breaks down. • The centriole divides and the two new daughter centrioles move to opposite poles (ends) of the cell. • Cytoskeleton protein (tubulin) threads form a spindle between these centrioles. The spindle has a 3D structure and is rather like lines of longitude on a virtual globe. In plant cells the tubulin threads form from the cytoplasm.
Metaphase chromosomes attached to spindle equator centromere · tubulin threads	▶ Digital 3D immunofluorescent light micrograph of a section through a mammalian cell during metaphase of mitosis. The tubulin microtubules of the spindle are coloured green, the chromosomes blue and the two centrioles are pink. [Magnification × 500.]	• The pairs of chromatids attach to the spindle threads at the equator region. • They attach by their centromeres.
Anaphase chromatids begin to be pulled apart	▶ Light micrograph of anaphase of mitosis in a bluebell cell	• The centromere of each pair of chromatids splits. • Motor proteins, walking along the tubulin threads, pull each sister chromatid of a pair, in opposite directions, towards opposite poles. • Because their centromere goes first, the chromatids, now called chromosomes, assume a V shape.
Telophase two sets of chromosomes new nuclear membranes form	▶ Digital 3D immunofluorescent light micrograph of anaphase of mitosis in a mammalian kidney cell. The chromatids (blue) of each chromsome have separated and been pulled to opposite ends of the cell.	• The separated chromosomes reach the poles. • A new nuclear envelope reforms around each set of chromosomes. • The cell now contains two nuclei, each genetically identical to each other and to the parent cell from which they arose.

Cytokinesis

Once mitosis is complete the cell splits into two, so that each new cell contains a nucleus.

- In animal cells, the plasma membrane folds inwards and 'nips in' the cytoplasm.
- In plant cells, an end plate forms where the equator of the spindle was and new plasma membrane and cellulose cell wall material is laid down either side along this end plate.

Two new genetically identical daughter cells are now formed. They are genetically identical to each other and to the parent cell. If the parent cell was diploid, then the daughter cells will also be diploid.

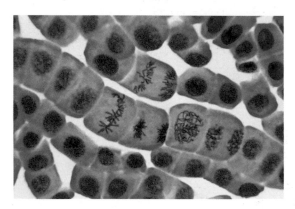

- Light micrograph of cells in root tip of *Allium sp.* (onion) undergoing mitosis. Can you identify cells in prophase, metaphase, anaphase and telophase? Mag × 1500.

Meiosis

Sexual reproduction increases genetic variation because it involves the combining of genetic material from two (usually) unrelated individuals, by the process of **fertilisation**. Genetic variation within a population increases its chances of survival when the environment changes, as some individuals will be adapted to the change.

In many organisms their body cells are diploid. For sexual reproduction to occur, they must produce haploid gametes so that when two gamete nuclei fuse during fertilisation, a diploid **zygote** is produced and the normal chromosome number is maintained through the generations.

Meiosis means 'reduction' and it is a type of nuclear division that occurs in diploid germ cells to produce haploid gametes. The diploid cells undergoing meiosis are in specialised organs called gonads – ovaries and testes in animals or ovaries and anthers in plants. These cells have been in interphase before they enter meiosis.

The main stages of meiosis

Here are the main stages of meiosis.

- In meiosis, the chromosomes pair up in their homologous pairs.
- There are two divisions in meiosis and in each division there are four stages.
- In the first meiotic division the four stages are: prophase 1, metaphase 1, anaphase1 and telophase 1.
- The cell may then enter a short interphase before embarking on the second meiotic division that also has four stages: prophase 2, metaphase 2, anaphase 2 and telophase 2.
- At the end of the second division cytokinesis may occur.
- During the S phase of interphase, each chromosome was duplicated as its DNA replicated, after which each chromosome consisted of two sister chromatids.

▸ **Table 11.3:** The first meiotic division (meiosis 1)

Stage of meiosis 1	Events during the stage
Prophase 1 crossing over paternal homologue chromosomes comprise two chromatids centrioles at opposite ends of cell nuclear envelope disintegrating maternal homologue Prophase 1 may last for days, months or years depending on the species and type of gamete (male or female) being formed sister chromatids a pair of homologous chromosomes in prophase 1 of meiosis crossing over point between non-sister chromatids centromere sister chromatids Crossing over	• The chromatin condenses and each chromosome supercoils. In this state they can take up stains and be seen with a light microscope. • The nuclear envelope breaks down and spindle threads of tubulin protein form from the centriole. • The chromosomes come together in their homologous pairs. • Each member of the pair consists of two chromatids. • Crossing over occurs where non-sister chromatids wrap around each other and may swap sections so that alleles are shuffled.
Metaphase 1 equator spindle fibres centriole crossing over Homologous pairs of chromosomes, still in their crossed-over state, on the equator of the spindle	• The pairs of homologous chromosomes attach along the equator of the spindle. • Each attaches to a spindle thread by its centromere. • The homologous pairs are arranged randomly with the members of each pair facing opposite poles of the cell. This arrangement leads to independent assortment – the way they line up in metaphase determines how they will segregate independently when pulled apart during anaphase.
Anaphase 1 exchange between non-sister chromatids has occurred during crossover in prophase I members of each homologous pair are pulled to opposite poles	• The crossed-over areas separate from each other, resulting in swapped areas of chromosome and allele shuffling. • The members of each pair of homologous chromosomes are pulled apart by motor proteins that drag them along the tubulin threads of the spindle. • The centromeres do not divide and each chromosome consist of two chromatids.
Telophase 1 nuclear envelope forming Xx Xx	• In most animal cells two new nuclear envelopes form around each set of chromosomes and the cell divides by cytokinesis. There is then a short interphase where the chromosomes uncoil. • Each new nucleus contains half the original number of chromosomes, but each chromosome consists of two chromatids. • In most plant cells the cell goes straight from anaphase 1 into prophase 2.

The second meiotic division takes place in a plane at right angles to that of meiosis 1.

Stage of meiosis 2	Events during the stage
Prophase 2 — centrioles replicate and move to poles; new spindle fibres form at right angles to previous spindle axis	• If the nuclear envelopes have reformed they now break down. • The chromatids of each chromosome are no longer identical due to crossing over in prophase 1. • Spindles form.
Metaphase 2 — chromosomes attached, by their centromeres, to the equator of the spindle	• The chromosomes attach, by their centromere, to the equator of the spindle. • The chromatids of each chromosome are randomly arranged. • The way they are arranged will determine how the chromatids separate (independent assortment) during anaphase.
Anaphase 2 — chromatid moving towards the pole	• The centromeres divide. • The chromatids of each chromosome are pulled apart by motor proteins that drag them along the tubulin threads of the spindle, towards opposite poles. • The chromatids are therefore randomly segregated.
Telophase 2 — haploid cells	• Nuclear envelopes form around each of the four haploid nuclei. • In animals the two cells now divide to give four haploid cells. • In plants, a tetrad of four haploid cells is formed.

How meiosis produces genetic variation

This is how meiosis produces genetic variation.

▶ Crossing over during prophase 1 shuffles alleles.

▶ Independent assortment of chromosomes in anaphase 1 leads to random distribution of maternal and paternal chromosomes of each pair.

▶ Independent assortment of chromatids in anaphase 2 leads to further random distribution (segregation) of genetic material.

▶ Production of haploid gametes that can combine their genetic material with that derived from another organism of the same species.

▶ Gametes fuse during fertilisation and this also leads to genetic variation.

▶ All the processes described above are normal and do not involve mutations.

Sex determination

In many organisms, including humans, sex determination is genetic. Females have two X sex chromosomes and males inherit one X and one Y sex chromosome. However, this is not universal. In birds and butterflies, males have matching sex chromosomes, ZZ, and females have non-homologous sex chromosomes, ZW.

In some organisms, e. g. crocodiles and turtles, sex determination is also environmental. The embryos develop as males or females according to their incubation temperature during a critical period. Cooler temperatures lead to males and warmer temperatures produce females.

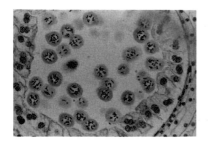

▶ Light micrograph showing transverse section through lily anther showing pollen mother cells in stages of meiosis

Ⅱ PAUSE POINT Compare mitosis and meiosis.

> **Hint** Depending on your style of learning, you could either make a table or write a short essay that may contain diagrams.

> **Extend** Explain (a) the importance of genetic variation (b) how sexual reproduction leads to genetic variation.

▶ False coloured scanning electron micrograph of a section through pollen mother cells in an anther of a Tradescantia flower, undergoing meiosis 1. Chromosomes are coloured yellow, cytoplasm violet and cell walls blue.

Assessment practice 11.2

B.P2 B.P3 B.M2 B.M3

The following activities will contribute to your practical portfolio.

Observe pre-prepared slides to observe the stages of mitosis and meiosis in plant tissue.

Demonstrate the preparation of microscope slides to observe the stages of mitosis and meiosis in plant tissue.

Make clear large labelled and annotated diagrams showing the stages of mitosis and meiosis and the behaviour of the chromosomes at each stage. Discuss and evaluate how the behaviour of chromosomes (i) during mitosis leads to genetically identical daughter cells and (ii) during meiosis leads to genetic variation.

Write a leaflet to explain the structure and function of chromosomes that could be available for members of the public while waiting to see a genetic counsellor.

Plan
- What is the task?
- How confident do I feel in my own abilities to complete the task?
- Are there any areas I think I may struggle with?

Do
- I know what it is I am doing and what I want to achieve.
- I can identify where I have gone wrong and can adjust my approach to get back on course.

Review
- I can explain what the task was and how I approached it.
- I can explain how I might approach the difficult parts differently next time.

C Explore the principles of inheritance and their application in predicting genetic traits

Principles of classical genetics

You will be aware that when you look at a group of people you can easily see that they are all humans but you can also see that they are each individual and unique. We are all similar to but also different from each other. The same can be said for any group of a particular animal or plant. Within any population there is variation. Some variation is genetic, some is caused by environmental factors and some is the result of a combination of both genetic and environmental differences.

Discontinuous and continuous variation

Discontinuous variation

This is where there are distinct **phenotypes** and no intermediates. Examples include height in pea plants and shape of ear lobes in humans. Often just one gene (with different alleles) is involved.

(a)
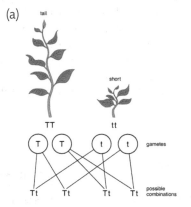

(b)
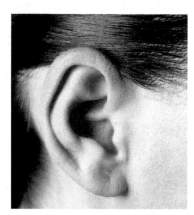
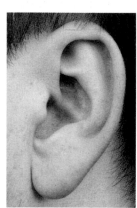

▶ (a) Tall and short-stemmed pea plants. (b) Attached and free ear lobes in humans.

Continuous variation

This is where there is a range of phenotypes. Examples include height, foot size, skin colour and intelligence in humans. Many genes are involved and environmental differences have a greater effect on continuous variation than they do on discontinuous variation.

(a)

(b)

▶ A group of humans showing differences in (a) height and (b) skin colour

Mendel's laws of inheritance

In 1866 Gregor Mendel published the results of his investigations into cross breeding pea plants. Although he did not know about chromosomes or meiosis he showed that units of inheritance existed and he predicted their behaviour during the formation of gametes (ova (eggs cells inside the ovules) and sperm (male gamete) inside the pollen). He worked with seven characteristics shown by pea plants: stem height, seed shape, seed colour, pod shape, pod colour, pod arrangement and flower colour.

He obtained accurate and quantitative records of his data that he analysed, and derived two laws.

Table 11.5 shows Mendel's two laws as he stated them, and their meaning in modern biological terms.

▶ **Table 11.5:** Mendel's laws

Mendel's laws	Modern interpretation
1 *The law of segregation* The factors of a pair of characters are segregated	In each diploid parent cell every gene has two alleles (one on each member of the homologous pair of chromosomes). At meiosis the chromosome number is halved and each gamete receives just one member of each pair of chromosomes and therefore only one allele of every gene.
2 *The law of independent assortment* The members of different pairs of factors assort independently	During meiosis 1 the chromosomes of each homologous pair assort independently and randomly according to how the chromosomes line up on the equator of the spindle during meiosis 1. During meiosis 2 the chromatids assort independently according to how they line up on the equator of the spindle at metaphase 2 and are subsequently pulled apart during anaphase 2. The result is that the members of the different pairs of alleles are assorted independently into the resulting gametes. Subsequent pairing of male and female gametes is random so any one of a pair of alleles for a gene in a female gamete can combine with any one of a pair of alleles for the same gene in the male gamete.

Monohybrid and dihybrid phenotypic ratios

A monohybrid cross

▶ This is a cross where both F_1 (first filial (offspring)) generation parents are heterozygous (having different alleles) for a particular **gene locus**.
▶ Heterozygous tall pea plants have the genotype Tt where T denotes tall and t denotes short.
▶ If two such F_1 parents are crossed and the seeds sown, about $\frac{3}{4}$ of the F_2 (second filial generation) offspring are tall and $\frac{1}{4}$ are short.
▶ Therefore, the phenotypic ratio of the F_2 generation is 3 tall:1 short.

Monohybrid crosses investigate just one gene and are therefore concerned with monogenic inheritance

A dihybrid cross

This looks at the simultaneous inheritance of two characteristics.
▶ Both F_1 parents are heterozygous at two gene loci.
▶ Heterozygous pea plants have green or yellow seeds (caused by two different alleles of the gene for seed colour) that may be round or wrinkled (caused by two different alleles of another gene that codes for seed coat texture).
▶ When Mendel crossed two pea plants, both having the phenotype yellow and round seeds, and the genotype Yy Rr he found the following phenotypes in the F_2 offspring:

315 of the total 556 (about $\frac{9}{16}$) had round and yellow seeds

108 of the total 556 (about $\frac{3}{16}$) had round green seeds

101 of the total 556 (about $\frac{3}{16}$) had yellow wrinkled seeds

32 of the total 556 (about $\frac{1}{16}$) had green wrinkled seeds.

▶ This equates to a phenotypic ratio of 9:3:3:1.

Key term

Gene locus – position of a gene on a chromosome.

Research

You can investigate such ratios using rapid cycling *Brassica* plants. Go to the Science and Plants for Schools website for more information about how to do this.

You can also carry out genetics investigations using fruit flies *Drosophila melanogaster*.

Using genetics diagrams and Punnett squares to interpret data from monohybrid and dihybrid crosses

Genetic inheritance diagrams

Genetic diagrams work best to explain the observed data of a monohybrid cross, or to predict the possible outcomes of a monohybrid cross.

- The first line shows the phenotypes of the parents.
- The next line shows possible genotypes of the parents.
- The third line shows the predicted genotypes of all the possible gametes.
- Lines join each female gamete to each male gamete to show the possible combinations resulting in the likely genotypes of the offspring.
- The last line shows the predicted phenotypes of the offspring – with ratios.

Figure 11.15 shows a genetic diagram.

P_1 cross

Parent phenotypes:	tall-stemmed	×	short-stemmed
Parent genotypes:	TT	×	tt
Gamete genotypes:	(T)	×	(t)
F_1 genotype:		Tt	
F_1 phenotype:		All tall-stemmed	

F_1 cross

F_1 phenotypes:	tall-stemmed	×	tall-stemmed
F_1 genotypes:	Tt	×	Tt
Gamete genotypes:	(T) (t)	×	(T) (t)
F_2 genotypes:	TT Tt		Tt tt
F_2 phenotypes:	tall tall		tall short
Phenotype ratio:		3 tall : 1 short	

Figure 11.15: Genetic diagram of Mendel's experiment on monohybrid inheritance – height in pea plants. There are two alleles for the height gene locus: T (**dominant**) and t (**recessive**). Where two heterozygous tall pea plants, genotype Tt, are crossed, among the offspring are tall and short plants in the ratio 3:1. Genotypes TT and Tt produce tall plants; genotype tt produces short plants.

Key terms

Dominant – allele whose expression is visible in the phenotype even if only one allele of the gene is present.

Recessive – allele whose expression is not visible in the phenotype if the dominant allele is also expressed.

Punnett squares

These were devised by the geneticist R. C. Punnett, and can be used to compute genotypic and phenotypic ratios of monohybrid and dihybrid crosses. They determine the possible genotypes of zygotes (fertilised ova), and subsequent phenotypes of the developed individual, produced by the fusion of gametes from the parents.

Figure 11.16 shows a Punnett square used for a monohybrid cross. Figure 11.17 shows a Punnett square for a dihybrid cross.

F_1 cross

F_1 phenotypes:	tall-stemmed $\times$ tall-stemmed
F_1 genotypes:	Tt $\times$ Tt
Gamete genotypes:	T t $\times$ T t

Punnett square:

F₂ generation:

♂ gametes / ♀ gametes	T	t
T	TT tall	Tt tall
t	Tt tall	tt short

Phenotype ratio: 3 tall : 1 short

▶ **Figure 11.16:** A Punnett square to explain the monohybrid cross between members of the F_1 generation, both heterozygous for the characteristic height. The phenotypic ratio is 3 tall : 1 short. The probability of obtaining phenotypically tall-stemmed plants is $\frac{3}{4}$ or 75%. The probability of obtaining homozygous tall-stemmed plants is $\frac{1}{4}$ or 25%. The probability of obtaining heterozygous tall-stemmed plants is $\frac{1}{2}$ or 50%. The probability of obtaining short-stemmed plants is $\frac{1}{4}$ or 25%.

P_1 cross

Parent phenotypes:	Yellow, round-seeded $\times$ Green, wrinkle-seeded
Parent genotypes:	YYRR $\times$ yyrr
Gamete genotypes:	YR $\times$ yr
F_1 genotype:	YyRr
F_1 phenotype:	Yellow round-seeded
F_1 cross:	YyRr $\times$ YyRr
Gamete genotypes:	YR Yr yR yr

Punnett square:

F₂ genotypes:

♂ gametes / ♀ gametes	YR	Yr	yR	yr
YR	YYRR	YYRr	YyRR	YyRr
Yr	YYRr	YYrr	YyRr	Yyrr
yR	YyRR	YyRr	yyRR	yyRr
yr	YyRr	Yyrr	yyRr	yyrr

Phenotypes:
9/16 yellow round-seeded
3/16 green round-seeded
3/16 yellow wrinkle-seeded
1/16 green wrinkle-seeded

▶ **Figure 11.17:** Punnett square showing the expected ratio of phenotypes in the F_2 generation of a dihybrid cross

Punnett squares can also be used to predict the outcome of a dihybrid cross.

Mendel crossed pea plants with yellow and round seed with pea plants having green and wrinkled seeds. All the F_1 (first filial) generation plants bore yellow and round seeds. These characteristics are dominant.

When he allowed members of the F_1 generation to self-fertilise and collected and grew the seeds, he obtained plants with yellow round seeds, green round seeds, yellow wrinkled seeds and green wrinkled seeds as shown above. They were in roughly a 9:3:3:1 ratio (see above).

Figure 11.17 shows a Punnett square predicting the outcome for this cross.

Meenakshi Rajkumar, genetic counsellor.

I inform prospective parents about any genetic conditions, brought about by mutations, present in them and potentially able to be inherited by their children. The GP or midwife dealing with a couple can refer that couple to me if the couple knows that there is a specific genetic disorder in one of both of the families. I can construct a family tree for that couple showing any individuals known to have or be affected by/suffer from the genetic disease. I can then use this family tree to show which parents were carriers, even though they themselves did not suffer from (they were non-sufferers/non-affected with) the disease. I may be able to tell from this family tree whether the members of this couple are carriers and calculate the risk or probability of their child inheriting the condition. I also give them information so that they can understand the options open to them. I do not advise them what to do, but give them the knowledge to make their own informed decision.

I also help patients who have a family member with a specific genetic condition that needs specialist diagnosis. Besides finding out about the symptoms of the family member I can arrange for genetic tests to be carried out that could confirm the presence of specific alleles associated with a specific genetic condition, such as certain types of breast cancer caused by specific mutations (*BRCA1* and *BRCA2*).

I graduated from the University of Liverpool with a degree in biology and I also studied some psychology. I then studied for an accredited MSc in Genetic Counselling at Manchester University. The Genetic Counselling Registration Board has approved this MSc course at Manchester University, as well as one at Cardiff University. Because I successfully graduated from one of these courses and successfully completed my registration portfolio, I was eligible to become a registered genetic counsellor in the UK. I am a non-medical health professional working in a clinical setting.

There are other non-accredited genetic counselling courses but graduates from them may find it hard to become registered and work as genetic counsellors in the UK. On my MSc course there were also students who had trained as nurses and midwives and then completed a counselling course and a genetics course before enrolling on the genetic counselling course.

Genetic counsellors need to have good people and communication skills so that they can help individuals or families. They need to build up good relationships with their clients and need to be able to:
- understand the information about the genetic condition in their family
- appreciate the inheritance pattern and risk of occurrence
- understand the options available
- make decisions appropriate to their situation.

1 Why do you think genetic counsellors should not directly advise their clients what to do?

2 Outline the possible routes to becoming qualified as a genetic counsellor.

Interpretation of Mendelian ratios from practical investigations

If you carry out genetics investigations using rapid cycling *Brassicas* or fruit flies you can see if your experimental data fit with your predicted phenotypic ratios. For example, fruit flies may be homozygous wild type (normal), genotype ++, with normal length wings or have short stumpy vestigial wings, genotype vv. Vestigial wings is a recessive characteristic; normal-length wings is a dominant characteristic.

When homozygous wild type parents are crossed with homozygous vestigial winged parents all the offspring are wild type – they have normal length wings. However, they are heterozygous: genotype +v.

If you carried out a monohybrid cross between members of the F_1 generation you would expect about $\frac{3}{4}$ of the offspring to have normal length wings and $\frac{1}{4}$ to have vestigial wings, a phenotypic ratio of 3:1.

It is unlikely that you will obtain data that gives exactly a 3:1 ratio, as:

▸ your sample may be small
▸ some gametes do not get fertilised
▸ which gametes actually fuse at fertilisation is a bit of a genetic lottery
▸ some fertilised eggs do not develop.

You can use a statistical test to see if the difference between your observed data and the data you expect is significant, in which case the inheritance pattern involved may be different from the one you think is involved, or is simply due to chance (for the reasons outlined above). If the difference is not significant and due to chance than you can assume that your ideas about the inheritance pattern are correct.

▸ Wild type Drosophila

Link

See *Unit 3: Science Investigation Skills*, Learning aim B, for more about statistical tests.

The chi-squared test

Consider the result Mendel obtained from his dihybrid cross, given above. You can apply the chi-squared test to these data because:

▸ the data are in categories (the different phenotypes)
▸ you have a strong biological theory you can use to predict the expected value
▸ the sample size is large
▸ the data gives raw counts
▸ there are no zero scores in the raw data.

Statistical tests cannot be used to test a **hypothesis** directly; instead they test a **null hypothesis**. If the null hypothesis is supported, then you accept it. If the difference between observed and expected data is due to chance, you can assume your model of the inheritance pattern is correct.

The null hypothesis states that 'There is no statistically significant difference between the observed and expected data. Any difference is due to chance.'

In Mendel's investigation:

▸ 315 of the total 556 had round and yellow seeds
▸ 108 of the total 556 had round green seeds
▸ 101 of the total 556 had yellow wrinkled seeds
▸ 32 of the total 556 had green wrinkled seeds.

These are the observed data

The expected data are as follows:

$\frac{9}{16}$ of 556 = 313; $\frac{3}{16}$ of 556 = 104; $\frac{1}{16}$ of 556 = 35

The formula for chi-squared (χ^2) is $\sum \dfrac{(O - E)^2}{E}$

or χ^2 = the sum of $\dfrac{(O - E)^2}{E}$

▸ Because differences between observed and expected data may be negative or positive, squaring prevents positive values cancelling out any positive values.
▸ Dividing by E takes into account the size of the numbers.
▸ The sum of sign (Σ) takes into account the number of comparisons being made.

Key terms

Hypothesis – a proposed testable explanation of a phenomenon or prediction based on that explanation that can be experimentally investigated.

Null hypothesis – a type of hypothesis used in statistical tests that proposes there is no significant difference between observed and expected data.

The procedure is as follows.

1 Calculate the value of χ^2.
2 Determine the number of degrees of freedom (df = number of categories – 1).
3 Determine the value of p from a distribution table.
4 Decide if the difference is significant or not at the p = 0.05 level of significance.
5 To calculate the value of χ^2, it is best to use a table, such as Table 11.6.

▶ **Table 11.6:** Calculation of χ^2

Category	Observed (O)	Expected (E)	$O - E$	$(O - E)^2$	$\dfrac{(O - E)^2}{E}$
yellow, round	315	313	2	4	4/313 = 0.0128
green, round	108	104	4	16	16/104 = 0.1539
yellow, wrinkled	101	104	–3	9	9/104 = 0.0865
green, wrinkled	32	35	–3	9	9/35 = 0.2571
				$\chi^2 =$	**0.5103**

The number of degrees of freedom = (number of categories – 1) = (4 – 1) = 3

If you look up the value of χ^2 in a distribution table (see Figure 11.18), you can see that there are p values ranging from 0.01 (1%) to 0.99 or 99%. The probability value of 0.05 means that this is the level that could occur by chance just 5 times in 100 (5%) or 1 in 20. You need to know that the probability of your deviation being the result of chance is greater than 5%.

Number of classes	Degrees of freedom	χ^2							
2	1	0.00	0.10	0.45	1.32	2.71	3.84	5.41	6.64
3	2	0.02	0.58	1.39	2.77	4.61	5.99	7.82	9.21
4	3	0.12	1.21	2.37	4.11	6.25	7.82	9.84	11.34
5	4	0.30	1.92	3.36	5.39	7.78	9.49	11.67	13.28
6	5	0.55	2.67	4.35	6.63	9.24	11.07	13.39	15.09
Probability that deviation is due to chance alone		0.99 (99%)	0.75 (75%)	0.50 (50%)	0.25 (25%)	0.10 (10%)	0.05 (5%)	0.02 (2%)	0.01 (1%)

Accept null hypothesis (any difference is due to chance and not significant)

Critical value of χ^2 0.05 p level; this is the level at which we are 95% certain the result is not due to chance, agreed on by statisticians as a cut-off point

Reject null hypothesis; accept experimental hypothesis (difference is significant, not due to chance)

▶ **Figure 11.18:** Part of a χ^2 (chi-squared) distribution table

You can see from Figure 11.18 that the critical value of χ^2 for three degrees of freedom is 7.82. Your value of χ^2 is smaller than this, therefore the difference between your observed and expected data is due to chance and is not significant. You accept the null hypothesis. If the value of χ^2 were greater than the critical value, then you would reject the null hypothesis and would need to think again about the inheritance pattern.

Worked example

A genetics research student carried out an investigation.

Homozygous male wild type fruit flies were mated with female fruit flies, homozygous for vestigial wings. All the F_1 offspring showed the phenotype wild type.

Members of this F_1 generation were allowed to interbreed and the parent generation separated from the pupae of the F_2 generation. As members of the F_2 generation hatched, their phenotypes were noted:
- 720 were wild type
- 260 had vestigial wings.

The student then carried out a chi-squared test.

Category	Observed (O)	Expected (E)	O – E	(O – E)²	$\frac{(O - E)^2}{E}$
Wild type	720	735	-15	225	0.306
Vestigial wings	260	245	15	225	0.918
				χ^2	1.224

Degrees of freedom = 1

At the $p = 0.05$ level of significance, for one degree of freedom, the critical value of chi-squared is 3.84.

The value here is less than that critical value.

Therefore, the difference between observed and expected data is not significant. You accept the null hypothesis: *There is no statistically significant difference between observed and expected data. Any difference is due to chance.*

In another investigation, tomato plants were grown. Bush type tomatoes are short and do not produce side shoots. They are homozygous for the mutant recessive allele s. Their genotype is ss.

Homozygous standard tomatoes are tall with side shoots and have the genotype SS.

The F_1 tomatoes obtained from crossing homozygous bush tomatoes with homozygous standard tomatoes are all tall, with genotype Ss.

When members of this F_1 generation interbreed, the F_2 offspring were as follows: 130 bush type and 360 standard type.

(a) Apply a chi-squared test to see if the difference between observed and expected data is significant or due to chance.

Observed ratio is 130 bush type and 360 tall; expected ratio is $\frac{1}{4}$ of 490 = 122.5 bush type and 367.5 tall.

Category	Observed (O)	Expected (E)	O – E	(O – E)²	$\frac{(O - E)^2}{E}$
Tall	360	367	-7	49	0.134
Bush type	130	123	7	49	0.398
				χ^2	0.532

Degrees of freedom = 1

At the $p = 0.05$ level of significance, for one degree of freedom, the critical value of chi-squared is 3.84.

The value here is less than that critical value.

Therefore the difference between observed and expected data is not significant. You accept the null hypothesis: *There is no statistically significant difference between observed and expected data. Any difference is due to chance.*

(b) In a family where both parents carry a recessive allele for cystic fibrosis, two of their four children have cystic fibrosis and two do not have any symptoms. Explain why you should not use a chi-squared test to examine these data.

You cannot use the chi-squared test for data relating to one family as the sample size is too small.

About 1 in 500 people in the UK inherit a condition, FHC (familial hypercholesterolaemia – high blood cholesterol). This is caused by a dominant allele. Affected individuals are **heterozygous**. Being **homozygous** for FHC is a lethal combination and the embryo does not survive. Construct genetics diagrams to predict possible offspring from (a) a couple, one of which has FHC and one who does not and (b) a couple, both of whom have FHC.

Hint ▸ Apply what you have learn about monohybrid inheritance together with the information in this question.

Extend ▸ Why do you think people with FHC are given medication to reduce their blood cholesterol levels rather than trying to reduce it by changing their diet alone?

Further genetics

There are several thousand genetic disorders known in humans. Some of them are extremely rare and some, such as cystic fibrosis and sickle cell anaemia, are more common and better known. Some are caused by single gene defects (mutations) and some are the result of chromosome abnormalities.

▸ Some gene defects have dominant inheritance patterns and some have recessive inheritance patterns.

▸ Some exhibit **co-dominance** at the molecular level but with a recessive inheritance pattern at the phenotypic level. Other characteristics, such as the ABO blood groups, exhibit co-dominance at the phenotypic level.

▸ It is important to realise that there are no genes *for* diseases – the gene in question codes for a protein or length of RNA that regulates the expression of another gene(s) and when this gene is mutated the normal protein or RNA is not made and a genetic disorder results.

In this section, you will examine the inheritance patterns and hence the ability to predict the inheritance of various conditions.

Huntington's disease

Huntington's disease is a progressive (it gradually gets worse) brain disorder that usually shows symptoms, in those affected, after the age of 35 years. It affects about 5 in every 100 000 people of European ancestry.

▸ It is caused by a mutation in a gene called *HTT* that codes for a protein named huntingtin, which is important for normal function of brain neurons.

▸ The mutation is an expanding triple nucleotide repeat, and is sometimes called a 'stutter'. In part of the *HTT* gene is a base triplet CAG that is normally repeated between 10 and 35 times. However, if this trinucleotide sequence repeats more than 40 times (in some cases the repeat occurs 120 times) then the altered, abnormally long protein is cut by enzymes in brain neurones into shorter, but toxic, fragments that bind together and interfere with brain neurone function. This leads eventually to death of brain neurones and to the symptoms.

▸ Early symptoms include depression, poor coordination and jerky involuntary twitching movements, known also as chorea. As the disease progresses these movements become more pronounced and sufferers have difficulty with walking, swallowing and talking. They also experience changes in personality and loss of thinking ability (dementia). Most sufferers die 15–20 years after the first symptoms appear.

▸ The *HTT* gene is on one of the autosomes. People with 27–35 CAG repeats do not develop the disease, but they are at risk of producing children with the disease since, at meiosis, the repeat sequence can expand.

Key terms

Heterozygous – having one or more pairs of dissimilar alleles for particular genes on homologous chromosomes.

Homozygous – having identical alleles at one or more gene loci on homologous chromosomes.

Co-dominance – the expression of both alleles of a gene is seen in the phenotype.

- People who have 36–40 or more CAG repeats may develop the disease later in life. By this time they have probably had children who may have inherited the mutated allele of the gene.
- Huntington's disease has an autosomal dominant inheritance pattern. The gene is question is on an autosome and sufferers only need to inherit one mutated allele for the disease to develop. Only one parent needs to be affected and at each pregnancy there is a 50% (1 in 2) chance of the child inheriting the disorder.

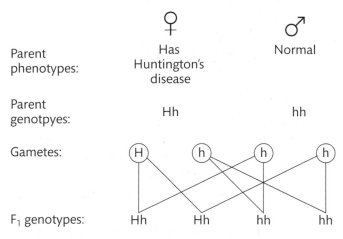

There is a 50:50 chance at each pregnancy that the child will inherit the allele for Huntington's disease.

▶ **Figure 11.19:** Genetic diagram illustrating the inheritance pattern for Huntington's disease

Research

Carry out an internet search to find out more about Huntington's disease.

Case study

Huntington's disease

David and John are brothers. David is 30 years old and John is 28. Their mother has been diagnosed with Huntington's disease. Their father does not have the disease and no one in his family is known to have suffered from it.

Both David and John have girlfriends they are planning to marry. They have both talked to their GP and been referred to a genetic counsellor who has told them that a test is available that will tell them if they have the faulty allele or not. They have a one in two chance of having inherited this allele from their mother. The counsellor has also told them that, at present, there

is no cure for this condition and no real effective treatment.

David and John are thinking about whether to take the test, whether to inform their partners and whether they should have children.

Check your knowledge

1 Explain why each brother has a 1 in 2 chance of having the faulty allele for Huntington's disease.

2 Explain why each brother has a 1 in 2 chance of developing the disease in later life.

3 Discuss the pros and cons of them taking the test for Huntington's disease.

Cystic fibrosis

This is the most common genetic disorder amongst Europeans. In the UK about 1 in 1600 live births are affected and about 1 in 20 people are symptomless carriers.

People who suffer from cystic fibrosis have a lot of thick sticky mucus in their lungs that leads to them suffering many chest infections, eventually leading to lung damage and possibly the need for a heart lung transplant. They also have excess mucus in their intestines and this reduces their ability to digest certain foods, especially fats, and to absorb the products of digestion. They may also be infertile as there is excess mucus in their reproductive tracts. They have to swallow encapsulated enzymes before meals, take antibiotics prophylactically to prevent the chest infections and have daily physiotherapy to loosen and expectorate (spit out) the mucus from their respiratory tract.

▶ The gene CFTR, identified in 1989, codes for a chloride ion channel protein (CFTR – cystic fibrosis transmembrane regulatory protein) in the cell surface membranes of epithelial (lining) cells in lungs, gut and reproductive tracts.

▶ There are several mutations but 70% of cases are caused by a mutation involving the deletion of a codon, leading to the loss of an amino acid (phenylalanine) at position 508 in the polypeptide chain.

▶ Symptomless carriers have one mutated allele and one normal allele of the CFTR gene. Sufferers have two mutated alleles of this gene, which is on an autosome – chromosome 7. They inherit one faulty allele from each parent.

▶ This disorder has an autosomal recessive inheritance pattern.

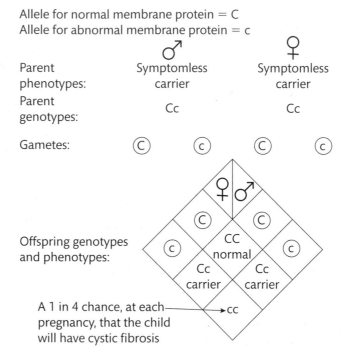

▶ **Figure 11.20:** Genetic diagram illustrating the inheritance pattern for cystic fibrosis

Tay-Sachs disease

This is also known as hexosaminidase A (an enzyme) deficiency. It is a rare autosomal recessive disorder that causes progressive deterioration of nerve cells and hence mental and physical abilities. Symptoms show at around the age of 7 months and sufferers usually die by the age of 4 years.

▶ Tay-Sachs results from a mutation in the *HEXA* gene on chromosome 15, which codes for a protein making up part of an enzyme.

▶ There are many different mutations including single base insertions or deletions and missense mutations, leading to non-functioning proteins.

▶ As a result, there is an accumulation of certain types of cell membrane lipids, called gangliosides, in brain neurones, causing early death of these cells.

▶ Sufferers have two mutated alleles of the HEXA gene, one copy from each carrier parent. The inheritance pattern is the same as for that of cystic fibrosis (see Figure 11.20).

▶ Although this is a rare genetic disorder it has a higher than usual incidence in genetically isolated populations such as Ashkenazi Jews (originally from Eastern Europe) and French Canadians of southeastern Quebec.

Sickle cell anaemia

Sickle cell anaemia (also called sickle cell disease, SCD) is an inherited disorder involving abnormal haemoglobin, leading to a severe form of anaemia.

▶ This disease has a recessive inheritance pattern and sufferers inherit two (one from each parent) mutated alleles of a gene, on chromosome 11, that codes for the β-globulin protein chains present in haemoglobin – the oxygen carrying molecule in red blood cells.

▶ The mutation is a single base substitution (GAG becomes GTG) leading to the amino acid glutamic acid being replaced by valine at position 6 in the protein chain.

▶ This causes the haemoglobin to collapse on itself in low oxygen concentrations and the red blood cells of sufferers assume a sickled shape.

▶ These erythrocytes can block capillaries, reducing oxygen delivery to organs and causing a painful crisis and other complications that eventually lead to death.

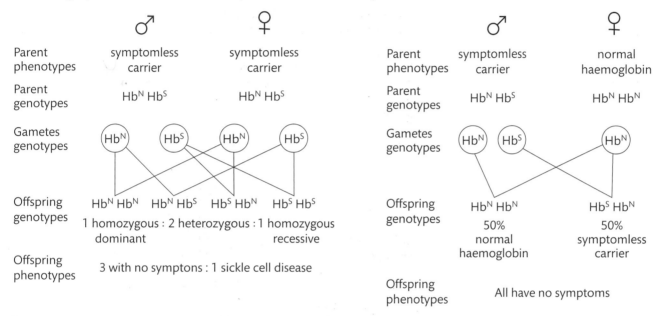

▶ **Figure 11.21:** Inheritance patterns for (a) parents both symptomless carriers of sickle cell disease (b) one carrier and one non-carrier

Co-dominance

Co-dominance is when both alleles of a gene locus are expressed in the heterozygote. This may occur at the molecular level and/or in the phenotype.

▶ With all three of the recessive conditions outlined above, both alleles (normal and mutated) are expressed at the molecular (biochemical) level in cells and tissues.

▶ However, because carriers, who are heterozygous, have one normal allele of the gene in question they produce enough of the normal protein to function and not suffer any symptoms.

▶ Therefore, in these cases, although both alleles are expressed (transcribed and translated to make protein) both alleles do not contribute to the visible characteristics of the phenotype and heterozygous people do not have symptoms of the disease. However, they may have only about 50% of the normal functioning protein in question.

In some cases, this can be an advantage:

▶ Carriers for Tay-Sachs (and possibly carriers of cystic fibrosis) may be less susceptible to TB (tuberculosis) and as this was a main cause of death for many hundreds of years, carriers of Tay-Sachs would have had a survival advantage over people homozygous for the normal allele of the gene.

▶ People heterozygous for sickle cell have some resistance to a severe form of malaria as roughly half their haemoglobin is abnormal. The malaria parasite spends part of its life cycle inside red blood cells of infected people.

Blood groups

Blood groups are an example of co-dominance, where both alleles in a heterozygous person contribute to the phenotype. The ABO blood group system also illustrates multiple alleles.

The four blood groups (phenotypes), A, B, AB and O are determined by three alleles of a single gene on chromosome 9. Although there are three alleles of this gene in the **gene pool**, any individual in the population will have only two of the three alleles within their **genome**.

▶ The gene encodes an **isoagglutinogen**, I, on the surface of erythrocytes.
▶ The three alleles present in the human gene pool are I^A, I^B and I^O.
▶ I^A and I^B are both dominant to I^O, which is recessive.
▶ I^A and I^B are co-dominant. If they are both present in the genotype they will both contribute to the phenotype.

Key term

Gene pool – all the alleles of genes within a population.

Genome – all the genes within a cell/organism.

Isoagglutinogen – a type of antigen on the surface of red blood cells.

Table 11.7 shows the blood group phenotypes produced by various genotypes.

A man of blood group A and a woman of blood group B, both heterozygous at this gene locus, could produce children of any of these four blood groups.

▶ **Table 11.7:** Genotypes and phenotypes

Genotypes	Phenotypes
I^AI^A I^AI^O	group A
I^BI^B I^BI^O	group B
I^AI^B	group AB
I^OI^O	group O

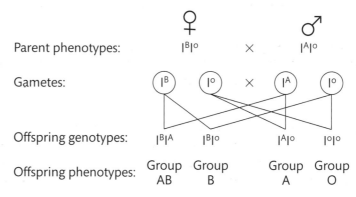

Parent phenotypes: ♀ I^BI^O × ♂ I^AI^O

Gametes: I^B I^O × I^A I^O

Offspring genotypes: I^BI^A I^BI^O I^AI^O I^OI^O

Offspring phenotypes: Group AB Group B Group A Group O

▶ **Figure 11.22:** Genetic diagram showing the probability of a heterozygous man of blood group A and a heterozygous woman of blood group B producing children of each blood group

---

Ⅱ PAUSE POINT

Two babies are mixed up in the maternity ward of a hospital. Baby 1 is blood group O and baby 2 is blood group A. Mr X is blood group A and Mrs X is blood group AB. Mr and Mrs Y are both blood group A. Which baby belongs to which couple?

Hint Write the possible genotypes of all the parents and the babies and then you can see which baby could be produced by which couple.

Extend Explain how the inheritance of ABO blood groups in humans shows multiple alleles and co-dominance.

Incomplete dominance

Incomplete, semi- or partial dominance is when intermediate phenotypes are observed in the offspring of parents with contrasting characteristics. For example, if snapdragon plants with red flowers are crossed with white-flowered snapdragon plants, all the offspring have pink flowers. This is because they have half the amount of red pigment as in the parent plants. It is possible to predict the outcome of crossing members of the F_1 generation as shown in Figure 11.23.

▶ Snapdragon flowers

Parent phenotypes:	red flowered plant (♀)	×	white flowered plant (♂)
Parent genotypes:	$C^R C^R$	×	$C^W C^W$
Gametes:	C^R	×	C^W
Offspring genotypes:		all $C^R C^W$	
Offspring phenotypes:		pink flowered plants	

▶ **Figure 11.23:** Inheritance pattern showing incomplete dominance

Sex linkage

You will recall that in humans, one pair of chromosomes, either XX or XY, is involved in determining our sex. These are the sex chromosomes.

▶ The human X chromosome contains over 1000 genes that are involved in determining many characteristics or metabolic functions, not concerned with sex-determination, and most of these have no partner alleles on the Y chromosome.

▶ If a female has one abnormal allele on one of her X chromosomes, she will probably have a functioning allele of the same gene on her other X-chromosome.

▶ However, if a male inherits, from his mother, an X chromosome with an abnormal allele for a particular gene, he will suffer from a genetic disease, as he will not have a functioning allele for that gene. Males are functionally haploid, or hemizygous, for X-linked genes. They cannot be heterozygous or homozygous for X-linked genes.

Haemophilia A

One of the genes on the non-homologous region of the X chromosome codes for a blood clotting protein called factor 8. A mutated form of the allele codes for non-functioning factor 8. A female with one abnormal allele and one functioning allele will produce enough factor 8 to enable her blood to clot normally when required.

If such a female passes the X chromosome containing the faulty allele to her son, because he has no functioning allele for this protein on his Y chromosome, he will suffer from haemophilia A. This is an inability to clot blood fast enough.

▶ He may internally haemorrhage after injuries sustained by, for example, falling over or receiving a knock.

▶ Injuries that involve cuts are easier to deal with, as the bleeding is visible and first aid can be given.

Genotypes representing sex-linked genes are represented by symbols that show they are situated on the X chromosome. The symbol H is used to represent the normal functioning allele for factor 8 and the symbol h represents the abnormal allele.

Figure 11.24 shows the inheritance pattern for haemophilia A where the father does not have haemophilia A and the mother is a symptomless carrier.

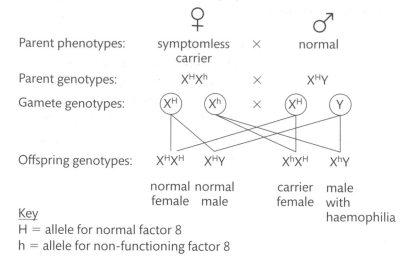

Key
H = allele for normal factor 8
h = allele for non-functioning factor 8

▶ **Figure 11.24:** Inheritance of haemophilia A

Colour blindness

One of the genes coding for a protein involved in colour vision is on the X chromosome but not on the Y chromosome.

▶ A mutated allele of this gene may result in colour blindness. This is an inability to distinguish between red and green.

▶ A female with one abnormal allele and one functioning allele will not suffer from colour blindness, but a male with an abnormal allele on his X chromosome will not have a functioning allele on his Y chromosome and will therefore suffer from red-green colour blindness.

▶ The inheritance pattern is the same as for haemophilia A – that of a recessive sex-linked disorder.

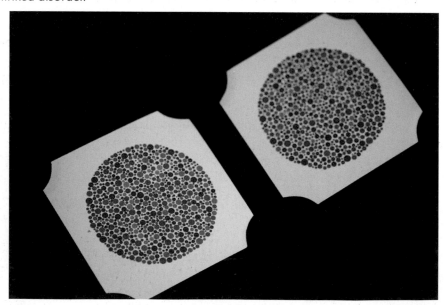

▶ Ishihara test cards used to diagnose colour blindness. People with red-green colour blindness cannot see the numbers.

It would appear that females have twice the number of X-linked genes, being expressed, as do males. However, there is a mechanism to prevent this disparity.

In every female cell nucleus one X chromosome is inactivated. Determination of which member of the pair of X chromosomes becomes inactivated is random and happens during early embryonic development.

PAUSE POINT

Tortoiseshell cats are ginger and black and are female. The gene is on the X chromosome and has two alleles C^B (black) and C^G (ginger).

Draw genetic diagrams to predict the outcome of the following crosses:
(i) tortoiseshell female and ginger male (ii) ginger female and black male
(iii) tortoiseshell female and black male.

Hint

This is a sex-linked inheritance. Remember females have two X chromosomes and males have only one.

Extend

Find out why tortoiseshell female have black patches and orange patches rather than individual hairs that are both ginger and black.

Chromosome mutations

Aneuploidy refers to a condition where the number of chromosomes in the cells of an individual is abnormal.

Sometimes, during meiosis, a pair of homologous chromosomes fails to separate (non-disjunction), resulting in one gamete with an extra chromosome or a gamete lacking a chromosome. If a gamete with an extra chromosome is fertilised by a normal haploid gamete, then the embryo will have three copies of a particular chromosome.

Down's syndrome

Also known as trisomy 21 (see Figure 11.15), Down's syndrome sufferers have an extra copy of chromosome 21. This leads to over expression of the genes on that chromosome and all the symptoms of Down's syndrome.

▸ This is a genetic condition but is not inherited.

▸ Neither parent usually has Down's syndrome and there are no symptomless carriers, since if you have an extra chromosome 21, you will have Down's syndrome.

▸ The condition arises spontaneously when either an egg or a sperm is produced.

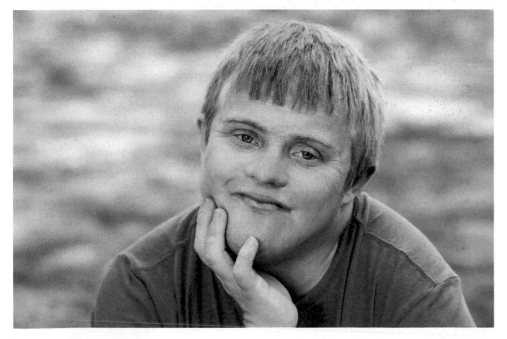

▸ Young man with Down's syndrome

Turner syndrome

This condition affects females who have only one functioning X chromosome. Part or all of one X chromosome is missing.

▶ Like Down's syndrome, Turner syndrome is not an inherited condition.

▶ An egg or a sperm without a sex chromosome is fertilised by a normal gamete with one X chromosome. Symptoms include webbed neck, short stature, underactive thyroid, heart defects, diabetes, vision and hearing problems, lack of menstrual periods and infertility. Intelligence is normal, although sufferers may experience some difficulty with aspects of maths.

▶ Growth hormone and thyroxine can be given.

Epistasis

In some cases different genes, at different loci on different chromosomes, interact to affect one phenotypic characteristic. When one gene masks or suppresses the expression of another gene this is termed epistasis.

In humans, there are many genes that determine skin colour but these genes, although present, are not expressed if an epistatic gene, A/a, has two recessive alleles present, genotype aa. At least one allele, A, is needed for any pigment to form in human skin. This condition has a recessive inheritance pattern. Albinism also affects pigment development in your eyes and sufferers are more susceptible to lens cataracts as well as being susceptible to sunburn and skin cancer.

Case study

Blood groups

Both of Jane's parents have blood group AB. When Jane became a blood donor, she found that her blood was group O. Jane understood that her parents could not produce a child of group O. She took a DNA test and it confirmed that her parents were her biological parents. She then researched and found that another epistatic gene is involved in the expression of blood group genes. The gene (H/h) controls the formation of a precursor molecule, H. The I^A and I^B alleles then govern the addition of sugar molecules to this precursor. Jane had inherited two rare recessive h alleles from her parents, so despite having an I^A and an I^B allele, the A and B agglutinogens cannot be made in her cells as she had no H molecules to modify. As Jane's red blood cells have no agglutinogens on their surface membranes, she is functionally group O but genetically group AB.

Check your knowledge

1 When Jane is donating blood, why should her blood be considered group O rather than group AB?

Discussion

Eilidh has red hair. Ten per cent of the population from her native Scotland also have red hair and geneticists say that about 35% of the Scots population carry the recessive allele for the gene *MC1R* present on chromosome 16. Red hair is also more common in Ireland than in many other areas of the world where only 1–2% of the population has red hair. Red hair contains less of the pigment eumelanin and more pheomelanin, a pigment that can direct UV light from sunlight onto the scalp of redheads, enabling them to synthesis more vitamin D.

Red hair has a recessive inheritance pattern and both parents of a red-haired child carry a recessive allele. However, there are other genes that interact so those with red hair exhibit a range of colours from pale red/blond to bright red to deep auburn.

Why do you think that red hair is more common among people of North West Europe than among the rest of the global population? By what process has this difference come about?

This assessment may form part of your practical portfolio.

Carry out investigations into monohybrid and dihybrid crosses, using fruit flies or rapid cycling *Brassica* plants, and record the data you obtain.

Show how the data you have obtained demonstrates Mendel's laws.

Analyse your data using the chi-squared test. Clearly state your null hypothesis and show all working plus your statement about accepting or rejecting your null hypothesis.

Produce posters with genetic diagrams, which could be used by a genetic counsellor, to predict and explain the genetic crosses between the following: (a) someone suffering from a recessive genetic disorder and a non-affected person; (b) a non-affected person and a symptomless carrier of a recessive genetic condition; (c) an affected person and a symptomless carrier for a particular recessive genetic condition; (d) someone suffering from a dominant genetic disorder and a non-affected person; (e) two people both affected by a dominant genetic disorder. *Hint: Research the disorder you choose because often inheriting a double dose of the faulty allele proves to be a lethal combination and individuals do not survive to birth.*

Plan
- What is the task?
- How confident do I feel in my own abilities to complete the task?
- Are there any areas I think I may struggle with?

Do
- I know what it is I am doing and what I want to achieve.
- I can identify where I have gone wrong and can adjust my approach to get back on course.

Review
- I can explain what the task was and how I approached it.
- I can explain how I might approach the difficult parts differently next time.

Explore basic DNA techniques and use of genetic engineering technologies

In this section, you will look at the principles and practical applications of the techniques and equipment in some aspects of DNA technology.

DNA extraction

Genomic DNA can be extracted from the bacterial chromosome and from nuclei of eukaryotic cells. Plasmids can be extracted from bacteria. Mitochondrial and chloroplast DNA can be extracted from eukaryotic cells that have been ground up and centrifuged differentially. The nuclei come down on the first spin, and if the supernatant is taken off and centrifuged at a slightly faster speed, mitochondria and chloroplasts will settle.

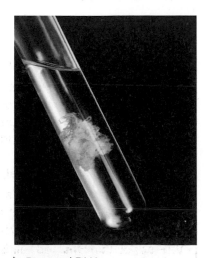

Gel electrophoresis

Electrophoresis is used to separate different-sized fragments of DNA. It can separate fragments that differ by only one base pair and is widely used in gene technology to separate DNA fragments for identification and analysis.

Restriction enzymes

A DNA sample can be cut into fragments by restriction enzymes. Each restriction enzyme has an active site that fits a specific recognition site on a length of DNA – a particular base pair sequence. The DNA sample is incubated with restriction enzymes at 35–40 °C for up to an hour.

▶ Extracted DNA

Electrophoresis

The technique uses an agarose gel plate covered by a buffer solution. The DNA fragments produced by the restriction enzyme activity are mixed with dense loading dye and placed into wells in the gel, covered with buffer solution. Electrodes are placed in each end of the tank so that, when connected to a power supply, an electric current can pass through the gel. DNA has an overall negative charge, due to its many phosphate groups.

The DNA fragments migrate towards the anode (positive electrode). Fragments of DNA all have a similar surface charge regardless of their size. Smaller DNA fragments pass more easily through the gel and travel further in a given time than larger fragments do.

You can carry out gel electrophoresis in your school or college. Figure 11.25 shows the apparatus.

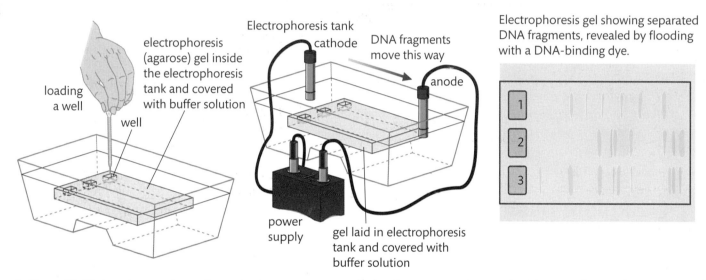

▶ **Figure 11.25:** Carrying out electrophoresis

DNA amplification

In 1983, Kary Mullis developed a technique, the polymerase chain reaction or PCR, to amplify DNA and this soon became incorporated into forensic DNA analysis and into the protocols for analysis of DNA for genetic diseases as well as many other functions in genetic engineering.

The polymerase chain reaction (PCR)

This is a cyclic reaction.

1 The sample of DNA is mixed with DNA nucleotides, primers, magnesium ions and the enzyme DNA polymerase.

2 The mixture is heated to around 95–98 °C to break the hydrogen bonds between complementary nucleotide base pairs and thus denature the double-stranded DNA into two single strands of DNA.

3 The temperature is cooled to around 65 °C so that the primers can anneal (bind by hydrogen bonding) to one end of each single strand of DNA. This gives a small section of double-stranded DNA at the end of each single-stranded molecule.

4 The DNA polymerase enzyme molecules can now bind to the end where there is double-stranded DNA.

5   The temperature is raised to 72 °C, which is the optimum temperature for the DNA polymerase used.

6   The DNA polymerase catalyses the addition of DNA nucleotides to the single-stranded DNA molecules, starting at the end with the primer and proceeding in the 5' to 3' direction.

7   When the DNA polymerase reaches the other end of the DNA molecule then a new double-strand of DNA has been generated.

8   The whole process begins again and is repeated for many cycles.

9   The amount of DNA increases exponentially, $1 \rightarrow 2 \rightarrow 4 \rightarrow 8 \rightarrow 16 \rightarrow 32 \rightarrow 64 \rightarrow 128$ and so on.

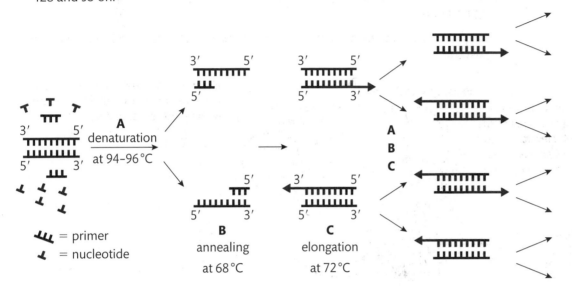

▶ **Figure 11.26:** Stages of the PCR

Research

Go to the Science and Plants for schools website. Research how you can isolate DNA from chloroplasts and amplify it using the PCR.

The purpose of using the PCR to amplify DNA

Amplifying DNA produces more of it for its analysis or cloning, and for genetic manipulation

DNA profiling

Modern DNA profiling can obtain results from as few as five cells. In theory, if a suspect had touched a surface and left a few cells behind, his or her DNA profile could be obtained. However, as we all leave skin cells behind when we touch a surface, a frequently touched surface could transfer innocent people's DNA on to the hands of a criminal and from there on to a crime scene.

Once a small amount of DNA has been amplified, it can be investigated using gel electrophoresis to give a banding pattern or DNA profile (DNA fingerprint).

Banding patterns produced when DNA of people is subject to electrophoresis can be compared. This can:

▶ find out if people are related to each other and demonstrate paternity or maternity

▶ show if a suspect's DNA matches that found at a crime scene.

DNA analysis was used to show that remains found under a Leicester car park were those of King Richard III as they were compared to the DNA of a known descendant. Susan Black, Professor of Anatomy and Forensic Anthropology at the University of Dundee, has used DNA profiling to identify remains of victims of the 2004 tsunami, and the genocide in Kosovo. Victims of plane crashes can also be identified from DNA profiles of tissues.

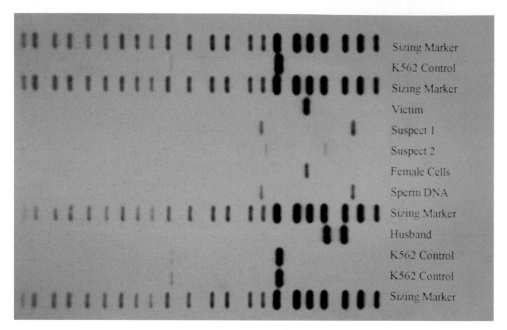

▶ DNA profiles

Cancer diagnosis

Many forms of cancer involve mutations to proto-oncogenes, involved in regulating the cell cycle and cell division, so these genes become oncogenes, promoting unrestricted cell division leading to tumour formation.

PCR-based tests can study these mutations. When specific mutations are detected, the therapy can be made specific for the patient, increasing the chance of a positive prognosis.

Tissue typing

Tissue typing is carried out to match a donor's tissue/organs to that of the recipient. Each person has a unique set of histocompatibility molecules that act as markers on the surface of all cells except erythrocytes. Your T cells can recognise these molecules, and if you receive a transplant that does not match your immune system, they will reject the new organ. Blood can be used for tissue typing. The white blood cells, lymphocytes, are obtained from the sample. The DNA can be obtained from these cells and amplified so that scientists can determine if your cells have the genes for specific surface antigens.

Pre-implantation genetic screening

If embryos are made via in vitro fertilisation (IVF), then when they are at the eight-cell stage, one cell can be removed without damaging the embryo. The DNA can be extracted from this cell, amplified and then tested for specific alleles that lead to genetic disorders, so that only healthy embryos are implanted into the mother's uterus. This may be carried out in families where there is a history of genetic disorders such as cystic fibrosis or haemophilia.

It can also be carried out on embryos before they are implanted, to see if they can be an umbilical cord stem cell donor for a sick sibling.

Link

See *Unit 12: Diseases and Infection* for more information on immunity.

PAUSE POINT   Explain how the PCR can be used for DNA profiling and pre-implantation genetic screening.

Hint   Explain what the PCR does and how it does it. Indicate why DNA needs to be amplified for profiling and pre-implantation genetic screening.

Extend   Compare the PCR reaction with normal DNA replication.

Transformation of cells

When a foreign gene is introduced into a cell the cell is transformed. Bacterial cells can be transformed by getting them to take up a **plasmid**, which is a **vector**, that has had a novel gene inserted. There is a kit available and you can transform *E. coli* cells by adding plasmids that contain the GFP gene obtained from the jellyfish *Aequorea victoria*. This gene codes for GFP, green fluorescent protein, that glows green when illuminated by ultraviolet light.

In the kit there are plasmids that contain the GFP gene and a gene for beta-lactamase, which gives resistance to the antibiotic ampicillin. This is a *marker gene* as only bacteria that have taken up a plasmid will have this gene and be able to grow in the special medium. When transformed bacteria are cultured in a medium also containing the sugar, arabinose, the GFP gene is switched on and expressed.

The plasmids were made in a lab by extracting them from bacteria, cutting them with a restriction enzyme, adding the GFP gene plus the marker genes and adding DNA ligase enzyme to catalyse the joining of these genes into the plasmid DNA.

There are also non-pathogenic *E. coli* bacteria in the kit and if you place them in a suspension of calcium chloride at 42 °C for 50 seconds and then place them on ice at 0 °C this produces heat shock and makes the bacterial walls and surface membranes more porous, enabling them to take up the plasmids.

The bacteria are then spread onto agar plates containing special nutrients. Some of the plates also contain arabinose sugar and the antibiotic ampicillin.

Plating out the bacteria onto the ampicillin-arabinose nutrient agar screens them and enables scientists to identify the bacteria that have been transformed.

Bacteria that have taken up a plasmid will be able to grow in the presence of ampicillin and the arabinose will switch on their newly acquired GFP gene.

Key terms

Plasmid – loop of DNA in prokaryotic cells.

Vector – agent that carries. In this context, it carries foreign DNA into a cell/organism.

(a)
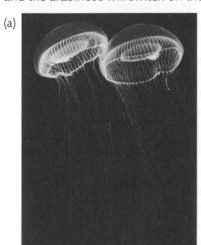

(b)
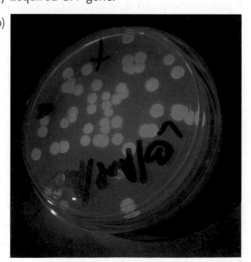

▶ (a) Jellyfish *Aequoria victoria*. (b) Transformed *E.coli* bacteria expressing the GFP gene making them glow when viewed under UV light.

Uses of genetic engineering

The GFP gene is used as a research tool. If it is inserted, along with another novel gene, into tissue of a transformed organism, scientists can see if the novel gene is in the correct location and if it is being expressed.

Genetic engineering (recombinant DNA technology) has many uses and potential applications.

Table 11.8 shows some uses of genetic engineering.

▶ **Table 11.8:** Types of genetic engineering

Type of genetic engineering	Examples
Genetically modified crops A gene for a desired characteristic is obtained from one species and inserted into another species. 	**Golden Rice™** is modified to contain a gene, obtained from daffodils, that causes rice plants to express a gene for making vitamin A in their seeds (rice grains). This staple food can be made available at no greater cost to consumers than standard rice and could prevent blindness and death due to vitamin A deficiency in many countries, e.g. India. **Potatoes**, an important staple crop, can be modified to be resistant to late blight and nematodes, reducing the need for pesticides. Low sugar potatoes can also be produced, reducing the generation of the carcinogen acrylamide when potatoes are heated to 120 °C, e.g. when fried.
Diagnostic tests Clinical geneticists use diagnostic tests to make diagnoses and try to tailor treatments to specific people's genetic makeup.	Diagnostic genetic tests are available to detect carriers of recessive disorders such as Tay-Sachs and cystic fibrosis; X-linked disorders such as haemophilia and fragile X syndrome; late-onset disorders such as Huntington disease; prenatal diagnosis, e.g. Down's syndrome, newborn screening, e.g. for phenylketonuria (PKU); pharmacogenomic testing, e.g. testing a patient's cytochrome 450 enzyme system to see how they will metabolise a specific drug such as warfarin.
Gene therapy This involves inserting functioning alleles of genes into a patient's cells to treat or prevent a genetic disease. Vectors such as liposomes (small aggregates of lipids with a gene inside) or modified harmless viruses are used to introduce the functioning alleles into living cells.	Some forms of blindness (e.g. retinitis pigmentosa – a form of retinal degeneration), SCID (severe combined immunodeficiency), Parkinson disease, cystic fibrosis and haemophilia may be able to be treated with gene therapy.
Genome editing **The CRISPR/Cas9** system is a prokaryotic immune system that protects bacteria and archaea from attack by phage viruses (viruses that infect bacteria) Scientists have isolated it and can use it as a gene editing tool.	The Cas9 protein and guide RNA molecules can target a particular DNA sequence, snip out (using an endonuclease enzyme) a faulty allele and insert a functioning allele in its place. This gives great potential to treat genetic disorders caused by a *dominant* allele. It may be able to be used to edit out pig viral genes associated with pig organs for xenotransplantation (using pig organs in humans).
Pharming Genes for human pharmaceutical proteins that are too large to be made in bacterial cells, are inserted into sheep or goats, near to a promoter for a beta-lactoglobulin gene, so that these genes are only expressed in mammary tissue so that the protein can be harvested from the animal's milk.	**Alpha-antitrypsin** can be used to prevent hereditary emphysema. The gene for the protein alpha-antitrypsin, which inhibits antitrypsin action in lungs causing breakdown of elastin in alveoli walls, is inserted into sheep embryos and the pharmaceutical is obtained from the sheep's milk. **Spider silk** fibres have tensile strength five times greater than steel of the same diameter. Spiders are impossible to farm, but genes for spider silk have been inserted into goats. These GM goats produce spider silk in their milk. Silk can be used for cables, sutures, artificial ligaments and bullet-proof vests.
Genetic testing This involves medical tests that identify changes in chromosomes, genes or proteins, which can rule out or confirm a suspected genetic condition or determine a person's chance of developing or passing on a genetic disorder to offspring. Some tests are available via the NHS to specific groups of people and some are only available privately, hence this is testing and not screening.	**Pre-implantation genetic testing** for chromosome abnormalities such as Down's syndrome (trisomy 21), Edwards syndrome (trisomy 13), Angelman or Prader-Willi syndromes. Testing to establish the possible cause of cancer within a family. **Predictive testing**, such as for the presence of BRCA1 or BRCA2 genes for breast cancer. Tests for bowel, prostate and ovarian cancer genes are also available on request, privately.

▶ **Table 11.8:** Types of genetic engineering – *Continued*

Type of genetic engineering	Examples
Stem cell therapy Regenerative medicine uses stem cells to replace and repair damaged tissue.	Potential to treat Parkinson'S disease, multiple sclerosis, some types of sight loss, leukaemia, arthritis and autism.
Xenotransplantation As techniques for organ transplantation have improved there is a shortage of organs for such operations. It may be possible to modify the organs of other animals.	It may be possible to genetically modify organs of animals, such as pigs, that have a very similar physiology to that of humans, so that the surface antigens on the organs are identical to those of humans. This would mean that these organs would not be rejected. However, scientists now know that all animals have viral genes in their genomes. Our viral genes do not harm us and pig viral genes do not harm pigs, but if pig viral genes were transplanted into humans along with the organs, these genes could pose problems.

PAUSE POINT Distinguish between gene therapy and gene editing

> Hint   You need to clearly understand the two processes and indicate how each may be used to treat genetic disorders.

> Extend   Discuss the pros and cons of GM crops.

Case study

The 2009 Welsh Grand National winner, Dream Alliance, owned by a village syndicate in Wales, achieved his victory despite having earlier damaged a tendon. He was the first horse to receive stem cell surgery (at Liverpool University) to repair the injury, and his surgery was paid for from his previous winnings!

Why do you think stem cells were needed to repair a tendon?

Assessment practice 11.4   D.P6 D.P7 D.M6 D.D4

Some of these activities can be part of your practical portfolio.

Extract DNA from plant material and amplify it using the PCR reaction.

You have been asked to help introduce information to stimulate debate, within your school/college about the potential uses of genetic engineering technologies in industry and medicine.

Produce a poster for an exhibition explaining the uses of genetic engineering technologies. Include some analysis of these technologies and evaluate possible future uses of genetic engineering technologies.

Plan
- What is the task?
- How confident do I feel in my own abilities to complete the task?
- Are there any areas I think I may struggle with?

Do
- I know what it is I am doing and what I want to achieve.
- I can identify where I have gone wrong and can adjust my approach to get back on course.

Review
- I can explain what the task was and how I approached it.
- I can explain how I might approach the difficult parts differently next time.

Further reading

Gelehrter, T. D., Collins, F. S. and Ginsburg, D. (1998) *Principles of Medical Genetics* (2nd edition). Williams and Williams.

Harper, P. S. and Clarke, A. J. (1999) *Genetics, Society and Clinical Practice.* Bios Scientific Publishers.

Klug, W. S. and Cummings, M. R. (2013) *Essentials of Genetics.* International edition. Pearson.

Kumar, D. and Eng, C. (2014) *Genomic Medicine Principles and Practice.* Oxford Monographs.

Snustad, P. and Simmons, J. (2009) *Principles of Genetics* (5th edition). ISV.

Websites

www.fightingblindness.ie gives information about using stem cells to treat some problems with the retina.

www.stemcells.ox.ac.uk has more information about stem cells.

www.yourgenome.org/facts/what-is-gene-therapy has information about gene therapy.

www.ghr.nlm.nih.gov provides a guide to understanding genetic conditions.

THINK ▶▶FUTURE

Sergei Filipov

public engagement
department worker
at Wellcome Genome
Campus

I have worked here for five years and am responsible for circulating information about the work done here to members of the general public. We often have visits from local school and college pupils and I show them around to see some of the labs and explain the work we do.

We support projects in many areas including theatre and film-making and exhibitions to help inform members of the public and enable them to take part in informed debates around contemporary issues with genetics.

The campus, set in the grounds of Hinxton Hall, Cambridge, is lovely and has grown substantially since its inception when Fred Sanger spotted the empty large house and set up the Sanger Institute. Many new buildings have been added for the Wellcome Trust Sanger Institute and the European Molecular Biology Laboratory's European Bioinformatics Institute. Much work on the Human Genome Project was carried out here and the bioinformatics department deals with the huge data sets generated by this research. Much more research on genomes is going on and we are also bar coding (investigating the sequences of specific lengths of DNA in various species to give an easily identifiable band pattern) various species. This can enable scientists to quickly identify, for example, if food products described as containing beef do, in fact, contain only beef.

People working in this department need to have a good understanding of the science of genetics as well as an understanding of how to involve the arts, media and education.

Questions

How could you use the science, drama and media studies departments and their students in your school or college to introduce a well-informed debate about genetic modification? Discuss your ideas in small groups and then share your ideas with others in your class.

Getting ready for assessment

Ganesh is studying for a BTEC National in Applied Science. He was given an assignment as part of his practical portfolio. He was asked to explain the underlying principles of inheritance of genetic characteristics and to investigate some practically. Ganesh shares some aspects of his experience below.

How I got started

I gathered all my notes on patterns of inheritance and also found some useful and relevant books in the college library and some useful websites.

Think about it

I asked myself: Have I thought about all the information I will need? Can I find enough information to carry out practical work to show similar patterns of inheritance to the examples in humans I have chosen? If needed, can I find some secondary data from websites and books, so I can analyse and explain those data? Does my school/college have all the necessary equipment for me to carry out some practical work? Do I understand how to analyse the data using a statistical test and can I write it up in my report in the proper way? I must remember to note down which information came from which source.

How I brought it all together

I chose interesting examples of inherited conditions – one involving a chromosome abnormality, one involving a recessive allele for a single gene defect and one involving a dominant faulty allele. I tried to find actual family pedigrees where these conditions were present and to show how these pedigrees can be used to

deduce whether a condition has a dominant or recessive inheritance pattern. Once I had worked that out in each case I showed, with genetic diagrams and Punnett squares, how we can predict possible outcomes from specific matings. I found out how to culture fruit flies and carry out cross mating experiments. I tried to use crosses that modelled the inheritance patterns I was investigating for humans – e.g. mutations resulting from recessive or dominant alleles. Using fruit flies enabled me to study the results of crosses as their life cycle takes about three weeks and they produce many offspring, which means I had enough data to apply the chi-squared test.

What I learned from the experience

I noted where all my sources of information came from but did not reference anything in the proper way. There are conventions to be followed and I need to get to grips with this as, when I progress to higher education, all reports must be correctly referenced. I forgot to include any information on sex-linked conditions or on epistasis. I also fell into the trap of forgetting that some mutations are neutral or useful – for example, paler skin among people who live in more temperate regions allowing them to make enough vitamin D so that enough calcium is deposited in the bones.

Glossary

Absolute zero: the lowest possible temperature which is 0 K on the Kelvin temperature scale and −273 °C on the Celsius temperature scale.

Absorbance: a measure of the amount of light of a particular wavelength which is absorbed by a sample.

Absorption spectrophotometry: the principle by which concentration of a chemical solution can be determined by the amount of light that it absorbs.

Acinus (plural acini): cluster of cells resembling a berry, for example, raspberry.

Accuracy: how close the readings are to the actual values.

Acrosome: a cap-like structure that covers the front section of the head of the sperm. It contains enzymes to break down the follicle cells and zona pellucida surrounding the oocyte.

Action potential: a sudden and rapid increase in the positive charge of a neuron caused when sodium and potassium ions move across the cell membrane.

Activation energy: the minimum energy required for collisions to break the bonds in the reactants and lead to a reaction.

Active site: the area of an enzyme that the substrate binds on to.

Active transport: movement of molecules into or out of cells against their concentration gradient. It uses carrier proteins in the cell surface membrane and energy from ATP.

Adsorption: the process by which atoms, molecules or ions from a gas or liquid adhere to a surface. The process is not permanent.

Aeration: introducing air into the soil.

Aerobic: requires oxygen.

Aerobic respiration: respiration with oxygen.

Afferent pathway: the route taken by impulses that travel away from the stimulus to the spinal cord.

Alkaline solution: a solution with a pH above 7.

Alkane: a hydrocarbon with the general formula C_nH_{2n+2}.

Allotropes: two or more different physical forms that an element can exist in, eg graphite and diamond are allotropes of carbon.

Alpha amino acid: a compound that contains a carboxyl group (COOH) and an amino group (NH_2) attached to a central carbon atom.

Alpha helix: a right-hand coiled formation in proteins.

Ambient temperature: the temperature of the surroundings.

Amino group: the group $-NH_2$ present in amino acids.

Amphoteric: substance that can act as both an acid and a base.

Amplitude: the maximum value of displacement in the oscillation cycle – always measured from the mean (rest) position.

Anabolic: reactions that produce a molecule.

Anaerobic: does not require oxygen.

Anaerobic respiration: respiration without the presence of oxygen.

Analogue signal: a signal with strength proportional to the quantity it is representing.

Analyte: a chemical solution or substance being analysed.

Anhydrous: a compound that contains no water, eg anhydrous copper sulfate, which is white and contains no water compared to blue copper sulfate, which contains water of crystallisation.

Anions: ions with a negative charge formed when an electron is gained by an atom.

Anomalous results: results that do not appear to fit the trend in the data.

Anomaly: a data point that does not fit the overall trend in the data.

Antigens: molecules, often proteins, on the surface of all cells, for example, on the surface of pathogens, and viruses.

Antinodes: points of maximum amplitude that occur halfway between each pair of nodes.

Appendicular skeleton: this is the bones forming the appendages (limbs) and the limb girdles that join your limbs to the axial skeleton.

Artery: blood vessel that carries blood away from the heart.

Atomic number: the number of protons in an atom. (This is the same as the number of electrons in an atom.)

ATP: adenosine triphosphate, an enzyme that transports chemical energy within cells for metabolism.

Atrioventricular node (AVN): specialised muscle cells in the junction of the atria and ventricles that receive impulses from the SAN and send impulses across the ventricle walls.

Autonomic nervous system: the part of the nervous system that controls bodily functions which are not consciously controlled such as the heartbeat and breathing.

Axial skeleton: this forms the longitudinal (lengthways) axis of the skeleton, which runs from your head to your feet. It consists of the cranium (top part of the skull) together with the mandible and maxilla (upper and lower jaw bones); the vertebral column (backbone) with its different types of vertebrae (cervical, thorax, lumbar and, between them, the intervertebral discs;); plus the rib cage and sternum (breast bone).

Axon terminal: the axon of a neuron ends in a swelling called the axon terminal. It contains mitochondria which provide energy for active transport and synaptic vesicles which release the neurotransmitter into the synaptic cleft.

Balanced diet: a diet that contains the correct amount of nutrients and energy to supply an individual's needs with respect to their age and activity level and to maintain their good health.

Baroreceptor: stretch receptors found in the blood vessels that respond to changes in blood pressure in the blood vessels.

Basic solution: solution containing a base that is a substance with react with acids and neutralise them.

Batch process: the production of materials in a small or limited number. The production does not go on all the time.

Beta pleated sheet: a flat flexible structure consisting of parallel polypeptide chains cross-linked found in proteins.

Biodiversity: the variety of life in a particular habitat. It includes all the plants, animals and microorganisms that live there.

Boiling point: the temperature at which a substance changes from a liquid to a gas.

Buffer solution: a solution that resists changes in pH when small quantities of an acid or an alkali are added to it.

Calibration: to adjust or correct the graduations of a measuring device when compared to a known value standard.

Calorimetry: the name given to science investigations using a calorimeter to measure changes of state, phase and chemical reactions in terms of the associated heat transferred.

Carbohydrate: a food source made up of the elements of carbon, hydrogen and oxygen.

Carboxyl group: consist of a carbon atom double bonded to an oxygen atom and single bonded to a hydroxyl group (–COOH).

Carboxylated: combined with carbon dioxide.

Carcinogen: an agent or substance that has been suspected of causing or increasing the risk of cancer.

Carcinogenic: causing cancer.

Cardiac output: heartbeat rate multiplied by the stroke volume.

Cardiovascular system: the heart and blood vessels.

Catabolic: reactions that involve the breakdown of a molecule.

Catalyse: to speed up or accelerate a reaction without itself being used up or changed.

Catalysts: substances that increases the rate of a chemical reaction but are unchanged at the end of the reaction.

Cations: ions with a positive charge formed when an electron is lost by an atom.

Central nervous system (CNS): consists of the brain and spinal cord.

Chemical indicator: any substance that gives a visible sign, usually by colour change, of the presence or absence of a chemical, such as an acid or an alkali in a solution.

Chemical plant: a place where industrial chemical processes are carried out on a large scale.

Chemiosmosis: the movement of ions across a semi-permeable membrane, down an electrochemical gradient.

Chlorophyll: the green pigment found in the leaves of plants, which is needed for photosynthesis.

Chloroplast: a plant organelle where the stages of photosynthesis take place, found in plant cells, photosynthetic bacteria and algae.

Chondroblasts: cells in cartilage that are actively dividing by mitosis. They give rise to chondrocytes – mature cells in cartilage.

Chromatogram: the resulting paper or plate produced showing the substance separation; the pattern of separated substances produced by chromatography (eg as seem on a TLC plate).

Chromatography: a method used to separate chemical mixtures for analysis.

Chyle: milky body fluid, consisting of lymph and emulsified fats and fatty acids, formed in the small intestine during digestion of fatty foods.

Chyme: semi-fluid mass of partly digested food formed in the stomach.

Ciliated cells: cells with tiny hair-like structures.

CLEAPSS: Consortium of Local Education Authorities for the Provision of Science Services.

Co-dominance: the expression of both alleles of a gene is seen in the phenotype.

Coherent: literally means 'sticking together' and is used to describe waves whose superposition gives a visible interference pattern. To be coherent, waves must share the same frequency and same wavelength and have a constant phase difference.

COMAH: Control of Major Accident Hazards.

Compact bone: one of the three layers of bone. Nearly 80% of a bone is this layer.

Complementary base pairing: the way in which nitrogenous bases in DNA pair with each other. Adenine (A) always bonds with Thymine (T) (or Uracil (U) in mRNA) and Guanine always bonds with Cytosine.

Concentration gradient: the change in concentration from an area of high concentration of molecules to an area of low concentration.

Condensation reaction: a chemical reaction involving the removal of a water molecule from two or more small molecules in order to form a larger molecule.

Conduction: the transfer of heat energy in a solid where there exists a difference in temperature.

Continuous process: production that occurs 24 hours a day, seven days a week. It is rarely shut down. Reactants are continually being added and products are being continually removed.

Convection: the transfer of heat by circulating currents from a region of high density to a region of less density in a gas or liquid.

COSHH: Control of Substances Hazardous to Health (legislation).

Cotransporter: a type of transport protein that transports two or more substances at the same time across a cell membrane.

Counter-current multiplier: a counter-current system (a system that maintains a concentration gradient along its length) that uses energy to actively transport substances across a membrane to create a diffusion gradient.

Covalent bonds: bonds formed when atoms share electrons; a chemical bond formed by the sharing of one of more electrons between atoms.

Critical angle: for a ray in the medium with a higher refractive index hitting the boundary with a less dense medium, this is the angle of incidence where the refracted angle would be at 90° – i.e. travelling along the boundary between the two media. So, at this and all higher angles of incidence, no reflected ray emerges.

Crystallisation: the process of forming crystals from a liquid or gas.

Cuvette: a small clear plastic, glass or quartz container usually rectangular in shape, used to contain a sample for spectroscopic analysis.

Data Protection Act: the Data Protection Act 1998 was passed by Parliament to control the way information is handled and to give legal rights to people who have data stored about them.

Delocalised electrons: electrons that are free to move. They are present in metals and are not associated with a single atom or covalent bond.

Denature: a change in the tertiary structure of a protein molecule.

Dendritic cells: antigen presenting cells; they process antigen material ad present the antigens to T cells.

Dendrons: extension of a nerve cell.

Depolarisation: when the axon is stimulated, channels in the axon membrane open. This allows sodium ions to diffuse into the axon. This creates a positive charge in the axon and causes the action potential.

Desiccator: a sealable jar containing substances that absorb water to keep a product dry.

Diffraction grating: a set of parallel, closely spaced slits which can separate light out into its specific colours because different wavelengths are diffracted (bent around the openings) at different angles.

Diffusion: random movement of molecules down their concentration gradient (from an area of high concentration to an area of low concentration). This may or may not be through a partially permeable membrane. It uses only the kinetic energy of the molecules, and does not use energy from ATP.

Digestion: break-down of large organic molecules to simpler soluble molecules that can be absorbed by a living organism/cell.

Digital signal: conveys in binary code a number that represents the size of the measured quantity.

Diploid: describes a cell that contains two sets of chromosomes; usually one set from the mother and the other from the father.

Dipole: separation of charges within a covalent module.

Disaccharide (double sugars): two monosaccharides bonded together by a gylcosidic bond.

Displacement: how far the quantity that is in oscillation has moved from its mean (rest) value at any given time. (Symbol and unit: varies according to the quantity that is oscillating.)

Disproportionation reaction: a type of redox reaction in which a reactant is simultaneously reduced and oxidised to form products.

Distal: situated away from the centre of the body or from the point of attachment.

Distillation: the action of purifying a liquid by a process of evaporation and condensation.

DNA: deoxyribonucleic acid, the hereditary material in cells.

Dominant: allele whose expression is visible in the phenotype even if only one allele of the gene is present.

Drifting: variations in the readings of the balance due to internal mechanical wear, for example.

DSEAR: The Dangerous Substances and Explosive Atmospheres Regulations 2002.

Duct: tube, canal or vessel that carries a body fluid, secretion or excretion.

Ductile: can be hammered thin or stretched into wires without breaking.

Dynamic equilibrium: when two processes take place at the same rate so there is no further change in concentration of the substances involved.

Effector: a muscle, organ or gland that is capable of responding to a nerve impulse.

Efferent pathway: the route taken by impulses that travel away from the spinal cord to the effectors (muscles or glands).

Ejaculation: the release of semen from the body via the urethra in the pelvis.

Electrolyte: a chemical compound that will conduct electricity in solutions.

Electromagnetic radiation: energy released by electrical and magnetic processes ranging from low to high frequency and short to long wavelength. It includes radio, microwaves, infra-red, visible light, ultra-violet, X-rays and gamma waves.

Electromagnetic spectrum: the range of energies produced by electrical/magnetic effects; the range of wavelengths over which electromagnetic radiation extends from gamma waves to radio waves.

Electron affinity: the change in energy when one mole of gaseous atom gains one mole of electrons to form a mole of negative ion. For example, for oxygen: $O(g) + e^- \rightarrow O^-(g)$

Electronegativity: the tendency of an atom to attract a bonding pair of electrons.

Electrostatic attraction: the force experienced by oppositely charged particles. It holds the particles strongly together.

Elodea: aquatic plant.

Eluting: extracting one substance from another using a solvent.

Endothermic reaction: a chemical reaction where heat energy is taken in from the surroundings.

End point: the point at which the indicator changes colour permanently.

Energy level: one of the fixed, allowed values of energy for an electron that is bound in an atom, or for a proton or neutron that is bound in a nucleus.

Enzyme: a biological catalyst.

Enzyme-substrate complex: a transition state where the enzyme and substrate are joined together, before the enzyme converts the substrate into a new producer or products.

Equivalence point: the point at which solutions have been mixed in exactly the right proportions relating to the chemical equation (stoichiometry).

Ester: an organic compound made by replacing the hydrogen of an acid by an alkyl or other organic group. It is the product of the condensation reaction between an alcohol and carboxylic acid.

Ester bond: the bond formed when the carboxyl group of a fatty acid combines with the hydroxyl group of glycerol.

Eukaryotic: an organism that contains the genetic information as linear chromosomes within the nucleus of the cells and numerous specialised organelles.

Evaluate: to make a judgement and determine the value, amount, quality or importance of something.

Evaporation: the change of state of liquid particles to gas near the upper most surface of a liquid, resulting in a drop in temperature of the remaining liquid; the process whereby a liquid turns into a vapour at a temperature below or at the boiling point of a liquid. It occurs at the surface of the liquid, where molecules with enough energy escape into the gas phase.

Exocytosis: process of vesicles fusing with plasma membrane and secreting contents.

Exothermic reaction: a chemical reaction where heat energy is given out to the surroundings.

Extracellular: taking place outside a cell.

Facilitated diffusion: diffusion that is enhanced by the presence of carriers or channels made of protein in the cell surface membrane.

Fermentation: the process by which glucose is converted into ethanol and carbon dioxide in the presence of yeast.

Fertilisation: the union of two gametes/gamete nuclei, to produce a zygote.

Filtration: technique to separate solids from the liquid in which they are suspended.

First ionisation energy: the energy needed for one mole of electrons to be removed from one mole of gaseous atom. For example, the equation shows one mole of potassium atoms losing one electron to become a mole of positive ion: $K(g) \rightarrow K^+(g) + e^-$.

Forcing frequency (or driving frequency): the frequency of wave energy from an external source that is coupled to a resonator. Efficient energy transfer into the resonator only occurs when this is close to one of the natural frequencies.

Fractional distillation: separation of a chemical mixture of liquids into fractions with different boiling point ranges.

Frequency: $f = \frac{1}{T}$ the number of whole cycles occurring in one second. (Symbol: f; SI unit: Hertz, Hz.) How often a particular value occurs in a set of values.

Fuel: a substance that undergoes combustion with oxygen to produce energy.

Functional group: specific part of a molecule that is responsible for particular characteristic chemical reactions.

Gamete: sex cells, eg sperm and ovum; one set of chromosomes compared to two sets in the parent cells.

Gametogenesis: the development of gametes (sex cells – sperm and ova) in the gonads (testes and ovaries).

Gas exchange: the diffusion of oxygen into cells and the diffusion of carbon dioxide out of the cells to enable respiration to take place.

Gene: length of DNA that codes for one or more proteins/polypeptides or codes for one or more length(s) of RNA that may regulate the expression of another gene/other genes.

Gene expression: the production of the product encoded by a gene; may be a protein or a length of RNA that regulates expression of another gene.

Gene locus: position of a gene on a chromosome.

Gene pool: all the alleles of genes within a population.

Genetic: related to heredity and variation.

Genome: all the genes within a cell/organism.

Genotype: genes/alleles present in an individual/cell, may refer to just one characteristic.

Giant ionic lattice: a regular arrangement of positive ions and negative ions, for example, in NaCl.

Glial cells: cells that provide support for neurons by carrying out processes such as manufacturing neuron cell components and digesting dead neurons.

Glycogen: many glucose molecules bonded together and stored in the liver and muscles.

Good Laboratory Practice (GLP): established set of principles that should be followed when working in a laboratory.

Ground state: the lowest energy state possible for a given bound particle.

Gut microbiota: all the microbes that live in the human gut.

H^+: positively charged hydrogen ion.

Habitat: a place with suitable conditions for a variety of different plants and animals to live in. There are many different types of habitat, eg woodland, tropical rainforest, freshwater ponds.

Haematopoietic stem cells: stem cells that divide and give rise to blood cells.

Haemoglobin: protein molecule in red blood cells. It carries oxygen from the lungs to other parts of the body and carbon dioxide back to the lungs.

Half equation: an equation that shows the loss or gain of electrons during a reaction.

Haploid: describes a cell that contains one of each type of chromosome.

Hazard: something which has the potential to cause harm.

hCG: human chorionic gonadotropin, a hormone produced by the chorion. It prevents the breakdown of the corpus luteum. This ensures that progesterone production continues and FSH production is inhibited.

Heterozygous: having one or more pairs of dissimilar alleles for particular genes on homologous chromosomes.

Histone proteins: proteins in nucleus of eukaryotic cells, around which the DNA is wound.

Homeostasis: the maintenance of a constant internal environment within an organism.

Homologous series: a group of organic compounds with similar chemical properties where one member of the series differs from the next by a CH_2 group.

Homozygous: having identical alleles at one or more gene loci on homologous chromosomes.

Hydrocarbon: a compound made up of only hydrogen and carbon atoms.

Hydrogen bond: a force of attraction between a very strongly de-shielded hydrogen atom's nucleus and the lone pair of electrons on an electronegative element on another molecule; a weak interaction that can occur between molecules that contain a slightly negatively charged atom and a slightly positively charged hydrogen.

Hydrolyse: a chemical reaction splits, by adding water, large molecules into smaller molecules.

Hydrolysis: a chemical reaction involving the addition of water molecules to break a covalent bond in order to break a larger molecule into smaller units.

Hydrophilic: has a tendency to mix with water.

Hydrophobic: does not mix with water.

Hypothesis: a prediction, based on scientific ideas, made as a starting point for further investigation; a proposed testable explanation of a phenomenon or prediction based on that explanation that can be experimentally investigated.

Immiscible: liquids that do not mix together, eg oil and water.

Incidence: the direction of the incoming ray.

Inorganic: does not contain carbon.

Intensity: (when related to light) the amount of light energy transmitted. Measured in photons (particles of light energy) per second.

Interference pattern: a stationary pattern that can result from a superposition of waves travelling in different directions, provided they are coherent.

Intermolecular forces: the attraction or repulsion between neighbouring molecules.

Internal reflection: when a wave that is already in an optically dense medium (eg glass) hits a boundary with a less dense medium (eg air or water) and energy is reflected back into the denser medium.

Interneuron: a type of nerve cell found inside the central nervous system that acts as a link between sensory neurons and motor neurons.

Interphase: the phase of cell cycle in which most cells spend most of their time. They synthesise molecules, grow and the organelles and DNA replicate prior to mitosis.

Intracellular: occurring within a cell.

Ion: electrically charged particle formed when an electron is lost or gained.

Ionic bonding: electrostatic attraction between two oppositely charged ions.

Isoagglutinogen: a type of antigen on the surface of red blood cells.

Isoelectronic: having the same number of electrons.

Isomerisation: the process of transforming one molecule into another which has exactly the same atoms, but they are arranged differently.

Kinetic model of matter: all matter is made up of very small particles (atoms, molecules or ions) which are in constant motion.

Kinetic theory: a theory describing the movement of particles in solids, liquids and gases.

Latent heat: the heat energy being taken in or given out when a substance changes state.

Limiting factor: a factor which limits the rate of the reaction or process.

Line of best fit: a straight line or smooth curve drawn to pass through as many data points as possible.

Lone pair: a non-binding pair of electrons.

Lumen: the space inside a structure.

Lymphocytes: white blood cells of three types: B cells (make antibodies), T cells (attack and kill infected and cancerous cells) and natural killer (NK) cells.

Macrophage: type of white blood cell that ingests foreign material; found in liver, spleen and connective tissues.

Magnification: the number of times larger the image appears compared to the actual size of the object being viewed.

Malleable: can be hammered into shape without breaking.

Mean: the sum of all the results divided by the number of results.

Meiosis: a type of cell division by which the amount of genetic material is precisely halved to produce a haploid gamete.

Melting point: the temperature at which a solid becomes a liquid.

Membrane-bound organelles: organelles surrounded by a phospholipid membrane. For example, lysosomes and Golgi apparatus.

Mesentery: double-layered extension of the peritoneum able to support organs within the abdominal cavity.

Metabolism: the chemical reactions that occur within the body to maintain life.

Mitochondria: an organelle where aerobic respiration takes place.

Mobile phase: the liquid that transports the substance mixture through the absorbing material which travels along the stationary phase or 'bed' and carries the substance components with it.

Mode: the data value that occurs most often.

Molar absorptivity: the Beer Lambert coefficient Standard International (SI) mol^{-1} dm^3 cm^{-1} also shown as m^2 mol^{-1}.

Molar mass: the mass of one mole of a substance.

Mole: a unit of substance equivalent to the number of atoms in 12g of carbon-12. One mole of a compound has a mass equal to its relative atomic mass expressed in grams. A standard scientific unit of measure for large quantities of atoms and molecules. One mole of a chemical substance has the same number of atoms as there are in 12g of Carbon-12. (This number is called 'Avogadro's Constant' and is 6.022×10^{23} mol^{-1}.)

Monomer: a single small molecule that can be joined with others to form a polymer.

Monosaccharide: a single carbohydrate molecule.

Mucous membranes: these line the body cavities that open to the exterior; consist of a layer of epithelial cells under connective tissue; cells in the mucous membrane secrete mucus that prevents the membranes from drying out.

Mutagen: an agent such as a chemical, ultraviolet light or a radioactive element that can induce or increase the risk of genetic mutation in an organism.

mV: millivolts, a small voltage/potential across a cell membrane.

Myofibril: basic rod-shaped unit of muscle cell.

Nanometres (nm): measure of wavelength which are $1\,000\,000\,000^{th}$ of a metre in length (1×10^{-9} m), one billionth of a metre.

Natural frequency: a resonator has a series of natural frequencies (or 'modes' or 'harmonics'), each of which corresponds to an exact number of half wavelengths fitting within its boundaries.

Neuron: a cell that transmits electrical impulses and is located in the nervous system.

Nodes: points along a stationary wave where the displacement amplitude is at a minimum (ideally zero).

Nodes of Ranvier: the gap in the myelin sheath of a nerve cell, between Schwann cells.

Non-polar molecule: a molecule where the electrons are distributed evenly throughout the molecule; molecules with an equal distribution of electrons, resulting in no observable electrical poles.

Normal line: a line at right angles to the surface of a transparent medium (eg glass or water) that passes through the point where a ray enters or exists that medium. The direction of rays is always described by measuring the angle between the ray and the normal line.

Nucleation: the initial process that occurs in the formation of a crystal when the dissolved substance starts to come out of the solution from a solution, a liquid or a vapour, in which a small number of ions, atoms or molecules become arranged in a crystalline solid, forming a site upon which additional particles are deposited as the crystal grows.

Nucleation sites: site on anti-bumping granules where small bubbles can form, preventing rapid boiling of a liquid during a reaction.

Nucleotide: the basic structural unit of nucleic acids; a nucleotide consists of a pentose (5-carbon) sugar, a phosphate and a nitrogenous base (adenine, thymine, cytosine, guanine or uracil). Nucleotides are the monomers of nucleic acids.

Nucleus: an organelle found inside the cell which contains genetic information.

Null hypothesis: a prediction which states that there is no relationship between two variables or no difference among groups; a type of hypothesis used in statistical tests that proposes there is no significant difference between observed and expected data.

OH^-: oxygen and hydrogen atoms held together by a covalent bond carrying a negative electric charge.

Ohm's Law: the law that states that the current through a conductor is proportional to the potential difference across it, provided the temperature remains constant.

Oogonia: ovum-producing cells in the germinal epithelium of the ovary.

Oogenesis: the process in which ova are formed in the ovaries.

Orbitals: regions where there is a 95% probability of locating an electron. An orbital can hold a maximum of two electrons.

Organelle: specialised structures found within a living cell.

Organic: derived from living things.

Organic compound: a compound that contains one or more carbons in a carbon chain; substance whose molecules contain one or more carbon atoms, with covalent linkages. They can be in the form of long carbon chains (including alkanes, alkenes and alcohols).

Organic molecule: a molecule that contains carbon.

Oscillation: a regularly repeating motion about a central value.

Osmosis: the movement of water from a region of high water potential to a region of low water potential across a partially permeable membrane.

Osteoblasts: cells that make bone.

Oxidation: loss of electrons from an atom/ion.

Oxidation state: the number assigned to an element in a chemical compound. It is a positive or negative number depending on how many electrons the element has lost or gained. (Also called oxidation number.)

Oxidising agents: substances that withdraw electrons from other atoms or ions.

Path difference: the difference in length between two (straight line) rays, eg one from a particular grating gap to a given point in space and the ray from the next-door grating gap to the same point.

Pathogen: a micro-organism that can cause disease.

Peptide bond: a covalent bond formed between two amino acid molecules when the carboxyl group of one molecule reacts with the amino group of the other molecule.

Peptide link: a functional group consisting of covalent chemical bonds formed between two amino acid molecules (CO–NH–).

Peptides: a chemical compound made of two or more amino acids.

Percentage yield: the actual amount of mass worked out as a percentage of the theoretical mass.

Periodicity: the repeating pattern seen by the elements in the periodic table.

Peripheral nervous system: consists of nerve cells linked to CNS with receptors and effectors.

Peristalsis: involuntary contraction and relaxation of smooth muscles of the intestine (and other canals in the body) creating wave-like movements that push forward the contents of the canal.

Peritoneum: membrane that lines the internal body cavity and organs within it.

Permeable: allowing movement of substance through it.

Personal development: improving yourself through a range of activities.

Phase difference: the difference in phase angle between two waves of the same frequency and wavelength where 360° (2π radians) represents a single whole cycle of the waveform.

pH calibration buffer: an aqueous solution of accurate pH used to set the pH meter levels.

pH curve: graphical shape describing how pH changes during acid-base titrations.

Phenotype: visible characteristics of a cell/organism.

Phosphorylation: production of ATP from ADP and P_i.

Photolysis: splitting of water in the presence of light.

Photomicrograph: photograph of an image seen using a light microscope.

Photon: a quantum of electromagnetic radiation. Photons have zero mass and zero charge, but a definite energy value linked to their frequency.

Photosynthesis: the process by which plants make food, using carbon dioxide, water and the energy from sunlight.

Photosynthometer: apparatus used to measure the rate of photosynthesis by collecting and measuring the volume of oxygen produced.

Plasmid: loop of DNA found in prokaryotic cells.

Point mutation: change in base sequence of DNA caused by a substitution of one base for another eg CGA becomes CCA.

Polarity: the property of molecules having an uneven distribution of electrons, so that one part is positive and the other part is negative.

Polar-molecule: a molecule with partial positive charge in one part of the molecule and similar negative charge in another part due to an uneven electron distribution; molecules without an equal distribution of electrons causing them to have opposite electrical poles.

Polymer: a single large molecule made from repeating units of monomers.

Polypeptides: a long chain of amino acids (and, therefore, peptides) bonded together producing proteins of a high molecular weight.

Polysaccharide (multiple sugars): polymers of monosaccharides. They are made up of thousands of monosaccharide monomers bonded together to form a large molecule.

Porosity: a measure of the volume of tiny holes (pores, from the Greek 'poros') in a material divided by the total volume of the material.

Postsynaptic membrane: the membrane of the cell body or dendrite of the neuron carrying the impulse away from the synapse. It contains a number of channels to allow ions to flow through and protein molecules which act as receptors for the neurotransmitter.

Power: the rate of doing work or the rate of transforming energy.

Precipitation reaction: a chemical reaction where a suspension of small solid particles, a precipitate, is produced from a liquid or a gas state.

Precision: how close two or more readings or measurements are to each other.

Presynaptic membrane: the axon terminal membrane of a neuron carrying the impulse to the synapse.

Primary oocyte: diploid cell formed by cell division in the oogonia. The primary oocyte starts to divide by meiosis but stops at prophase 1.

Primary spermatocyte: diploid cell formed by cell division in the spermatogonia.

Primary structure (of a protein): the sequence of amino acids within the protein chain.

Prokaryotic cell: a cell with no true nucleus or nuclear membrane.

Proximal: situated nearer to the centre of the body or to the point of the attachment.

Purity: freedom of a substance from other matter of different chemical composition. In chemistry, elements and compounds are pure, a mixture is not.

Qualitative analysis: practical experiment producing observational results such as colour, odour, transparency (quantities are not measurable).

Qualitative data: observations made without using numbers.

Quantitative analysis: practical experiment producing numerical results (quantities are measurable).

Quantitative data: data which involves using numbers.

Quantum: the smallest unit that

can exist independently. A quantum has clearly defined values of energy, mass, charge and other physical quantities.

Quantum theory: combines ideas from wave motion and particle mechanics theories to create a new 'wave mechanics'. At the sub-atomic level all the particles – protons, neutrons, electrons, photons, etc. – also behave like waves (eg they can be diffracted). When they are bound into an atom or molecule, these particles behave like stationary waves with a fixed wavelength and energy.

Radiation: the transfer of energy, such as heat from a source to its surroundings.

Receptor: a specialised cell or group of cells that respond to changes in the surrounding environment.

Recessive: allele whose expression is not visible in the phenotype if the dominant allele is also expressed.

Recrystallisation: a technique used to purify a chemical by dissolving both the chemical and the impurity in a solvent and warming the solution. Separation is possible due to the product and the impurity having different solubilities in hot and cold solvents.

Redox: the transfer of electrons during chemical reactions.

Redox reactions: reactions in which atoms have their oxidation state changed.

Reduction: when an atom/ion gains electrons. The phase OIL RIG will help you remember the difference between oxidation and reduction. **O**xidation **I**s **L**oss (of electrons), **R**eduction **I**s **G**ain (of electrons).

Reflection: wave energy that bounces off a surface and has its direction of travel altered by more than 180°.

Reflux: a method involving heating a reaction mixture to the boiling point temperature of the reaction solvent and using a condenser to recondense the vapours back into the reaction flask. This allows a longer reaction time so that the reaction can complete.

Refraction: means bending of the direction of travel, so it describes the direction of an outgoing ray after bending.

Refractive index: of a transparent medium is the ratio of the speed of light in vacuum to its speed in the medium.

Refractory period: the brief period following an impulse before another impulse can be generated.

Reliability: how trustworthy the data is.

Resolution: the ability to distinguish between objects that are close together.

Resonance: the storing of energy in an oscillation or a stationary wave, the energy coming from an external source of appropriately matched frequency.

Respiration: the process by which glucose in living cells is converted into carbon dioxide and water, releasing energy; a series of oxidation reactions that take place in all living cells to produce ATP, carbon dioxide and water from organic compounds such as glucose.

Retention factor (R_f): distance moved by the solute/distance moved by the solvent on chromatography paper or plate.

Reticular fibres: fibres made of collagen and coated with glycoprotein. They form a network around fat cells, nerve cells, muscle cells and in the walls of the blood vessels.

Reversible reaction: a reaction where the reactants react to form products and the product simultaneously react to re-form the reactants, for example, in NaCl.

RIDDOR: Reporting of Injuries, Diseases and Dangerous Occurrences Regulations 2013.

Risk: the harm that could be caused by a hazard and the chances of it happening.

RNA: ribonucleic acid, a molecule with long chains of nucleotides.

Saltatory conduction: (from the Latin verb *saltus*, which means *to leap*) in myelinated neurons the impulse appears to jump along the axon between nodes. The action potentials are propagated from one node of Ranvier to the next node, which increases the conduction velocity of action potentials.

Sarcolemma: cell membrane of a striated muscle cell.

Saturated solution: a solution in which the maximum amount of solute has been dissolved.

Secondary oocyte: cell formed when the primary oocyte completes the first meiotic division. The second meiotic division takes place after fertilisation.

Secondary spermatocyte: cell formed when the primary spermatocytes divides by meiosis.

Semi conservation replication: mode of replication where two new molecules are formed, each identical to the other and to the parent molecules and each consisting of one old strand and one new strand of DNA.

Semi-permeable membrane: a membrane that will allow small molecules such as water, carbon dioxide and oxygen to pass through it, but will not allow large molecules to pass through it.

Significance level or confidence level (*p*): This is used in hypothesis testing. It is a figure used to reject or accept the null hypothesis. Scientists usually use figures ranging from 1% (0.01) to 5% (0.05) significance levels.

Sinoatrial node (SAN): specialised muscle cells in the right atrium that start the cardiac cycle by sending impulses across the atria walls. This is often called the heart's pacemaker as these cells control the speed of the cardiac cycle.

SI units: a system of units that have been agreed internationally.

Skills: the abilities required to do something well or expertly.

Sodium-potassium pump: carrier proteins in the cell membrane that transport sodium ions and potassium ions in opposite directions across the cell membrane.

Solute: the substance dissolved in a solvent to form a solution; a substance which is dissolved in another substance and is usually the lesser amount.

Solution: a liquid mixture where a solute is dissolved in a solvent; the resulting liquid which has the solute dissolved in a solvent.

Solvent: a liquid that dissolves another substance to form a solution; the liquid in which a solute dissolves.

Somatic nervous system: the part of the nervous system that brings about the voluntary movements of muscles as well as involuntary movements such as reflex actions.

Specific heat capacity: the energy required to raise the temperature of 1 g of a substance by 1 °C.

Spectroscopic analysis: analysis of a spectrum to determine characteristics of a substance, eg its composition.

Spermatogenesis: the process of sperm formation in the testes.

Spermatogonia: sperm producing cells in the germinal epithelium of the seminiferous tubules.

Sphincter muscle: circular muscle that surrounds an opening and acts as a valve.

Spongy bone: one of the layers of bone. Only 20% of the mass in bone is spongy bone but the surface area is ten times that of compact bone tissues.

Standard deviation: a measure of how far data values are from the mean value.

Standard form: a way of writing down small and large numbers easily using powers of ten.

Standard Operating Procedures (SOPs): established procedures or methods for the completion of a routine operation.

Standard solution: a solution of known concentration used in volumetric analysis.

Stationary phase: the solid material that absorbs the mixture flowing through it.

Stationary waves (or standing waves): wave motions that store energy rather than transferring energy to other locations.

Stoichiometry: involves using the relationships between the reactants and the products in a chemical reaction to work out how much product will be produced from given amounts of reactants.

Stroke volume: the volume of blood pumped out of the heart with each contraction.

Substrate: the molecule that is affected by the action of an enzyme.

Superposition: the adding together of wave displacements that occurs when waves from two or more separate sources overlap at any given locations in space. The displacements simply add mathematically.

Supersaturation: the difference between the actual concentration and the solubility concentration at a given temperature.

Teratogen: a drug or other substance capable of interfering with the development of the fetus (unborn child), therefore causing birth defects.

Tertiary structures (of a protein): the 3D shape of a protein, caused by its folding.

Theoretical mass: the expected amount of product from a reaction calculated from the balanced equation.

Thermal equilibrium: point at which there is no temperature change due to heat energy being used to break molecular forces at phase change.

Threshold level: the point at which increasing stimuli trigger the generation of an electrical impulse.

Titration: a method of volumetric analysis used to calculate the concentration of a solution; the process of determining the concentration of an unknown solution using a solution of known concentration.

Tolerance: the acceptable upper or lower limits for a measurement.

Total internal reflection: all the wave energy is internally reflected. None of it is lost as a refracted ray. This happens for all angles of incidence larger than the critical angle.

Transcription factors: proteins that activate or suppress the expression of genes.

Transduction: the conversion of a signal from outside the cell to a functional change within the cell, eg odour to electrochemical signals.

Transmission: wave energy passing through an object, eg a diffraction grating, and mostly continuing forward in the original direction, though some energy will be diffracted through angles of less than 90°.

Transmittance: a measure of the amount of light of a particular wavelength which passes through a sample.

Transpiration: evaporation of water from the surface of the leaves of plants.

Turgor: rigidity of plant cells due to pressure of cell contents in the cell wall.

Van der Waals forces: all intermolecular attractions are van der Waals forces.

Variables: factors which can change or be changed in an investigation.

Vector: agent that carries. In this context, it carries foreign DNA into a cell/organism.

Ventilation centre: groups of nerve cells located in the brain that control the pattern and rate of breathing.

Viscosity: a measure of how easily a liquid flows. The thicker and less runny the liquid, the more viscous it is.

Voluntary response: a conscious action taken in response to a stimulus (change in the environment).

Water potential: a measure of the ability of water molecules to move in a solution.

Wavelength: the distance along the wave in its direction of travel (propagation) between consecutive points where the oscillations are in phase.

Yield: the amount produced.

Zona pellucida: the membrane that forms around a secondary oocyte as it develops.

Zygote: diploid cell produced by the union of two gametes.

Index